Calculus and Linear Algebra

Aldo G. S. Ventre

Calculus and Linear Algebra

Fundamentals and Applications

 Springer

Aldo G. S. Ventre
Napoli, Italy

ISBN 978-3-031-20551-4   ISBN 978-3-031-20549-1 (eBook)
https://doi.org/10.1007/978-3-031-20549-1

To
Diletta
Lorenzo
Giuseppe
Rebecca

Preface

Some university courses envisage that mathematics coexists with diverse disciplines, such as history, arts, economics, natural sciences, architecture and design. Mathematics should be related to these disciplines, and the student should not detect objectives and contents in mathematics that are decentralized from his or her interests and from his or her world. Actually, for centuries and even a couple of millennia, and not only in the Western world, mathematics has also directed culture, philosophical and scientific speculation, has lived in company of arts, philosophy, architecture and music, in one rich and fertile environment full of perspectives.

In fact, mathematics is the place of durable goods, rationality, wisdom, and emotions. The interested reader is the student who desires clarity. However, presenting concepts with simplicity is a demanding task: in every context there is a threshold beyond which simplification alters the meaning.

The book is designed for students enrolled in first level courses, for which the knowledge of mathematics is functional to the entire educational programme (e.g., chemical or biological sciences, material science, information technology, and various engineering disciplines). In addition, the text can be a valid support for students enrolled in courses oriented to social sciences (e.g., economy and finance, marketing, management).

This book is the result of my experience gained in the years of teaching Mathematics, Geometry, and Calculus at the University of Naples Federico II, University of Salerno, University of Trento and University of Campania, as well as in the context of editing, assembling and publishing texts for university courses. The structure of this book is compliant to what is usually utilized in undergraduate courses of a British/American system: the results, theorems, statements, and exercises are proposed by illustrating the unifying principles of Mathematics in various practical contexts and applications. Therefore, on the one hand attention is focused on the construction of solid and robust fundamentals through the presentation of the theoretical bases and the demonstrations of the theorems; on the other hand, wide selections of examples, problems and exercises, both posed and solved, corroborate and finalize the theoretical framework.

Clarity and appropriate examples aid the formal setting in the general presentation. Some theorems, whose proof is trivial, are simply stated and commented on. Also, regarding some demonstrations where rigor contrasts with clarity, I propose a reasoning by analogy or an intuitive presentation. My main concern was to make the book understandable for young students.

The order of the chapters can be altered during the delivery of the course. For example, it is possible to premise the differential calculus to the linear algebra by anticipating the chapters from 17 to 22 before chapter 11.

The presentation is accompanied by a set of examples and exercises with increasing complexity. With the exclusion of the first three chapters in which the discussion rests on logical and intuitive bases, each chapter contains sections dedicated to exercises and problems both solved and proposed.

Advances in mathematics, since ancient times, are presented in their historical framework as a consequence of the relations of thoughts and their development. Scientific achievements and struggles are no coincidence, but a need for constructing a coherent world. Each chapter mentions some names linked to a mathematical fact or to a society and its time.

Along this line, the past struggles and the triumphs of those who have contributed most to the development of mathematics become cornerstones that highlight the evolution of mathematics and science in general. The idea is to place the student in a dynamic context that, together with the fundamental formative elements of the subject, makes him or her to assume the role of an active participant in the evolution of ideas rather than a passive observer of the results, thus enabling him or her to develop an analytical as well as critical mindset.

Napoli, Italy   Aldo G. S. Ventre

Acknowledgements The completion of this work would not have been possible without the precious help of my wife Vanda. During the composition of the manuscript, when difficulties arose, she feels my state of doubtful anxiety that leads me to close and be absorbed, thinking of nothing but a resolving sentence. In those moments the matter is out of control, but she puts all the pieces together and transmits calm and trust. I must thank Vanda for her kind support and understanding.

Contents

Chapter 1
Language. Sets

1.1 Language

Let us observe the language of a child learning to speak. Any progress in his way of expressing himself reveals his thinking and behavior (Adler 1961; p. 57).

Let us accept without question the legend that the first word a baby learns is *Mama*. He soon learns his own name, say *Peter*, and identifies *Daddy* and other members of the household. He also learns words like *spoon, cup, table*, and so on. Before long, he discovers that the words *Mama* and *table* are used in different ways. The word Mama is applied only to one object in his experience, the warm, gentle woman who feeds him, bathes him, dresses him, and hugs him. But the word table applies to several things. He may use it for the table in the kitchen, the table in the dining room, or the little table in his own bedroom. The word table is the name of a class of things, and is applied to any member of the class. Similarly, spoon and cup are class names, while Peter and Daddy refer to individuals only. In grammar, we distinguish between class names and the names of individuals by using different labels for them. We call a class name a common noun. We call an individual name a proper noun.

Language soon develops rapidly beyond all foreseeable limits (Chomsky 1970; p. 7).

Most of our linguistic experience, both as speakers and hearers, is with new sentences; once we have mastered a language, the class of sentences with which we can operate fluently and without difficulty or hesitation is so vast that for all practical purposes (and, obviously, for all theoretical purposes), we may regard it as infinite. Normal mastery of a language involves not only the ability to understand immediately an indefinite number of entirely new sentences, but also the ability to identify deviant sentences and, on occasion, to impose an interpretation on them.

Thus, the matter grows on which language develops and also the level of depth and precision of language. To communicate ideas, concepts, opinions, emotions we choose a suitable language. For various purposes, one is to thin out ambiguities, there are different languages.

© The Author(s), under exclusive license to Springer Nature Switzerland AG 2023
A. G. S. Ventre, *Calculus and Linear Algebra*,
https://doi.org/10.1007/978-3-031-20549-1_1

Sailors use seafaring language, sports reporters draw on the language of games and competitions, lawyers use legal language. And then there is the language of medicine or music, mathematics, and many others. Each group of professionals, artists, scientists has its own language and the need to communicate with those who are not proficient with these specificities.

A suitable language is required to deal with specific concepts relevant to a topic. As for us, what is being asked of the reader? Simply an attitude open to understanding. There is no need to dress the concepts hand by hand introduced with fictional complications, nor to cling to fragments of memory.

The words and concepts of mathematical language will be defined, at least described, before being used. In this sense the book is self-sufficient. The symbols can simplify communication: using them during repetitions, carrying out exercises or proving theorems, helps in understanding and memorization.

1.2 Sets

The concept of *set* is widespread in common language, but to explain it in mathematics, at a first approach, some synonyms can be used: a *set is a collection of distinct objects* or a set is an *aggregate of things having a shared quality*; a *class*, a *collection*, a *list* are sets. We recognize that this is not an exhaustive way to exhibit the concept of set. However, for our purposes, we remedy by considering as intuitive or *primitive* the concept of set.

Usually the sets are denoted by capital letters, $A, B, C, \ldots$.

In the same way the concept of *membership* of an element to a set, is considered primitive. If a is an element of the set A, we write $a \in A$, or $A \ni a$, and we read "a is an element of A", or "a belongs to A". If the element b does not belong to the set A, we write "$b \notin A$" and we read "b is not an element of A".

A set can be identified by the list of the elements that belong to it: all the elements belonging to the set are those mentioned in the list and only the elements of the list belong to the set. The list of the elements of a set is usually enclosed in curly brackets. This way of identifying a set is called an *extensive formulation*, or *extensive definition*, of the set.

For example, the writing $A = \{l, m, n\}$ denotes the set A whose elements are, all and alone, the letters of the alphabet l, m, n. The elements can be named in any order. The set A formed by only the element a is denoted with $A = \{a\}$, i. e., there is a distinction between the element a and the set $\{a\}$.

A set can sometimes also be identified by a property owned by the elements of the set and only by them, elements having a shared quality which is a *characteristic property* of the set. Such an identification of the set is called an *intensive formulation*, or *intensive definition*, of the set.

The set $A = \{l, m, n\}$ is also identified as the set of consonants of the word "lemon" and indicated in symbols by means of the writing $A = \{x \mid x$ is a consonant of the word "lemon"$\}$, meaning that x is the generic element of A and has the property indicated after the slash; a colon can be used instead of the slash, so we can also write $A = \{x : x$ is a consonant of the word "lemon"$\}$.

If A and B are sets such that each element of A is also an element of B, we say that A is a *subset* or a *part* of B, or A is *contained* or *included* in B, denoted $A \subseteq B$, or B *contains* or *includes* A, denoted $B \supseteq A$. The symbol $\subseteq$ is called the symbol of *inclusion*. If $A \subseteq B$ and also $B \subseteq A$, i. e., if each element of A is also an element of B and each element of B is also an element of A, then the sets A and B are said to be *equal*, and we write $A = B$; for example, the sets $A = \{l, m, n\}$ and $B = \{m, n, l\}$ are equal because every element of A is an element of B and every element of B is an element of A.

If every element of A is in B and there exists at least one element of B not belonging to A, then A is said to be a *proper part,* or a *proper subset* of B, or A is *properly included* or *properly contained* in B, in symbols $A \subset B$ or $B \supset A$.

Sometimes to indicate a set an oval curve is drawn to isolate a region in which all the elements of the set are imagined to be contained (Fig. 1.1).

Graphs such as in Fig. 1.1 are called *Euler-Venn diagrams*.

Given the sets A, B and C, if A is included in B, and if B is included in C, then A is included in C. Naturally, if $A = B$ and $B = C$, then $A = C$.

Starting from two sets, other sets can be constructed. Let us give some examples.

The *intersection* of the sets A and B is, by definition, the set whose elements belong both to A and B. The intersection of A and B is denoted $A \cap B$ (Fig. 1.2).

The *union* of the sets A and B is, by definition, the set whose elements belong to A or B, i. e., any element of the union of A and B belongs to at least one of the two sets. The union of A and B is denoted $A \cup B$ (Fig. 1.3).

Fig. 1.1 a, b, c are the elements of the set A

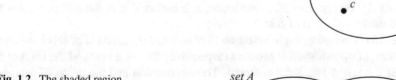

Fig. 1.2 The shaded region common to A and B represents the intersection A ∩ B of the sets A and B

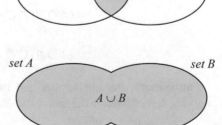

Fig. 1.3 The shaded region inside the closed curve represents the union $A \cup B$ of sets A and B

Fig. 1.4 $A - B$ is the
difference of the sets A and B

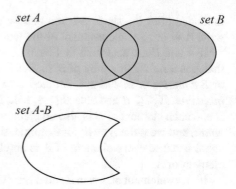

Observe that if an element c belongs both to set A and set B, the element c must be counted only once in the union set. For example, if $A = \{a, c, d, e\}$ and $B = \{c, l, m, n, d\}$, then $A \cup B = \{a, c, d, e, l, m, n\}$.

The existence of a set without elements, called *empty set* denoted $\emptyset$, is assumed.

It is postulated that the empty set is unique and it is defined by a false property: the empty set is the set of common points to two distinct and parallel lines, it is the set of donkeys that fly, it is the set of camels that pass through the eye of a needle, it is the set of politicians who before the elections promise to raise taxes for poor people, is the set of triangles with two parallel sides.

If A and B are sets with no elements in common, then $A \cap B = \emptyset$: in this case the sets are said to be *disjoint*. For example, the set A of the ruminants and the set B of the bipeds are disjoint: it is known that the ruminants are all quadrupeds.

Furthermore, the following property is assumed:

(P) The empty set is contained in any set.

Given the sets A and B, the set whose elements are the elements of A that do not belong to B is called the *difference* of A and B. The set difference of A and B is denoted by $A - B$ (Fig. 1.4). If, in particular, B is included in A, the difference $A - B$ is named the *complement* of B in A.

The symbols denoting the intersection $\cap$ of sets and the union $\cup$ of sets have also the meaning of operations that produce respectively the set $A \cap B$ and the set $A \cup B$ whenever the sets A and B are assigned. The intersection and the union of sets are operations that satisfy the *associative property*, i. e., whatever the sets A, B and C are, the following equalities are verified:

$$(A \cap B) \cap C = A \cap (B \cap C)$$

$$(A \cup B) \cup C = A \cup (B \cup C)$$

Furthermore, the intersection and the union of sets satisfy the *commutative property*, i. e., whatever the sets A and B are, the following equalities are verified:

$$A \cap B = B \cap A$$

$$A \cup B = B \cup A$$

Remark 1.1 As the operations of intersection and union between sets satisfy the previous equalities, the intersection of a finite number of sets and the union of a finite number of sets can be performed by taking the sets in any order and grouping them in an arbitrary way. For example, whatever the sets A, B, C and D are, it can be checked, possibly using the Euler-Venn diagrams, that the following equalities hold:

$$A \cup (B \cup (C \cup D)) = (B \cup (D \cup C)) \cup A = ((D \cup C) \cup B) \cup A$$

Therefore, the calculation of the union of the sets A, B, C and D, can be performed by grouping the sets in an arbitrary way denoted $A \cup B \cup C \cup D$.

References

Adler, I.: Thinking Machines. The New American Library, New York (1961)
Chomsky, N.: Current Issues in Linguistic Theory. Mouton, The Hague (1970)

Bibliography

Seifert, H.: Heinführung in die Mathematik, Zahlen und Mengen. C. H. Beck'sche Verlagsbuchhandlung, München (1973)

Chapter 2
Numbers and Propositions

2.1 The Natural Numbers

From early childhood we learned to count, 1, 2, 3, 4, 5, ..., with and without fingers. Who left his country as a child and had to move to a country where another language was spoken, now thinks and speaks in this one, but continues to count in the language of childhood.

These numbers, so persistent from the first learning, are called *natural numbers* and, taken together, they form the *set of natural numbers*, in fact. Therefore, we denote with

$$\mathbf{N} = \{1, 2, 3, 4, 5, \ldots\}$$

the set of natural numbers.

When we count we not only call natural numbers by name, but we say them in a precise order and by this we mean: the number 1 precedes 2; 1 and 2 precede 3; 1, 2, 3 precede the number 4, and so on. The number 1 is not preceded by any natural number.

The fact that the natural number m precedes n means that m is smaller than n. If m is smaller than n, we write $m < n$ and read "m is less than n". This also means that n follows m, i. e., n is larger than m, and we write $n > m$, which reads "n is greater than m".

We assume that the natural number $n > 1$ follows the numbers 1, 2, 3, 4, 5, ..., $n - 1$.

If the natural number m does not follow n, then one of two cases occurs:

1. m precedes n, $m < n$;
2. m is equal to n, $m = n$.

The notation $m \leq n$ includes both cases and reads "m is less than or equal to n".

© The Author(s), under exclusive license to Springer Nature Switzerland AG 2023
A. G. S. Ventre, *Calculus and Linear Algebra*,
https://doi.org/10.1007/978-3-031-20549-1_2

The natural numbers can be added and multiplied with each other and the result of the addition or multiplication of two natural numbers, whatever they are, it is still a natural number.

The number 0, zero, is not a natural number. The symbol N_0 denotes the set $\{0, 1, 2, 3, 4, 5, \ldots\}$:

$$N_0 = \{0, 1, 2, 3, 4, 5, \ldots\}.$$

A natural number n is said to be a *multiple* of the natural number m if n is equal to the product of m by a natural number q; then we write $n = m \times q$, or $n = mq$. For example, 51 is a multiple of 3 because $51 = 3 \times 17$. So, if $n = 51$ and $m = 3$, then $q = 17$. Of course, 51 is also a multiple of 17. If n is a multiple of m, then we say that n is *divisible* by m, or m is a *submultiple*, or *divisor* of n, or m *divides* n. The number 17 is a divisor of 51.

The numbers of N_0 multiples of 2 are called *even numbers*; those that are not multiples of 2 are called *odd numbers*. Zero is even.

Let us now consider the natural numbers n and m and let $m \le n$. Then it is always possible to find a natural number q and a number r of the set N_0, such that the two conditions are met:

(1) $r < m$;
(2) the number n is equal to the sum of a multiple of m, mq, plus the number r:

$$n = mq + r$$

For example, if $n = 38$ and $m = 5$, conditions (1) and (2) are verified by the numbers $q = 7$ and $r = 3$. Indeed, $38 = 5 \times 7 + 3$.

The procedure described is called the *Euclidean division* of n by m, and the numbers q and r are called the *quotient* and the *remainder* of the Euclidean division, respectively. Euclidean division is known also as *division with remainder*.

If n is a multiple of m, then the remainder of the Euclidean division is 0, indeed $n = m \times q$. For example, $51 = 3 \times 17 + 0$.

If we perform the Euclidean division of 52 by 3, we obtain the quotient 17 and the remainder 1, i. e., $52 = 3 \times 17 + 1$. This equality tells us that 52 is not a multiple of 3, but it is equal to a multiple of 3 plus 1. The number 28 is also equal to a multiple of 3 plus 1, i. e., $28 = 3 \times 9 + 1$, while we obtain 53 adding 2 to a multiple of 3 and the same goes for 29, indeed $53 = 3 \times 17 + 2$ and $29 = 3 \times 9 + 2$.

The remainder of the Euclidean division of the natural number n by 3 is a number r that cannot take other values than 0, or 1, or 2. Then in the set N of natural numbers there are:

(a) the multiples of 3. The division by 3 of every number that is multiple of 3 has remainder 0;
(b) the numbers which are the sum of a multiple of 3 plus 1;
(c) the numbers which are the sum of a multiple of 3 plus 2;

and there are no natural numbers greater than 2 that do not satisfy one of the conditions (a), (b), (c).

Therefore, the following subsets are defined:

$$\{3, 6, 9, 12, \ldots, 3q, \ldots\}$$
$$\{4, 7, 10, 13, \ldots, 3q + 1, \ldots\}$$
$$\{5, 8, 11, 14, \ldots, 3q + 2, \ldots\}$$

q being a natural number.

Let us join the number 0 to the first subset and get a subset we call *remainder class* [0]; let us join the number 1 to the second subset and we get a subset that we call *remainder class* [1]; let us join the number 2 to the third subset and we get a subset that we call *remainder class* [2]. The union of the three classes is equal to the set $\mathbf{N_0}$.

Therefore, $\mathbf{N_0} = [0] \cup [1] \cup [2]$ and the intersection of any two different classes of the three is the empty set, i. e., the classes [0], [1] and [2] are two by two disjoint (Sect. 1.2).

2.1.1 Counting Problems

Nail stuff. A problem to be solved with natural numbers

Ms. Julia has to solve a problem concerning ten sacks, placed in a row next to each other and full of nails: there is a first sack, a second, a third, and so on. Julia knows that each sack contains nails of the same weight: nine sacks contain only nails weighing ten grams each and one sack contains only nails of nine grams: Julia has to individuate this sack.

To solve the problem Julia has at her disposal a scale with which she can carry out only one weighing, then she must be able to identify the sack containing nine-gram nails.

Julia tries to clean up the problem of contingent data. Then she numbers the sacks: $S_1, S_2, \ldots, S_{10}$. Therefore, each sack is identified by its numbered position. Free to weigh nails in the quantity she deems appropriate, Julia has an intuition. She takes one nail from the sack S_1, two nails from the sack S_2, three from S_3, ..., i nails from S_i, and so on, and finally from the sack S_{10} she takes ten nails.

Julia collects all these nails in a handful and weighs them. Well, if the sack with the nine-gram nails is S_7, then, in the handful of selected nails, the nine-gram nails weigh $9 \times 7 = 63$ g, while the ten-gram nails, which are in number of $1 + 2 + 3 + 4 + 5 + 6 + 8 + 9 + 10 = 48$, weigh 480 g. So, the weight of the whole handful of nails is $480 + 63 = 543$ g. Since the handful will always consist of $1 + 2 + 3 + 4 + 5 + 6 + 7 + 8 + 9 + 10 = 55$ nails, it is not difficult, at this point, to find the sack of nails of nine grams, which is the sack S_i, such that

$$(55 - i)10 + 9i = 550 - 10i + 9i = 550 - i$$

Then, the number i of grams missing to 550 is the number of the sack with nine-gram nails. If the handful weighs 543, then $i = 7$.

Baby Gauss counts

Another intelligent count, most famous, was made by Gauss. Carl Friedrich Gauss (1777–1855) was the greatest mathematician of his time. As a child he attended the Braunschweig school. One day, Carl was ten years old, the teacher assigned an exercise to the class, planning to keep it busy for quite a while: adding all the natural numbers from one to one hundred. The teacher asked each to write the result on their own slate to place on the table as soon as he had finished the calculation. Soon after, Carl handed over the slate saying "That's it", while the other students diligently scrambled to do the calculation.

When everyone had finished, the master examined the results and found that Gauss's slate was the only one that presented the exact result, 5050. How did he do it?

He performed the sum $1 + 2 + 3 + \cdots + 98 + 99 + 100$ by observing that the sum of the two extreme addends and the sum of two addends equidistant from the extremes is always 101:

$$1 + 100 = 2 + 99 = 3 + 98 = \ldots = 50 + 51 = 101$$

Then the sum of the natural numbers from one to one hundred is 50 times 101, that is 50 times 100 plus 50 times 1, just 5050.

We will see that, whatever the natural number n is, the sum of the natural numbers not greater than n is equal to the half of $n(n + 1)$.

2.2 Prime Numbers

Recall (Sect. 2.1) that if a, b and c are natural numbers and

$$c = a \times b \tag{2.1}$$

a, b are *factors*, or *divisors* of c. The right-hand side of (2.1) is said to be a *factorization* of c, and c is *factorized* into $a \times b$. The following equalities provide examples of factorizations:

$$6 = 2 \times 3, 9 = 3 \times 3, 30 = 2 \times 15, 30 = 3 \times 10, 51 = 1 \times 51,$$
$$51 = 3 \times 17, 108 = 2 \times 2 \times 3 \times 3 \times 3, 210 = 2 \times 3 \times 5 \times 7$$

Every natural number c has the factorization $c = 1 \times c$, called *trivial factorization*. Among the examples above there is only one trivial factorization, $51 = 1 \times 51$. Each of the other factorizations is a product of two or more smaller factors than the factored number. If the natural number c distinct from 1 has a non-trivial factorization, then c is called a *composite* natural number. If the natural number c different from 1 has only the trivial factorization, i. e., c is divisible only by itself and by 1, then c is called *prime number*. As a result, 1 is not a prime number. The prime numbers less than 200 are:

$$2, 3, 5, 7, 11, 13, 17, 19, 23, 29, 31, 37, 41, 43, 47,$$
$$53, 59, 61, 67, 71, 73, 79, 83, 89, 97,$$
$$101, 103, 107, 109, 113, 127, 131, 137, 139, 149,$$
$$151, 157, 163, 167, 173, 179, 181, 191, 193, 197, 199$$

The only even prime number is 2. Since 1 is not a prime number, the factorization of a natural number into prime factors is unique.

It is the case to know better the prime numbers. Let us consider a prime number p greater than 2. Then the number $p + 1$ is not a prime number because p is odd and p $+ 1$ is even, i. e., divisible by 2. We now add 1 to the product of two, or more, prime numbers, for instance,

$$2 \times 3 \times 5 + 1 = 31$$
$$3 \times 5 + 1 = 16 = 2 \times 2 \times 2 \times 2$$

We observe that 31 is a prime number while 16 is not a prime number and 3 and 5 are not factors of 16. Again,

$$2 \times 23 + 1 = 47$$
$$7 \times 13 + 1 = 92 = 2 \times 2 \times 23$$

The number 47 is a prime number while 92 is not a prime number and 7 and 13 are not factors of 92. These observations give rise to an important property of the prime numbers, proved by Euclid: *there exist infinitely many prime numbers.* Euclid made this reasoning. Suppose we know a certain list of k prime numbers:

$$p_1, p_2, p_3, \ldots, p_k$$

Let us consider the number $q = p_1 \times p_2 \times \ldots \times p_k + 1$, which is obtained by adding 1 to the product $p_1 \times p_2 \times \ldots \times p_k$. Each number $p_1, p_2, p_3, \ldots, p_k$ is not a divisor of q, so either q is a prime number or q has a prime divisor different from each of the numbers $p_1, p_2, p_3, \ldots, p_k$. In any case, there exists a prime number other than $p_1, p_2, p_3, \ldots, p_k$ and this prime number is either q or a divisor of it. In other words, it has been shown that at least one additional prime number not in the known list exists.

2.2.1 Codes and Decoding

Codes used to protect military or industrial secrets are based on systems that are inviolable in practice, even if not in principle. One of the systems consists in using operations that are simple to perform in one direction, but very difficult in the reverse.

Consider, for instance, two very large prime numbers, each made up of hundreds of digits, if we want the product of them a computer will multiply them in less than a second.

But if we give the computer the product of two unknown prime numbers and ask to find its prime factors, then the fastest computer could take years to give us the answer. Let us consider an example of coding (Barrow 2002).

Suppose I want to send Larry a secret message. My way of *coding* it is to put it in a briefcase and close it with a padlock.

The *decoding* corresponds to the opening of the padlock. Of course, when Larry receives the briefcase, he will need the key of the padlock to open it. But I cannot also send the key to Larry and thus risk that "the enemy" takes possession of the message.

How to do?

Let us activate a sequence of actions:

First step. I put the message in the briefcase, I lock it with the padlock and keep the key with me. Then I send the closed case to Larry.

Comment. Larry receives the briefcase, but he cannot open the lock and take the message.

Second step. Larry closes the briefcase with another padlock and keeps the key with him. Larry sends the briefcase back to me.

Comment. The briefcase is now closed by two padlocks and nobody can take the message.

Third step. I use my key to open my lock, I remove it and keep it with me and I send the briefcase back to Larry.

Comment. The only lock on the briefcase belongs to Larry, who has the key.

Fourth step. Larry removes his lock from the briefcase and he can take the message.

Comment. Neither of us, neither Larry nor I, needed the other's key.

Let's move on to a "symbolic" encoding and decoding: we use numbers as keys. I transform my message into a many-digit number (or numeric string) S and multiply it by my many-digit secret prime number p, which is my key, thus obtaining the product Sp. I transmit Sp to Larry who multiply by its key, its secret prime number q, to get the new number Spq. Larry sends me Spq, a number which I divide by p (known to me), obtaining Sq which I send to Larry. He divides the number Sq by q (known to him) and gets S, which is my message. I never need to know q and

Larry doesn't need to know p. If some ill-intentioned guy gets in the way and wants to decipher the message, he will have to factorize into prime factors numbers with many digits, operations that are burdensome or impractical.

Remark 2.1 Encoding information, in order to make it incomprehensible, uses prime numbers. Applications include pay-TV decoders, internet transactions, money movements, etc. One might think that with the evolution of computers it will be more and more simple to break these protections. But that's not the case: the magnitudes of the numbers involved will increase. It is recent the discovery of a "monster" prime number of 17,425,170 digits (www.isthe.com/chongo/tech/math/digit/m57885161/huge-prime-c.html. To download this number 25 Mb are needed). Euclid's property (Sect. 2.2) assures us that there are even more large primes.

2.3 Integer Numbers

Starting from the set $\mathbf{N_0}$, the set $\mathbf{Z}$ of *integer numbers*, called also *integers*, or *relative integers*, is defined, whose elements are the negative integers, zero, the positive integers:

$$\mathbf{Z} = \{\ldots, -3, -2, -1, 0, 1, 2, 3, \ldots\}$$

In the set $\mathbf{Z}$ we find the elements of $\mathbf{N}$, but also other elements which are the *negative integers* and the *zero*: it is said that $\mathbf{Z}$ is an extension of both $\mathbf{N}$ and $\mathbf{N_0}$. $\mathbf{N_0}$ is called the set of *non-negative integers*. The result of the subtraction of an integer n from an integer m, even when m is smaller than n, is still an integer.

For example,

$$7 - 5 = 2$$
$$5 - 7 = -2$$
$$6 - 6 = 0$$

2.4 Rational Numbers

We have seen (Sect. 2.1) that the Euclidean division of the natural number n by the natural number m is an operation that gives rise to a quotient q and a reminder less than m such that $n = mq + r$. If $r = 0$, then m is a factor of n and the number q, the quotient, tells us by which natural number we need to multiply m to get n; as we say "the number of times m enters n". And so the 3 enters 9 times the 27, while the 3 enters 9 times the 28, but with the remainder of 1. The number $n - r$ is a multiple of m, in fact, $n - r = mq$; therefore $n - r$ results divided into q equal parts, each of which is worth m. If we do not want to remain undivided r of the n units that form the number n, we must introduce the existence of other numbers that are not integers. For instance, the quotient of the division 27: 3 is worth 9, while the quotient of 28: 3 is worth more than 9, but less than 10. Are there numbers between 9 and 10? Or can we define numbers between 9 and 10? And, if so, how to define the number q, the quotient of the division 28: 3, such that $28 = 3q$?

If we want to introduce the possibility of performing the division $n:m$ of two integers n and m, with zero remainder, whatever they are, with m different from zero, we must extend the set of integers **Z**.

A *rational number* is defined by a fraction:

$$\frac{n}{m}$$

with n and m integers and $m \neq 0$. The fraction is called a *fractional representation* of the rational number.

Hence the rational number q such that $28 = 3q$ is identified with the fraction $\frac{28}{3}$. The number n is called *numerator*, the number m *denominator*. The set of rational numbers is denoted **Q**. The fractional representation $\frac{n}{m}$ of the rational number is not unique: let p be any non-null integer, the fractions $\frac{n}{m}$ and $\frac{pn}{pm}$ represent the same rational number. For example, fractions $\frac{2}{5}$ and $\frac{4}{10}$ represent the same rational number.

A fraction whose numerator and denominator have no common integer factors, other than 1 or -1, is said to be *simplified* or *irreducible* or *reduced to lowest terms*. For example, the fractions $-\frac{2}{5}$ and $\frac{2}{5}$ are irreducible.

Positive rational numbers are represented by fractions having numerator and denominator of the same sign (i. e., both positive or both negative); negative rational numbers are represented by fractions having the numerator and denominator of opposite signs. Fractions with null numerator represent zero.

Sometimes it will be appropriate to define the fractional representation $\frac{n}{m}$ with integer n and natural m, so that the sign of the fraction depends on the sign of the numerator.

A fraction with denominator equal to 1 is identified with its numerator: therefore, the integers are particular rational numbers.

Therefore: the set of rational numbers contains the set of integers and the set of integers contains the set of natural numbers.

The *decimal representation* of a fractional number $\frac{n}{m}$ is the quotient of the division $n:m$ with n, m integers, and $m \neq 0$.

The decimal representation of the number $-\frac{n}{m}$ is obtained by the decimal representation of fractional number $\frac{n}{m}$ preceded by the minus sign. The quotient of the decimal representation of the fractional number $\frac{5}{4}$, that means 5:4, is 1.25 because $5 = 4 \times 1.25$.

Let us set:

Fig. 2.1 The cake divided into 5 slices and one smaller slice

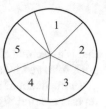

$$\frac{5}{4} = 1.25$$

which is a division with zero remainder.

A way for cutting a cake

We have a cake that weighs one kilogram and 600 g, or 16 hectograms. There are also 5 greedy diners. If we cut the cake into 5 slices of 3 hectograms each, a smaller slice of 1 hectogram remains.

If the diners want nothing left, the remainder of 1 hectogram must be divided into five parts, five small slices of 20 g to each cake lover. Therefore, $16 = 5 \times 3.20$ and the division without the remainder gives the result $16 : 5 = 3.20$, that is a rational quotient (Fig. 2.1).

Examples

$$\frac{1}{2} = 1 : 2 = 0.5 \quad \frac{-5}{4} = -5 : 4 = -1.25$$
$$\frac{18}{6} = 18 : 6 = 3 \quad \frac{5}{8} = 0.625$$
$$\frac{1}{1000000} = 1 : 1000000 = 0.000001$$
$$\frac{1}{3} = 0.3333\ldots \quad \frac{723844}{99900} = 723844 : 99900 =$$
$$7.24568568568\ldots$$

In the last two examples the division does not stop, in the sense that some digits after the dot, that are the *decimals*, are repeated indefinitely. On a closer inspection, this circumstance is also true in the other examples, if we agree to add zeros. For example,

$$\frac{5}{8} = 0.6250000\ldots$$

2.4.1 Representations of Rational Numbers

We have dealt with the *fractional representation* and *decimal representation* of rational numbers: by performing the division $\frac{n}{m}$ we get the *decimal representation* of the rational number $\frac{n}{m}$

$$\frac{n}{m} = p.c_1c_2c_3\dots$$

that is a decimal alignment made of an integer number p, which is called the *integer part*, followed by a dot and some digits which form the *decimal part*.

In the decimal representation of a rational number a set of digits, called *period*, repeats indefinitely in the decimal part. Such an alignment which is *generated* by a fraction is called a *recurring decimal* and provides the decimal representation of a rational number. Observe that the decimal representation of a rational number may include some digits in the decimal part, between the dot and the period, that form the *anti-period* (see the number 7.24568568568... among the examples above, where the anti-period is 24 and the period is 568); in the decimal representation the period is marked by a bar. For example, $7.24568568568\dots = 7.24\overline{568}$ and $0.333\dots = 0.\overline{3}$.

If the period consists of the only digit 0, it is left out; for example, instead of $-5.48\overline{0}$ we write -5.48. Of course, $6.\overline{0} = 6$.

It is always possible to carry out the passage from any decimal alignment to its fractional expression: the procedure is performed as illustrated by the example:

$$7.24\overline{568} = \frac{724568 - 724}{99900} = \frac{723844}{99900}$$

where the denominator is the integer number formed by as many nines as there are digits of the period followed by as many zeros as there are digits of the anti-period.

Let us observe that in the passage from fractional to decimal representation of a rational number not all possible decimal alignments are obtained: indeed, alignments whose recurring part is the single digit 9 are excluded.

For example, there is no fraction from which we get the decimal alignment $7.56\overline{9}$. It is assumed to identify $7.56\overline{9}$ with 7.570. In general, each decimal alignment with period 9 is identified with the alignment obtained by replacing the period 9 with zero and increasing by one unit the last digit preceding it. For example, 1 is identified with $0.\overline{9}$.

2.4.2 The Numeration

In the ancient world, small objects or balls were used to count animals returned to the stable or those to be sold at the market. The ancient Romans and the Greeks had systems of representing numbers different from ours. The Romans used base five numbering (they counted on the fingers of one hand); our usual numbering system is in base ten.

Another system uses the base twenty; we see traces of it in the French idiom in which eighty is said *quatre-vingt* and seventy and ninety are rendered with compound words, *soixante-dix* and *quatre-vingt-dix*, respectively. The Mesopotamians were interested in astronomy and used *sexagesimal* notation, that is, they represented numbers in the base sixty, which is still useful today in dividing time and calculating with angles. What is the advantage of the sexagesimal system? The Mesopotamians, and not only them, were convinced by the fact that the number 60 is relatively small and has a large number of divisors, which comes in handy in measuring time and in astronomy. (The day is divided into 24 h and the hours into 60 min.)

Let us mention also the *dozen* and the *gross*, i. e., twelve dozens.

The bases two and sixteen numbering systems find applications in computer science.

If we had eight fingers, we would have found it natural to use the number eight as the basis of the numbering system. In this system, only eight distinct symbols are needed, 0, 1, 2, 3, 4, 5, 6, 7, to be attributed to each non-negative integer not greater than seven. To designate the number eight we write 10, a symbol that indicates a group of eight elements plus zero units. Then 11 means a group of eight plus one, that is nine, 12 means ten, and so on. In this system where eight is the base, 100 means eight times eight, or sixty-four, 1000 means eight times eight times eight, or five hundred and twelve (Adler 1961).

In short, the number has its own "personality" independent of the way of representing it.

2.5 The Real Numbers

There are other numbers. We refer to the set R of *real numbers*, which contains all the previous numerical sets, natural, integer, rational numbers, but also other numbers, called *irrational numbers*, because they cannot be expressed as fractions but are decimal non periodical numbers.

For example, the famous number *pi*, $\pi = 3.14159\ldots$, the ratio of the circumference to its diameter is an irrational number. This number cannot be put in the form of a fraction. But what does it mean that π is an irrational number? Well, a stretch of road 57 m long can be measured by placing the meter consecutively 57 times. We can measure a stretch of road 23.5 m long in the same way, but after counting 23 m there remains a piece that is measured by half a meter.

Again, a straight segment measures 8.564 m: it means that 8 full meters are needed to measure the segment, but it is not enough, the part of the segment that remains is measured placing one thousandth of a meter and reporting it 564 times. But you can proceed differently: taking the millimeter and placing it 8564 times consecutively: now the same number 8.564 has been thought in the form $\frac{8564}{1000}$. Beyond the physical fatigue which emerges at the mere thought of performing these measurements, a conceptual fact stands: after a finite number of times in which you have placed consecutively the millimeter you come to know the precise measurement of the segment. In the same way you measure a segment of 7.358243 m, but now you have to connect 7,358,243 times the millionth part of the meter. It is always a finite number of times.

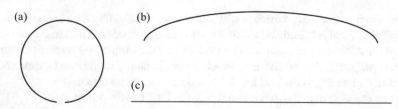

(a)   (b)

(c)

Fig. 2.2 Rectification of a circumference

Let us now consider the circumference which has the length of one meter and imagine to rectify it, which means:

(a) cut it;
(b) let's take it for the two ends;
(c) spread it on the straight line, as if it were a string (Fig. 2.2).

There is no segment, no matter how small, whose length is a submultiple of 1 (meter), which can be juxtaposed consecutively a finite and integer number (even very high) of times on the whole circumference and a finite and integer number of times on its diameter. Irrational numbers are generated when we measure *incommensurable magnitudes*, which do not have a common submultiple. The circumference and its diameter are incommensurable magnitudes.

Homogeneous magnitudes

We can compare two segments by establishing whether they have the same length or one is greater than the other; moreover, addition and subtraction operations can be defined between segments. The same can be said with regard to the angles, using their amplitudes, and polygons, with regard to their extensions. We will then say that the set of segments, as well as the set of angles and the set of polygons, are sets of *homogeneous magnitudes*.

Postulate of Eudoxus-Archimedes. Given two homogeneous magnitudes A and B, there exists a non-negative integer m such that the multiple mA of A is greater than B.

The postulate of Eudoxus-Archimedes also involves numerical magnitudes, such as the real numbers. In this case the postulate means that no matter how large the positive real number a is and no matter how small the positive number b is, successive repetitions of b will eventually exceed a.

2.5.1 Density

Between two distinct rational numbers there exists a rational number distinct from the first two and also an irrational number is there. In other words, taking any two distinct rational numbers x, y, with $x < y$, there exists a rational number z and an irrational number w, such that $x < z < y$ and $x < w < y$. We can deduce that between two distinct rational numbers there are infinite rational numbers and infinite irrational numbers. For this reason, the set of rational numbers and the set of irrational numbers are said to be *dense* in the set of real numbers.

2.5.2 Closure of a Set with Respect to an Operations

We know that the sum of two natural numbers is a natural number. Then, we say that the set of natural numbers is *closed* with respect to the operation of addition. The set of relative integers is closed with respect to addition, but it is also closed with respect to subtraction since subtracting an integer from another results in an integer. And so the set of rational numbers is closed with respect to addition, subtraction, multiplication. The set $\{-1, 1\}$ is closed with respect to multiplication, but not with respect to addition.

2.6 Abbreviated Notations

We have introduced (Chap. 1) some abbreviated notations, sometimes useful in formulating mathematical concepts. Some other notation, or symbol, which we define below, may help us:

> the symbol $\forall$ means for any, whatever they are
> the symbol $\exists$ means there is at least one, exists a
> the symbol : means such that

For example, the proposition: "for every real number x, there exists a real number y which is greater than x" may be written, using these symbols, and the membership symbol $\in$, introduced in Chap. 1:

$$\forall x \in \mathbf{R}, \exists y \in \mathbf{R} : y > x$$

We stress that the use of these symbols simply allows us to report some concepts in shorthand form.

2.6.1 There is at Least One ...

When we say "there is a student who is 18 in this classroom" we mean that there is at least one student who is 18 and so there can be several students who are 18 in the classroom. If, on the other hand, we want to express the fact that there is an 18-year-old student in the classroom and there are no other students of this age, we must say "there is an 18-year-old student in the classroom and this student is unique", or "in the classroom there is exactly one student who is 18", or "in the classroom there is one and only one student who is 18". In short, in everyday language we sometimes say "one" instead of "exactly one".

2.7 The Implication

We came across some examples of "mathematical reasoning", for example, in (Sects. 2.1.1, 2.2 and 2.2.1). We want to see how a mathematical reasoning develops.

2.7.1 Implication and Logical Equivalence

Among the concepts introduced so far, we mention the union of sets, the intersection, the complement, the numbers. We have given the definitions on which to reason, make observations, find further properties, which are expressed by propositions, such as the one seen above, "for every real number x, there exists a real number y greater than x". Propositions are the elements of reasoning.

A proposition (or assertion, affirmation, statement) is an expression of the language for which one can decide whether it is true or false.

For example, *the expressions*:

"Archimedes is a mathematician of the third century B.C."
"The whale is a reptile"
"Man is mortal"

are propositions. The first expression is a true proposition, the second expression is a false proposition. While the expressions: "What time is it?", "Hi, Becky!" are not propositions.

What is meant by "reasoning" in mathematics? In this regard, we illustrate the process that leads to the deduction of a proposition from another and we describe the fundamental construction that animates mathematics: the *theorem*.

The deductive process consists in giving rise to a statement, a proposition, from another; in finding a property, starting from other known properties, to which it is coherently, rationally, connected. This activity is essential to the mathematical method.

Let us start by introducing two symbols that are linked precisely to the deductive process. The symbol $\Rightarrow$ called the *implication symbol*, means "implies". If from proposition P it follows, we deduce, proposition Q, then we write $P \Rightarrow Q$.

Example 2.1 Let us consider the two propositions

$$P = \text{the battery is flat}$$
$$Q = \text{the phone does not work}$$

If we use the implication symbol, and write $P \Rightarrow Q$, we mean that the proposition "the battery is flat" implies the proposition "the phone does not work". The implication $P \Rightarrow Q$ also reads "if P, then Q", if the battery is flat, then the phone does not work.

Let's talk now about the $\Leftrightarrow$ symbol. It, placed between two propositions, indicates that from the first we deduce the second, and, moreover, from this we deduce the first. The symbol $\Leftrightarrow$ expresses the logical equivalence of two propositions and it is called the symbol of *double implication*.

Example 2.2 It is true that (we will realize this shortly) if a nonnegative integer number n is even, then its square n^2 is even, but (and we will soon realize this) the vice versa also holds: if n^2 is even, then n is even. Hence the logical equivalence of the two propositions, and this means:

$$n \text{ is even} \Rightarrow n^2 \text{ is even and } n^2 \text{ is even} \Rightarrow n \text{ is even}$$

and, therefore,

$$n \text{ is even} \Leftrightarrow n^2 \text{ is even},$$

which reads: n is even if and only if n^2 is even.

Coming back to the example of the phone, we cannot say that $Q \Rightarrow P$, i. e., if the phone does not work then the battery is flat, because there are various causes of a phone malfunction.

Therefore, we cannot affirm the logical equivalence of the two propositions, that is, it is not true that $P \Leftrightarrow Q$.

Given two generic propositions P and Q, the implication $P \Rightarrow Q$ is a proposition because it is an expression of the language for which one can decide whether it is true or false. In the case that $P = $ "the battery is flat", $Q = $ "the phone does not work" the implication $P \Rightarrow Q$ is a true proposition; the implication $Q \Rightarrow P$ is still a proposition, but false.

Another example about the implication.

Example 2.3 It is night, it is pitch-dark. Two cars proceed along a street on opposite lanes, one towards the other, with their headlights on. The cars are not immediately close, they are, for now, about half a mile far. Each driver sees nothing but the lights of the car in front of him and a bit of the road. So, the cars proceed towards each other and each sees the headlights of the other approaching. Can we deduce that the stretch of road that separates the two cars is straight? No. It may be that the stretch of road contains a curve (Fig. 2.3).

Fig. 2.3 Cars in the night

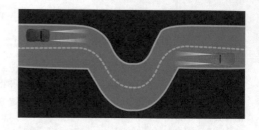

2.7.2 The Theorem

We have seen that implication, as well as double implication, establish a transition from a proposition to another: from a given proposition P one reaches proposition Q.

Again, it seems obvious to admit that if a proposition implies a second and this implies a third, then the first proposition implies the third. Let's say that the implication between propositions is *transitive*.

For example, the proposition $P_1 =$ "it's raining" implies the proposition $P_2 =$ "the streets are wet". In turn, proposition P_2 implies proposition $P_3 =$ "whoever walks on the street gets his/her own shoes wet". From these two implications we deduce, for transitivity, the third "if it is raining, then whoever walks on the street gets his shoes wet"

$$\text{Summing up}: \text{ if } P_1 \Rightarrow P_2 \text{and } P_2 \Rightarrow P_3, \text{ then } P_1 \Rightarrow P_3.$$

A *theorem* connects a proposition P with a proposition Q by means of a chain of implications: $P \Rightarrow P_1 \Rightarrow P_2 \Rightarrow P_3 \Rightarrow \ldots \Rightarrow P_n \Rightarrow Q$.

The proposition P is called the *hypothesis* of the theorem, Q the *thesis*, the chain of intermediate implications constitutes the *proof* of the theorem; building this chain means *to prove the theorem*. Since the implication is transitive Q is deduced from P, and P and Q are therefore related by the implication $P \Rightarrow Q$. The implication $P \Rightarrow Q$ is a proposition called the *statement* of the theorem.

In Sect. 2.2 we showed the proof of the theorem of existence of infinite prime numbers.

We need to look closely at a new theorem to recognize its parts and the formal setting. In the previous Example 2.2 we find the *statement* of a theorem (which we will prove): "n is even" $\Rightarrow$ "n^2 is even", having supposed that n is an element of $\mathbf{N_0}$, i. e., the set of nonnegative integers.

Let n and k be elements of $\mathbf{N_0}$. Recall (Sect. 2.1) that an even number n has the form $n = 2\,k$, and an odd number n has the form $n = 2\,k + 1$.

We now exhibit the statement and the proof of the following.

Theorem 2.1 *If n is an even element of* $\mathbf{N_0}$*, then n^2 is even.*

Proof If n is even, then $n = 2k$, being k a nonnegative integer.

The square of n is

$$n^2 = (2k)^2 = 4k^2 = 2(2k^2)$$

which is an even natural number.

So, we have built the chain that links the hypothesis and the thesis and the proof ends. $\square$

(The symbol $\square$ indicates the end of the proof.)

Remark 2.2 Add 1 to an even number we get an odd number and adding 1 to an odd number we get an even number. The addition of a unit to an integer alters its *parity*, i. e., the property of the number of being even or odd.

We now prove the following.

Theorem 2.2 *If n is an odd element of $\mathbf{N_0}$, then n^2 is odd.*

Proof If n is odd, then $n = 2k + 1$, being k a nonnegative integer. The square of n is

$$n^2 = (2k + 1)^2 = 4k^2 + 4k + 1 = 2(2k^2 + 2k) + 1$$

which is an odd number. $\square$

Remark 2.3 The word "hypothesis" exists in Greek and Latin languages. While in classical Greek "hypothesis" is the foundation, the base, in Latin it is the topic of a speech or a judicial discussion, in short, something to discuss or questionable. Today's current language uses the word in a sense close to the Latin meaning. In short, the hypothesis is a possibility, it must be verified, it is subjective, it is not a certainty. In the mathematical language the hypothesis retains the meaning of the "foundation of the reasoning", which it has in Greek.

Before returning to Theorems 2.1 and 2.2, let's add some considerations on theorem proving techniques. If P is a proposition, the symbol *non*P denotes the proposition which is the negation of the proposition P, i. e., the proposition "it is not true that P". For example, if P = "the battery is flat", then *non*P = "it is not true that the battery is flat".

Not always, in order to prove a theorem we will build a chain of implications: P $\Rightarrow$ P$_1$ $\Rightarrow$ P$_2$ $\Rightarrow$ P$_3$ $\Rightarrow$ ... $\Rightarrow$ Pn $\Rightarrow$ Q, that from the hypothesis P leads to the thesis Q.

Sometimes we will use a different technique to prove a theorem: we build a chain that has as hypothesis the initial hypothesis P and, in addition, the negation of the thesis, *non*Q. With the new hypothesis, i. e., the proposition P and *non*Q, a new chain starts.

If at some point in this new chain of reasoning we yield a proposition that contradicts the initial hypothesis P, then we must deny the negation of the thesis: therefore, it is not true that Q is not true, *non*(*non*Q) = Q, and the thesis Q is true, i. e., it is achieved as the last link in the chain.

We have described an indirect way of proving a theorem, named *proof by contradiction*, or *reductio ad absurdum*. The absurdity consists in the coexistence of a proposition and its negation.

We have so far acquired two techniques of proving: the *direct* proof method and the proof by contradiction.

2.7.3 *Tertium Non Datur*

The proof by contradiction ends with "it is not true that Q is not true ", *non(non*Q), which means, without a shadow of doubt, "Q is true". The conclusion, two negations claim, is reasonable. It is a principle that we accept. The method of proof by contradiction was loved by the ancient Greeks. In fact, Aristotelian logic has among its cornerstones the principle that only one of the two propositions Q, *non*Q is true; in other words, a proposition can only take one and only one of the two states: true, false. *Tertium non datur*, the third is excluded, the Aristotelian logicians say.

In the course of the twentieth century, for a more realistic description of the phenomena of physical reality, and also for theoretical reasons, alongside the true and false states it was supposed that a proposition can also have an indeterminate state, undecidable, uncertain, may it be neither true nor false. Theories of probability, quantum mechanics and other theories related to imprecision and uncertainty have been developed.

These studies, which are fueled by the needs of efficient communication and language adaptation, seem to be in contrast with what we have built up to now. In reality these studies, in their variety, are dealt with as mathematical procedures.

We attribute the value 1 to the "true" state of a proposition and the value 0 to the "false" state. To the "indeterminate" or "uncertain" state of a proposition we attribute a degree of indeterminateness or uncertainty that can be a value between 0 and 1. These values are called *truth values* of the propositions. For example, the truth value of the proposition "man is mortal" is 1; the truth value of the proposition "the whale is a reptile" is 0. If I have to predict the outcome of the toss of a coin, I don't want to commit myself, and then I say that the truth value of the proposition "At the next launch heads will come out" is probably 0.5.

We can assign truth value to the description of a phenomenon because the description can be more or less clear. An object can satisfy a property to a certain extent: what is the borderline that makes us to distinguish the tree from the shrub?

A motivation for the progress of studies about uncertainty is due to the development of complex systems; indeed,

> ...as the complexity of a system increases, our ability to make precise and yet significant statements about its behavior diminishes until a threshold is reached beyond which precision and significance (or relevance) become almost mutually exclusive characteristics (Zadeh 1973)

Among the highly complex systems let's mention human systems, real world societies, political and economic systems.

The theories of probability, fuzzy logic (Zadeh 1965; Kosko 1993) are among the subjects which provide methods and techniques for describing and looking into complex systems.

2.7.4 Proofs in Science

Pythagoras and his school studied and demonstrated important links between nature and numbers and between forms and numbers, but the Pythagorean theorem is the most important result of the thought of that school because it offers us an equality valid for all triangles rectangles; indeed, it defines the right angle.

From the right angle arises the definition of perpendicular line, the passage to dimension two and therefore to dimension three. The right angle is the basis of understanding the space in which we live. The statement of the Pythagorean theorem is relatively simple. Take a right triangle, that is a triangle with a right angle, measure the two shorter sides that form the right angle, the *catheti*; let a and b be the lengths of the catheti, take the squares a^2 and b^2, add them, $a^2 + b^2$. Well, the length c of the third side, the hypothenuse, satisfies the equality $a^2 + b^2 = c^2$ that is true for all right triangles. The Pythagorean theorem is a universal law that you can rely on whenever you come across a right triangle. Figure 2.4 represents a right triangle with catheti $a = 3$ and $b = 4$ and hypothenuse $c = 5$.

The concept of mathematical proof is stronger than that of proof in everyday life. Proofs are fundamental for law, physics, chemistry, natural sciences. In these

Fig. 2.4 $a^2 + b^2 = c^2$

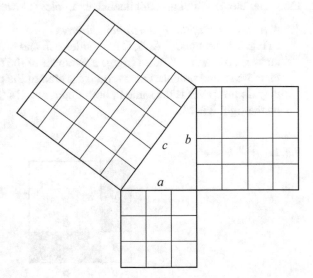

sciences a hypothesis is put forward to explain a phenomenon. If the characteristics of the phenomenon agree with the hypothesis, this circumstance becomes a clue in favor of the hypothesis (Singh 1997).

Experiments can be performed to verify the predictive power of the hypothesis and, if it proves effective again, then there are further clues to support it. In the end, the weight of the clues can be overwhelming and the hypothesis is accepted as a theory.

Think a physics theory or a criminal trial. A mathematical proof, as we have said, starts from a true proposition or from an axiom, which is a proposition assumed to be true; then, after a certain number of steps, we reach the thesis, the conclusion. If the starting point, that is the hypothesis, be it a proposition or an axiom, is true, then the conclusion is true without any doubt.

Therefore, two kinds of proofs exist: the mathematical proof and the empirical proof of the sciences. But, we observe, they are not comparable, they belong to different worlds. Mathematical proof is the discovery of consequences, the proof of science is the discovery of causes. The former belongs to the deductive method, the latter to the inductive method.

Science of induction is statistics. The Pythagorean theorem, true from the sixth century B.C., will always be true, while the theoretical physicist, Nobel laureate, knows that his theory may turn out to be false tomorrow. No emotion.

2.7.5 Visual Proofs

Imagination, intuition, often ignite the spark that guides scientific investigation, the development of reasoning and the resolution of problems. Hence, the importance of the visual element. Let us exhibit some examples (Gowers 2002).

1. *A visual proof of the Pythagorean theorem*

   In Fig. 2.5 the squares A, B, C have sides a, b and c, respectively, and, therefore, areas a^2, b^2 and c^2. Since a (rigid) movement of the four triangles does not alter their areas, and produces no overlap, the area of the part of the large square that they do not cover is the same in both diagrams. On the left this area is $a^2 + b^2$, on the right it is c^2.

Fig. 2.5 $a^2 + b^2 = c^2$

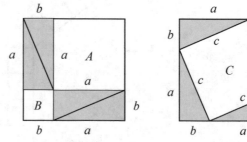

Fig. 2.6 The mutilated
chessboard

2. *The mutilated chessboard*

   A chessboard is made up of 64 squares, the chess. With all 32 tiles of the set of
   dominoes, each as big as two adjacent squares, you can exactly cover the whole
   board. Let us remove the two white squares at opposite corners from the board,
   as in Fig. 2.6.

Question: can the mutilated board be covered by 31 tiles? Just think: each tile occupies
two chess squares, twice 31 is 62. You could do it. Then take the tiles and arrange
them to cover the board.

You are unable to put all the tiles in place to cover the board exactly, even after
many attempts. You find there are many clues that covering cannot be done. However,
you cannot be sure that the covering cannot be done: among the thousands of possible
combinations you may have missed one.

Why can't the covering be done? Is it because you do not have enough time to
solve the problem or does the impossibility arise as a consequence of a specific
reasoning, i.e., a proof?

If we proceed with a mathematical method, we entrust to the chain the proof of
the proposition Q = "the covering cannot be done". The thesis is established.

The hypothesis is: P = "we have a mutilated chessboard".

Let's describe the mutilated chessboard: it has 32 black and 30 white squares and
two contiguous squares do not have the same color.

Then we have:

P_1 = "30 tiles cover 60 chess, 30 white and 30 black"
P_2 = "two black squares remain uncovered"
P_3 = "two black squares are not adjacent"
P_4 = "the 31st tile cannot cover these two black squares"

and now we deduce:

Q = "the covering cannot be done"

The chain is complete: $P \Rightarrow P_1 \Rightarrow P_2 \Rightarrow P_3 \Rightarrow P_4 \Rightarrow Q$.   □

Remark 2.4 The proof of the previous theorem, P (the chessboard is mutilated) $\Rightarrow$
Q (the covering cannot be done), is generalized to mutilated grids n times n, whatever

the natural number n is. Therefore, a mutilated grid 1000×1000 cannot be covered with dominoes.

Remark 2.5 The theorem proved above asserts that, under a certain hypothesis, a certain operation cannot be done. Theorems like this are called *impossibility theorems*.

2.7.6   The Inverse Theorem

We will now see that Theorems 2.1 and 2.2 can be inverted and new theorems are obtained by interchanging the hypothesis with the thesis. This circumstance does not occur for all theorems, because, as we know, if $P \Rightarrow Q$, it does not mean that $Q \Rightarrow P$ too.

Let us prove the inverse of Theorem 2.1.

Theorem 2.3 *If n^2 is even, then n is even.*

Proof We prove the theorem by contradiction: we deny the thesis, i. e., we add to the hypothesis "n^2 is even" the denied thesis "it is not true that n is even", which implies that n is odd. If n is odd, then n^2 is odd by Theorem 2.2, against the hypothesis "n^2 is even": we got a result that contradicts the hypothesis. Therefore, n is even.   □

Similarly, it can be proven the following.

Theorem 2.4 *If n^2 is odd, then n is odd.*

Proof If n were even, by Theorem 2.1, n^2 would be even, contrary to the hypothesis. Then n is odd.   □

Now let us summarize the statements of Theorems 2.1–2.4 in the following form:

Theorem 2.5 *The element n of $\mathbf{N_0}$ is even if and only if its square is even. The element n of $\mathbf{N_0}$ is odd if and only if its square is odd.*

2.7.7   Irrationality of $\sqrt{2}$

According to the Pythagorean theorem, given a square with side 1, the square constructed on the diagonal must have area 2, because $1^2 + 1^2 = 2$. Therefore, the length d of the diagonal must be a number whose square is 2, $d^2 = 2$.

We ask ourselves: given the ease of constructing the diagonal, is it reasonable to think that, fixed any non-zero segment as the unit of measurement of the lengths, we can measure both the side and the diagonal of the square? The answer is: yes, it is reasonable to think all of this. However, as we will see, the side and the diagonal of a square are not commensurable.

This means (Sect. 2.5) that there is no a submultiple segment u common to the side and the diagonal of the square; i. e., there is no segment that can be arranged consecutively an integer number of times n on the side and an integer number of times m on the diagonal, such that $1 = nu$ and $\sqrt{2} = mu$.

Passing to the ratios, we will see that it is not true that

$$\sqrt{2} = \frac{\sqrt{2}}{1} = \frac{mu}{nu} = \frac{m}{n}$$

Therefore, we prove the.

Theorem 2.6 *If the square of the real number a equals the number 2, then the number a is not a rational number.*

Proof Let us prove the theorem by contradiction. Suppose then that a is rational, i. e., a is equal to a fraction

$$a = \frac{m}{n} \tag{2.2}$$

(with m and n integers and n distinct from 0) and let the fraction be irreducible. From the hypothesis

$$a^2 = 2$$

and from (2.2) we have

$$\left(\frac{m}{n}\right)^2 = 2$$

that implies

$$m^2 = 2n^2 \tag{2.3}$$

Then m^2 is the double of a positive integer and therefore it is even. By Theorem 2.3, m is also even and takes the form $m = 2\,h$, with h integer. By (2.2), we have $(2h)^2 = 2n^2$, i. e., $4h^2 = 2n^2$, and therefore $2h^2 = n^2$, i. e., n^2 is even and, by Theorem 2.3, n is also even.

Since m is even, it cannot also be n even since we have supposed $\frac{m}{n}$ irreducible.

Therefore, (2.3) is not true, and neither (2.2) is. So a is not a rational number, i. e., $\sqrt{2}$ is an irrational number. $\qquad\qquad\square$

2.7.8 The Pythagorean School

Pythagoras of Samos lived in the sixth century B.C.. In his youth he traveled a lot: he
was in Egypt, Babylon, India. In his maturity he settled in Kroton, Magna Graecia,
where he founded his scientific and religious school. Pythagoras conceived the geometry in its etymological meaning of measuring the earth, a meaning that he intended
to project to the whole universe to explain it numerically. The fundamental principle
of the Pythagorean school held that every number is the quotient of two natural
numbers, what is equivalent to say that any two quantities or magnitudes must be
commensurable. For the Pythagorean school it was unfounded, indeed "heretical",
to think of the existence of non-rational numbers. The discovery of the incommensu-
rability of the diagonal with respect to the side of the square, which legend attributes
to Ippasos of Metapontum, a pupil of Pythagoras, caused the crisis of Pythagorean
thought and the separation of arithmetic from geometry. The legend ends tragically:
Ippasos is expelled from the school for blasphemy, his colleagues abandon him in a
boat, which Zeus wrecks.

2.7.9 Socrates and the Diagonal of the Square

In the dialogue *Menon*, Plato relates that Socrates knows that the square of double area
of the square of area 1 has side $\sqrt{2}$. By asking appropriate questions, Socrates extracts
this knowledge from the virgin mind of the servant of Menon, thus demonstrating
that the knowledge is within us and we have nothing to do but extract it. This is the
maieutic process, that is extractive, the cognitive process according to Socrates.

2.7.9.1 More on the Pythagorean Theorem

We have discussed about the Pythagorean theorem, meaning that its statement is: if
a right triangle has catheti a and b and hypotenuse c, then $a^2 + b^2 = c^2$.

But Pythagoras asserted that the inverse theorem is also true: if the triangle with
sides a, b, c satisfies the condition $a^2 + b^2 = c^2$, then the angle opposite to the side
c is right, and therefore the triangle is a right triangle. Furthermore, the theorem has
a practical content that has contributed largely to its fame. In fact, the theorem was
primarily used to determine perpendicularity. Let's think of a Greek architect who
was about to check if two walls were perpendicular. First, he took a tool to measure
lengths, for example a rope with knots at equal intervals, then he marked three units
on one wall and four on the other: the walls were perpendicular if there were five
units between the two marked ends, $5^2 = 3^2 + 4^2$. It was a brilliant way of reducing
an angular problem to an easier relationship between lengths (Alsina 2011).

2.8   The Inductive Method and the Induction Principle

We dealt with the deductive method and the inductive method, which are the two fundamental scientific methods (Sect. 2.7.4).

We know that the inductive method is typical of research in statistics, physics, economics, medicine, psychology, the natural and social sciences, in short, the experimental sciences.

It is the method used to discover a law of nature, to be able to make predictions.

If, by observing a phenomenon, we intuit the conditions that determine it, we repeat the observation several times by recreating the environment and the situations in which the phenomenon occurred: if the intuition is confirmed, as observed in the different attempts, we formulate a law. This law has a statistical value, in the sense that, under assigned conditions, it is (very) probable that the phenomenon will occur. The inductive method cannot be applied to the mathematical construction. The following example is discussed in (Lombardo Radice and Mancini Proia 1979, p. 160). Let us consider the polynomial: $x^2 - x + 41$:

> when in this polynomial we replace x with zero, we get the number 41, if we replace x with 1 we get the number 1. If in the polynomial, we put $x = 2$ we get 43, if we put $x = 3$ we get 47. Well, if we continue the sequence of operations, replacing x with the natural numbers 4, 5, ..., up to 40, the values taken by the polynomial are always prime numbers. However, if $x = 41$ we get a square, the number 41^2, which is not prime, and for x greater than 41 we obtain prime or composite numbers (Sect. 2.2), depending on the values of x. Therefore, if using the induction, we had stated, as it is usual in the experimental research after thirty tests and even less, that the polynomial takes, for any natural x, a prime value, we would have made a gross mistake. Induction, based on the extrapolation of a law after a good number of checks, or tests, which is the main method of the experimental sciences, is therefore not good for mathematics.

Another example of induction that seems possible, and then turns out not to work, is built by Richard Guy. Let us inscribe a polygon with n vertices and all its diagonals in a circumference (Fig. 2.7).

Let us see what happens with the first six polygons:

if $n = 1$, then the polygon is reduced to a point
if $n = 2$, then the polygon is reduced to a segment
if $n = 3$, then the polygon is a triangle
if $n = 4$, then the polygon is a quadrilateral
if $n = 5$, then the polygon is a pentagon
if $n = 6$, then the polygon is a hexagon

The maximum number p of regions of the circle determined by the diagonals that meet is 1, 2, 4, 8, 16 when n is equal to 1, 2, 3, 4, 5, respectively. If $n = 6$ then $p = 31$, while we would have expected 32.

Fig. 2.7 Subdivisions of the circle

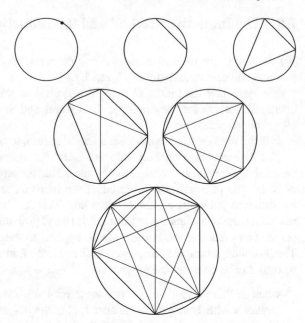

The examples above reveal that the inductive method does not work in mathematics as it is. The mathematician Giuseppe Peano defined an induction method suitable for mathematics, the *method of complete induction* (or, *mathematical induction*), also called the *induction principle*, defined as follows.

Let P(n) be a proposition concerning a non-negative integer n and suppose we know that:

1. there exists a non-negative integer $\bar{n}$ such that the proposition P($\bar{n}$) is true;
2. for every $n \geq \bar{n}$, supposed P(n) true, then P($n+1$) is proved to be true. In symbols: P(n) $\Rightarrow$ P($n+1$).

Well, the induction principle states that if the hypotheses 1 and 2 are verified, then the proposition P(n) is true for every non-negative integer $n \geq \bar{n}$. The number $\bar{n}$ is called the *basis* of induction.

A proof obtained by applying the induction principle is said to be obtained by *induction* or *recurrence*.

The induction principle provides another technique of proving, in addition to those known to us, namely the construction of a chain of implications and the proof by contradiction.

We prove the following result already mentioned (see Sect. 2.1.1).

Theorem 2.7 *If n is a non-negative integer, then the sum of the non-negative integers not greater than n is $\frac{n(n+1)}{2}$.*

Proof Let P(n) denote the following proposition, concerning the non-negative integer n:

$P(n)$ = the sum of non-negative integers not greater than n is $\frac{n(n+1)}{2}$, i. e.,

$$P(n) = 0 + 1 + 2 + \ldots + n = \frac{n(n+1)}{2} \qquad (2.4)$$

which reads: "the sum of non-negative integers less than or equal to n is equal to the product of n by the next of n divided by two". Let us formulate Properties 1 and 2, that define induction principle, in the case of the present theorem:

1. the proposition $P(0)$ is true. In fact, the two sides in the Eq. (2.4), when $n = 0$, reduce to $P(0) = 0$. The basis if induction principle holds $= \overline{n}0$,
2. suppose true $P(n)$, $n \geq 0$, and prove that $P(n + 1)$ is also true. Indeed, if $P(n)$ is true, it is:

$$1 + 2 + \ldots + n = \frac{n(n+1)}{2}$$

Then, adding $n + 1$ to both sides, we have:

$$(1 + 2 + \ldots + n) + n + 1 = \frac{n(n+1)}{2} + n + 1 = \frac{(n+1)(n+2)}{2}$$

This means that proposition $P(n + 1)$ is true.

Therefore, proposition $P(n)$ is true for any non-negative integer n. □

Let us quote a consideration on induction:

–   The purpose of the induction method is to verify whether something that we have guessed through a certain number of successful tests, is true. However, it does not construct the result: we must already 'suspect' the result! (Lombardo Radice and Mancini Proia, ibid, p. 161).

2.8.1 Necessary Condition. Sufficient Condition

We have considered (Sect. 2.7.1) the propositions:

P = "the battery is flat"
Q = "the phone does not work"

and we have observed that the following implication holds: if the battery is flat, then the phone does not work. In other words, the fact that the battery is flat is a sufficient condition for the phone to fail. We have observed that the proposition cannot be reversed, i. e., if the phone does not work, the battery is not necessarily flat. In other words, the fact that the battery is flat is not a necessary condition for the phone to malfunction.

In the case of the cars in the night, we think in a similar way. If the street is straight, then the lighthouses are opposite: the straight way is a sufficient condition for the lighthouses to be in opposite positions, but if the lighthouses are opposite, the street is not necessarily straight.

It is true (Sect. 2.7.2) that if n is even, then n^2 is even: the fact that n is even is a sufficient condition for n^2 to be even. It is also true that if n^2 is even, then n is even: the fact that n^2 is even is a sufficient condition for n to be even and if n^2 is even, then it necessarily occurs that n is even. It is concluded that being n even is a sufficient and necessary condition for n^2 to be even.

An Olympic athlete must have good health: a necessary condition for a person to be an Olympic champion athlete is that she is in good health. The converse is not true: the condition of "enjoying good health" is not enough to make every person in good health an Olympic champion in athletics.

The political elections ended and the results are official. No party has achieved the absolute majority what would have been a necessary condition to form the government. Some political commentators observe that the sum of the elected representatives in the Yellow party and in the Red party forms an absolute majority: this circumstance is a necessary condition for giving rise to a new government, but it is not sufficient because the Red party does not share the Yellow political program: this prevents the coalition of the two parties.

Let us go back to the phone. We have established that its failure does not necessarily depend on the discharged battery. While the implication

$$(\text{battery is flat}) \Rightarrow (\text{phone does not work})$$

is a deduction, tracing the reasons of the failure activates an induction procedure, the search for the causes of the occurrence of an event. Which means making a diagnosis. As mentioned (Sect. 2.7.4), statistics is concerned with making diagnoses.

2.9 Intuition

The construction of mathematics is made up of definitions and theorems and develops on intuition and creativity. Theoretical results often apply to solving questions, problems, using reasoning and calculation. But what produces the progress of science, in general, is the creative drive of intuition.

Let us quote some comments on intuition in mathematics:

- if the logic is the hygiene of the mathematician, it does not provide him with any food (André Weil 1950);
- […] a significant discovery or an illuminating insight is rarely obtained by an exclusively axiomatic procedure. Constructive thinking, guided by the intuition, is the true source of mathematical dynamics. Although the axiomatic form is an ideal, it is dangerous fallacy to believe that axiomatics constitutes *the* essence

of mathematics. The constructive intuition of the mathematician brings to mathematics a non-deductive and irrational element which makes it comparable to music and art (Courant and Robbins 1978, p. 216).

The *conjecture* belongs to the world of intuition. Supported by reasonable clues, the scholar, or the amateur, launches a conjecture.

Encouraged by all the verifications he can make, he conjectures that every natural even number greater than 4 is the sum of two prime numbers (remind that 1 is not prime): 12 is the sum of 5 and 7; 26 is the sum of 7 and 19 (but also 13 and 13); 102 is the sum of 61 and 41 (but also 13 and 89). So we go on, having fun or not, we verify that, however we fix an even natural number greater than 4, there are two prime numbers whose sum is the fixed number.

But many successful checks are not a proof. Though the verification is performed for a lot of cases, it cannot be done for all even natural numbers.

Up to now no one has proved that the proposition "every even natural number greater than 4 is the sum of two prime numbers" is true or false. Therefore, we cannot state the theorem:

"If n is an even natural number greater than 4, then there are two prime numbers whose sum is n",
nor the other:
"It is not true that if n is an even natural number greater than 4, then there exist two prime numbers whose sum is n".

And this last proposition is formulated in the equivalent form:

"There exists an even natural number greater than 4 which is not the sum of two prime numbers".

The statement

"Every even natural number greater than 4 is the sum of two prime numbers"

is known as the *Goldbach conjecture*, after the Prussian mathematician Christian Goldbach, who lived in the eighteenth century.

2.10 Mathematics and Culture

2.10.1 On Education

Mathematics is an aspect of thinking, mathematics is culture. At one time, the unity of culture was a universally recognized, practiced and accepted fact. The arts and sciences originate from a single source, they are inspired by the same reality.

Here we quote a thought of Albert Einstein on the unity of culture and education (from a speech in Albany NY on October 15, 1936, published in 1950).

For this reason I am not at all anxious to take sides in the struggle between the followers of the classical philologic-historical education and the education more devoted to natural science. On the other hand, I want to oppose the idea that the school has to teach directly that special knowledge and those accomplishments which one has to use later directly in life. The demands of life are much too manifold to let such a specialized training in school appear possible. Apart from that, it seems to me, moreover, objectionable to treat the individual like a dead tool. The school should always have as its aim that the young man leave it as a harmonious personality, not as a specialist. This in my opinion is true in a certain sense even for technical schools, whose students will devote themselves to a quite definite profession. The development of general ability for independent thinking and judgement should always be placed foremost, not the acquisition of special knowledge.

2.10.2 Individual Study and Work

Lucio Lombardo Radice (1916–1982) was a mathematician engaged in culture, society and politics. The formation of young people has always been an important goal in his life. Let us quote a reflection from "The education of the mind" (1962).

Intellectual development and the acquisition of a serious and effective cultural heritage require a systematic effort: they are a job. Every serious job, even the one we love the most, even the one we have freely chosen and that we would not abandon for anything in the world, has its different phases, has complex needs. Work is not a succession of joys, achievements, creations. Joy, conquest, creation are the tiring result of a daily, humble, dark, boring effort. In every job, even in that of the poet and the creative scientist, there are technical problems, there is the need to devote a lot of time to the acquisition of notions, of systematic knowledge, to the possession of tools, to the premises of true creative work. [...] The beautiful poetry was born after long and patient linguistic, literary and historical studies; scientific discovery is the result of a patient installation of equipment, of an inflexible intellectual tenacity aimed at understanding theories and experiments that others have laboriously constructed. The genius-magician is a deceptive and educationally harmful romantic myth: the genius, poet or scientist, Leopardi or Fermi, is above all a tireless worker.

References

Adler, I.: Thinking Machines. The New American Library, New York (1961)
Alsina, C.: La Secta de Los Números. RBA, Barcelona (2011)
Barrow, J.D.: Perché il mondo è matematico? Editori Laterza, Roma-Bari (2002)
Courant, R., Robbins, H.: What is Mathematics? Oxford University Press (1978)
Einstein, A.: On education. Excerpts from an address to the State University of New York at Albany, 15th October, 1931 (1950)
Gowers, T.: Mathematics: a very short introduction. Oxford University Press (2002)
Kosko, B.: Fuzzy Thinking. Hyperion Publishing, New York, The new science of fuzzy logic (1993)
Lombardo Radice, L.: L'educazione della mente. Editori Riuniti, Roma (1962)
Lombardo Radice, L., Mancini Proia, L.: Il metodo matematico III. Principato Editore, Milano (1979)
Singh, S.: Fermat's last theorem. © Simon Singh (1997)
Weil, A.: The future of mathematics. Am. Math. Month. **57**, 295–306 (1950)

Zadeh, L.A.: Fuzzy sets. Inf. Control **8**, 338–353 (1965)
Zadeh, L.A.: Outline of a new approach to the analysis of complex systems and decision processes. IEEE Trans. Syst. Man Cybernet. SMC3-28–44 (1973)

Bibliography

Roberts, S.: King of infinite space. © Siobhan Roberts (2006)
Seifert, H.: Heinführung in die Mathematik, Zahlen und Mengen. C. H. Beck'sche Verlagsbuchhandlung, München (1973)

Chapter 3
Relations

3.1 Introduction

Connections, links between people, images, words, memories, are a large part of everyday life. Relations between people, or within a community, are an important object of study in all sciences, from economics to sociology, from psychology and biology to medicine and the sciences of individual and collective behavior. The individual determines which choices to make in a set of alternative actions in order to make decisions: to do this he or she relates objects, people, pieces of information and knowledge.

Let us quote from M. Buchanan (2002):

In the 1960s, an American psychologist named Stanley Milgram tried to form a picture of the web of interpersonal connections that link people into a community. To do so, he sent letters to a random selection of people living in Nebraska and Kansas, asking each of them to forward the letter to a stockbroker friend of his living in Boston, but he did not give them the address. To forward the letter, he asked them to send it only to someone they knew personally and whom they thought might be socially "closer" to the stockbroker. Most of the letters eventually made it to his friend in Boston. Far more startling, however, was how quickly they did so, not in hundreds of mailings but typically in just six or so. The result seems incredible, as there are hundreds of millions of people in the United States, and both Nebraska and Kansas would seem a rather long way away – in the social universe – from Boston. Milgram's findings became famous and passed into popular folklore in the phrase "six degrees of separation." As the writer John Guare expressed the idea in a recent play of the same name: "Everybody on this planet is separated by only six other people.... The president of the United States. A gondolier in Venice.... It's not just the big names. It's anyone. A native in the rain forest. A Tierra del Fuegan. An Eskimo. I am bound to everyone on this planet by a trail of six people"(J. Guare 1990).

Relations permeate the world: characteristics of the food chain of an ecosystem, electrical and communication networks, neural nets, the web, economic and social relations.

These diffuse and chaotic relations can somehow be handled mathematically.

We will study the properties of the relations between sets, by starting with some considerations about schematic examples.

© The Author(s), under exclusive license to Springer Nature Switzerland AG 2023   39
A. G. S. Ventre, *Calculus and Linear Algebra*,
https://doi.org/10.1007/978-3-031-20549-1_3

3.2 Cartesian Product of Sets. Relations

Let A and B be non-empty sets. The *cartesian product* of the sets A and B, denoted $A \times B$, is defined as the set of *ordered pairs* (a, b), with a element of A and b element of B; in symbols,

$$A \times B = \{(a, b) : a \in A \text{ and } b \in B\}$$

Two pairs (a, b) and (a', b') are equal if and only if $a = a'$ and $b = b'$. The product $A \times B$ is not equal to the product $B \times A$, unless A is equal to B. The Cartesian product $A \times A$ is also denoted A^2 and is the set of ordered couples of elements of A. If $A = \{2, 4, 6, 8\}$,, then $A^2 = \{(2, 2), (2, 4), (2, 6), (2, 8), (4, 2), (4, 4), (4, 6), (4, 8), (6, 2), (6, 4), (6, 6), (6, 8), (8, 2), (8, 4), (8, 6), (8, 8)\}$.

The cartesian product extends to any number of sets. For example, given the sets $A_1, A_2, \ldots, A_n$, with $n \geq 2$, the Cartesian product $A_1 \times A_2 \times \ldots \times A_n$ is the set of the n-tuples $(a_1, a_2, \ldots, a_n)$, with $a_1 \in A_1, a_2 \in A_2, \ldots, a_n \in A_n$.. The n-tuples are ordered, by definition. The element a_1 is called the *first component* of the n-tuple $(a_1, a_2, \ldots, a_n)$, a_2 the *second component*, and so on. Two n-tuples $(a_1, a_2, \ldots, a_n)$, $(b_1, b_2, \ldots, b_n)$ are defined *equal*, and we write $(a_1, a_2, \ldots, a_n) = (b_1, b_2, \ldots, b_n)$, if the equality of the components of the same place is verified, that is, $a_1 = b_1, a_2 = b_2, \ldots, a_n = b_n$. The cartesian product $A \times A \times \ldots \times A$ of n sets, each equal to A, can be denoted A^n; it is the set of the ordered n-tuples of elements of A. For example, if $A = \{1, 2, 3\}$, the triples $(1, 1, 3), (1, 3, 1), (3, 1, 1)$ are distinct elements of A^3. Remark that 2-tuples and 3-tuples are usually called *couples* and *triples*, respectively.

Again, for example, let us consider the set $F = \{0, 1\}$, the 4-tuples $(1, 0, 0, 0), (0, 1, 0, 0), (0, 0, 1, 0), (0, 0, 0, 1)$ are two by two distinct elements of F^4.

If $n \geq 1$, a choice $a_1, a_2, \ldots, a_n$ of elements not necessarily distinct of the set A is called a *system* of n elements of A, or a unordered n-tuple of elements of A, and it is denoted $[a_1, a_2, \ldots, a_n]$. Thus, the symbols $[1, 0, 0, 0], [0, 1, 0, 0], [0, 0, 1, 0], [0, 0, 0, 1]$ identify the same system of four elements of F. Then, with the elements of the system $[1, 0, 0, 0]$ we construct four 4-tuples of F^4 and a single system of four elements of F.

Let us define the concept of relation between two sets. If A and B are non-empty sets, a *relation*, or *correspondence*, between A and B, taken in this order, is defined as a subset of the cartesian product $A \times B$. Then, denoted R the relation between A and B, from the definition it follows $R \subseteq A \times B$.

Example 3.1 The weather report relates regions with atmospheric perturbations, the minimum or maximum temperature in a day, and so on. The couples (London, 23), (Rome, 27), (Madrid, 33), (Paris, 25) are elements of the relation R_1 between a certain set A of European capitals and a set B of numbers, say between -60 and $+60$, which associates to each city of A the number that expresses the maximum temperature in centigrade degrees, measured in a precise day.

It makes sense to define a relation between A and A, which is a relation between a set and itself, i. e., a relation in $A \times A$. A relation in $A \times A$ is called a *relation defined in A*, or a *binary relation* in A.

Example 3.2 The set of the couples $(n, n + 1)$ whose first element is the natural number n and the second is the next defines a binary relation in the set of natural numbers; $(5, 6)$ is a couple in the relation, while $(6, 5)$ does not belong to the relation.

If the couple (a, b) belongs to the relation R, we say that a is in the relation R with b, and then we use the notation $a\text{R}b$. If it is not true that a is in the relation R with b, we write $non(a\text{R}b)$.

Example 3.3 Suppose real numbers represent amounts of money: positive numbers stand for income, negative numbers stand for expenses. In the set of real numbers we define the binary relation R_2 between income and consumption in this way: income i is related to consumption c, $i\,R_2\,c$ if the two quantities i and c verify the equality $c = 60 + 0.8i$: the equality expresses a link between i and c, establishing how one of the quantities expressed in euros varies as the other varies. Therefore, if in the equality we set wage $= i = 1000$, we find the consumption $c = 60 + (0.8)1000 = 860$; and if $i = 300$, then $c = 300$. Hence the couples $(1000, 860)$ and $(300, 300)$ are in the relation R_2. Consumption $c = 60 + 0.8 \times 0 = 60$ corresponds to zero income, which means that an individual needs a guaranteed minimum to live. The couple $(300, 300)$ tells us that an income of 300 corresponds to a consumption of 300, i. e., what is earned all is spent.

Example 3.4 Fixed a point P in the plane, a binary relation R_3 in the set of the lines of the plane is formed by the couples (a, a') of lines, where a' is the line passing through P and parallel to a.

Example 3.5 Let A be the set of cars sold in 2020 in Europe. Let group the elements of A by manufacturer, model and engine displacement. The following binary relations in A are defined: two hatchback cars a and b are in the relation R_H, $a\text{R}_H b$; the cars c and d having the same engine displacement 1.8 L are in the relation $R_{1.8}$, $c\text{R}_{1.8}d$; the sedan cars e and f are in the relation R_S, $e\text{R}_S f$. Of course, it makes sense to consider the intersection relations $R_{F \cap 1.8}$, $R_{F \cap S}$, $R_{1.8 \cap S}$, $R_{F \cap 1.8 \cap S}$.

3.3 Binary Relations

A binary relation R in a set A can satisfy particular properties (Fishburn 1970):

P1. if $a\text{R}a$, for every $a \in A$, then the relation is called *reflexive*;

P2. if $non(a\text{R}a)$, for every $a \in A$, then the relation is called *irreflexive*;

P3. if $a\text{R}b$ implies $b\text{R}a$, for every $a, b \in A$, then R is called *symmetric*;

P4. if, for every $a, b, c \in A$, aRb and bRc imply aRc, then R is called *transitive*;

P5. if, for every $a, b, c \in A$, *non*(aRb) and *non*(bRc) imply *non*(aRc), then R is called *negatively transitive*;

P6. if, for every $a, b \in A$, aRb implies *non*(bRa), then R is called *asymmetric*;

P7. if, for every $a, b \in A$, aRb and bRa imply $a = b$, then R is called *antisymmetric*;

P8. if, for every $a, b \in A$, aRb or bRa, then R is called *total*, or *connected*, or *complete* (this means that either aRb or bRa or both are true);

P9. if aRb or bRa, for every a different from b, then the relation R is called *weakly connected*.

Example 3.6 The binary relation R_2, between income and consumption, is irreflective and antisymmetric (in fact, when $iR_2 c$ and $cR_2 i$ happens, and this occurs only when $300R_2300$, we have $i = c = 300$), is not transitive, not reflexive (only in the case $300R_2300$ it occurs $i R_2 c$ and $i = c$), not symmetric (e.g., $1000R_2860$, but it is not true that $860R_21000$).

Example 3.7 The parallelism relation between lines of a plane is a reflexive relation (each line is parallel to itself), symmetric (if the line r is parallel to the line s, then s is parallel to r), transitive (if r is parallel to s and if s is parallel to the line t, then r is parallel to t) and is not antisymmetric.

Example 3.8 The perpendicularity relation between lines of the plane is not reflexive, it is symmetric, it is not transitive, it is not antisymmetric, nor negatively transitive (because if r is not perpendicular to s and s is not perpendicular to t, it does not mean that r is not perpendicular to t) (Fig. 3.1).

Example 3.9 The implication is a reflexive and transitive relation in a suitable set of propositions.

Theorem 3.1 *If a binary relation R is irreflexive and transitive, then it is asymmetric.*

Proof If by contradiction R were not asymmetric, then the proposition P6 should not occur, i. e. from being aRb it should not follow *non*(bRa); this means if aRb, then bRa, and therefore for P4, aRa, which contradicts the hypothesis P2. So, property P6 follows from P2 and P4. □

Fig. 3.1 Parallelism and perpendicularity relations

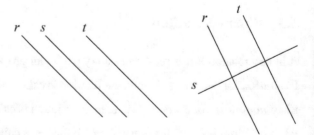

A binary relation cannot be simultaneously reflexive and irreflexive nor symmetric and asymmetric.

Theorem 3.2 *If the binary relation R in A has the properties P6 asymmetric and P5 negatively transitive, then R fulfills the transitive property.*

Proof We show that if, for every $a, b, c \in A$, aRb and bRc, then aRc. In fact, let us suppose by contradiction, that the thesis is not verified:

$$non(aRc) \tag{3.1}$$

The hypothesis bRc and the asymmetry of R imply

$$non\,(cRb) \tag{3.2}$$

From (3.1) and (3.2) it follows, from negatively transitivity of R, $non(aRb)$, against the hypothesis $aR\,b$. We then conclude that R is transitive. □

3.3.1 Orderings

A binary relation R in the set A, antisymmetric and negatively transitive is called a (relation of) *weak order*; a binary relation R in A, antisymmetric, negatively transitive and weakly connected is called a (relation of) *strict order*; a binary relation R in A, reflexive, transitive and antisymmetric is called an *order relation*, or an *ordering* of A. Let us remark that weak order, strict order and order relations are generically named *orderings*.

Example 3.10 The relation R in the set of real numbers: $R = \{(a, b) : a < b$, with a, b real numbers$\}$, for which, instead of the notation aRb, we employ the usual $a < b$, is a weak order because it is asymmetric (if $a < b$, then it is not true that $b < a$), and negatively transitive (for example, if it is not true that $5 < 4$ and $4 < 3$, then it is not true that $5 < 3$). The relation R, being also weakly connected, is also a strict order because if $a \neq b$, then $a < b$ or $b < a$. The notation $a < b$ is equivalent to $b > a$.

Example 3.11 If a and b are real numbers, the notation $a \leq b$ means $a < b$ or $a = b$. The relation $R = \{(a, b) : a \leq b\}$, in which, instead of aRb we write $a \leq b$, is an order relation; in fact, it is reflexive ($a \leq a$, for every a) transitive (e. g., if $5 \leq 7$ and $7 \leq 8$, then $5 \leq 8$) and antisymmetric (if $a \leq b$ and $b \leq a$, then $a = b$). The notation $a \leq b$ is equivalent to $b \geq a$.

3.3.2 The Power Set

Consider a non-empty set A having $n \in \mathbf{N}$ elements. The *power set*, or *the set of the parts*, of A is, by definition, the set whose elements are all the subsets of A (Sect. 1.2). The power set of A is denoted by $p(A)$. Let's recall (Sect. 1.2 (P)) that the empty set $\emptyset$ belongs to $p(A)$, whatever A is. For example, the set of parts of the set $A = \{a, b, c\}$ is

$$p(A) = \{\emptyset, \{a\}, \{b\}, \{c\}, \{a, b\}, \{a, c\}, \{b, c\}, \{a, b, c\}\}$$

Observe that the number of the elements of the set of the parts of A is 8. If a set contains $n \in \mathbf{N}$ elements, then the power set contains 2^n elements.

Let us consider two subsets A_1, A_2 of the set A, and define a relation between them in this way: we say that A_1 is in the relation R with A_2 if A_1 is contained in A_2. This relation is called the *relation of inclusion* $\subseteq$, and it is a binary relation in the set $p(A)$ of the parts of A. The relation of inclusion is an order relation since it is reflexive (every set is contained in itself), transitive (if $B \subseteq C$ and $C \subseteq D$, then $B \subseteq D$, for all B, C, D subsets of A) and antisymmetric (if $B \subseteq C$ and $C \subseteq B$, then $B = C$,, for every B and C subsets of A).

3.3.3 Total Order

Let R be a binary relation in the set A, precisely let R be one of the orderings defined in (Sect. 3.3.1). Then, if aRb, we say that "a precedes b", or "b follows a", in the relation R. Furthermore, two elements a, b of A are said to be *comparable* if aRb or bRa.

If R is reflexive, transitive and antisymmetric, i. e., R is an *order relation* and furthermore if R satisfies property P8, i. e., R is total and therefore, fixed any two elements a and b in A, they are comparable, then R is defined as a *relation of total order* in A.

For example, whatever the real numbers a, b are, they are comparable in the usual ordering and therefore the ordering $\leq$ in the set of real numbers is a total order in the set of real numbers.

We observe that it is not certain that, taken two subsets of the set A they are comparable in the relation of inclusion; in fact, two subsets are comparable in the relation of inclusion if and only if one of them is contained in the other.

3.4 Preferences

Each individual establishes preferences between objects, actions, individuals, situations, etc. .... The preference is usually established by comparing two elements. The individual places a couple in relation and the relation is the expression of his/her tastes, of his/her convenience, of how he/she sees the world: each individual has his own system of preferences. If element b is preferred to element a, we write $a \angle b$.

We can suppose that *preference* is a weak order relation; in fact, asymmetry seems an obvious requirement for preference: if I prefer b to a, I cannot simultaneously prefer a to b. As shown above, asymmetry and negative transitivity imply transitivity, which seems a reasonable criterion for the consistency of an individual's preferences: if I prefer an interesting job in a place thirty-five minutes from home, to a job less interesting, but in a more easily accessible location and if I prefer this to a job, even interesting, that forces me to an exhausting commute, then I should prefer the first to the third, following common sense.

However, the transitivity of preferences, is a non-trivial and much debated topic. A simple example is enough to introduce the seed of doubt: I prefer milk to tea, tea to coffee and coffee to milk. These preferences are not transitive. To restore the transitivity, I should prefer milk over coffee. Yet, I am not inconsistent as this is the taste I possess. These preferences are said to form a *cycle*.

Personal preferences also transcend transitivity. If Becky loves Tom and Tom loves Diletta, then Becky doesn't necessarily love Diletta, nor Diletta loves Tom.

3.4.1 Indifference

Next to the concept of preference there is that of *indifference*, as the absence of preference. Indifference has various origins.

The individual does not perceive any difference between object a and object b: she may wish to get one rather than the other, or vice versa, indifferently. Or she is uncertain about which of the two elements a, b, to prefer, or does not feel like it because she has insufficient information and therefore judges a and b to be equally desirable, or undesirable.

Again, the individual expresses indifference towards a and b if she considers these elements not comparable. In each of these situations, in which the individual affirms neither $a \angle b$, nor $b \angle a$, the notation $a \sim b$ is used.

3.5 Equivalence Relations

A binary relation R in A with reflexive, symmetric and transitive properties is called (a relation of) *equivalence*.

Example 3.12 The parallelism relation P in the set of lines of the plane, where aPb means "line a is parallel to line b", is an equivalence relation between the lines of the plane. The relation P is reflexive, a is parallel to a; the relation P is symmetric, if a is parallel to b, then b is parallel to a; the relation P is transitive, if a is parallel to b, and b is parallel to the line c, then a is parallel to c.

Example 3.13 The binary relation S, in the population of Northampton, where aSb means "a has the same high school qualification as b", is an equivalence relation.

Example 3.14 The relations described in the Example 3.5 are equivalences.

Sometimes the symbol $a \approx b$ instead of aRb is used when R is an equivalence relation. We will say that a and b are *equivalent* or also that a is *equivalent* to b, in (or, with respect to) the relation $\approx$.

If $\approx$ is an equivalence relation in A and x is an element of A, we call *equivalence class* (with respect to the relation $\approx$) determined by x, or *equivalence class of x*, the set of the elements of A that are equivalent to x. The equivalence class of x is denoted by $[x]$. In symbols:

$$[x] = \{a \in A : a \approx x\}$$

Example 3.15 If P denotes the parallelism relation in the set of the lines of the plane and r is a line, the equivalence class $[r]$ is the set of the lines that lie in the plane and are parallel to r, including the line r.

Example 3.16 Let us consider the individuals of the same age in a population. We have thus defined a relation C that is an equivalence relation. Relation C divides the population into groups of people of the same age: the kids who are not yet one year old, those who are one year (and not yet two), those who are two, and so on, up to individuals of one hundred and fifty years, hopefully. How many equivalence classes are there with respect to C? There are at most one hundred and fifty-one.

Theorem 3.3 *If b belongs to the equivalence class $[a]$ with respect to the relation $\approx$, then $[a] = [b]$ (i. e., the equivalence class of a is equal to the equivalence class of b).*

Proof If we prove that the set $[a]$ is contained in $[b]$ and also $[b]$ is contained in $[a]$, then we will have proved the theorem (see Sect. 1.2). We begin by proving that $[a] \subseteq [b]$, i. e., if an element belongs to the class $[a]$, then it belongs to $[b]$. Indeed, if $c \in [a]$, then $c \approx a \approx b$ and therefore by the transitivity of the relation $\approx$, we have $c \approx b$, i. e., $c \in [b]$. Similarly, we show that if an element belongs to $[b]$, then it is also in $[a]$.   □

Theorem 3.4 *If two classes are not equal, then they are disjoint.*

Proof We prove that if two classes are not equal, then they do not have elements in common. Let c be an element of the class $[a]$ that does not belong to the class $[b]$. If, by contradiction, there were an element d in common to the classes $[a]$ and $[b]$, then $c \approx a \approx d \approx b$ and by transitivity c would be equivalent to b, which contradicts the hypothesis that c is not in $[b]$. $\qquad\qquad\qquad\qquad\qquad\qquad\qquad\qquad\qquad\qquad\qquad\square$

3.5.1 Partitions of a Set

Definition 3.1 Given a non-empty set A whose subsets $A_1, A_2, \ldots, A_m$ are such that:

1. their union is equal to A $\quad A = A_1 \cup A_2 \cup \ldots \cup A_m$;
2. the intersection of any two distinct subsets is empty, $A_i \cap A_j = \emptyset$, for every i, j $= 1, 2, \ldots, m$ and $i \neq j$;
   then we say that the sets $A_1, A_2, \ldots, A_m$ form a *partition* of A, and each set A_i is called *an element of the partition*.

From Theorems 3.3 and 3.4 we deduce:

Theorem 3.5 *Each partition $A_1, A_2, \ldots, A_m$ of a non-empty set A, defines an equivalence relation in A, in which any two elements of A_i are said to be equivalent. Conversely, each equivalence relation in A defines a partition of A and the elements of the partition are the equivalence classes.*

3.5.2 Remainder Classes

We saw some properties (Sect. 2.1) of the Euclidean division. In particular, we found that the Euclidean division with divisor 3 operates a partition of the set N_0 into classes $[0], [1], [2]$, of the numbers that divided by 3 have remainder 0, 1, 2, respectively.

Then, by the previous theorem, Euclidean division defines an equivalence relation in N_0. Precisely we say *equivalent modulo* 3 two natural numbers which, divided by 3, have the same remainder; the classes are called *modulo* 3 *remainder classes*.

The remainder classes modulo 4, 5,..., n are defined similarly. Then the numbers 61 and 96 belong to the same remainder class modulo 7, because $61 = 8 \times 7 + 5$ and $96 = 13 \times 7 + 5$. If two non-negative integers s and t belong to the same remainder class modulo n, we write $s \equiv t \pmod{n}$. Therefore, $61 \equiv 96 \pmod 7$; i. e., 61 and 96 belong to the remainder class $[5]$ modulo 7.

References

Buchanan, M.: Nexus: Small Worlds and the Groundbreaking Science of Networks. W. W. Norton &
  Co., New York and London (2002)
Fishburn, P.C.: Utility Theory for Decision Making. John Wiley & Sons, New York and London
  (1970)
Guare, J.: Six Degrees of Separation. Random House, New York (1990)

Chapter 4
Euclidean Geometry

4.1 Introduction

That the geometry of the space in which we live were the only conceivable geometry was universally accepted until the early nineteenth century. Geometry, i. e., *Euclidean geometry*, developed from composite, sensory and psychological experiences subjected to processes of abstraction and schematization.

Euclidean geometry is a tactile geometry: by touching the top of a table, or its edge, we develop sensations that lead us back to the plane, to a straight line, to a polygon, to the point. The geometry of the eye, introduced and experimented by Renaissance painters, Piero della Francesca, Mantegna, Albrecht Dürer, studied by philosophers and mathematicians, Blaise Pascal, Girard Desargues, Jakob Steiner, called *projective geometry*, has among its objects even infinity, into which the eye wants to penetrate.

Euclidean geometry and projective geometry are each founded on its own system of axioms which underlies the properties and theorems that form the body of the resulting geometry.

Projective geometry does not admit the axiom of the unique parallel and admits that two coplanar lines have a point, possibly at infinity, in common.

Among the mathematicians who introduced non-Euclidean geometries we mention Carl Friedrich Gauss, Nikolai Ivanovich Lobačewskij and János Bolyai: from their studies the conviction that there are infinite geometries. We will study Euclidean geometry in the most interesting intrinsic synthetic aspects and for the analytical purposes. The study of other geometries has important theoretical and practical implications (Beutelspacher and Rosenbaum 1998).

© The Author(s), under exclusive license to Springer Nature Switzerland AG 2023   49
A. G. S. Ventre, *Calculus and Linear Algebra*,
https://doi.org/10.1007/978-3-031-20549-1_4

4.2   First Axioms

A set of points in the line, or plane, or space, is called a *geometric figure*, or simply a *figure*. Obviously, the nomenclature and the operations defined between the sets are valid for the figures because figures are particular sets. Euclidean geometry, also called *elementary geometry*, states properties and relations between figures in the space. Euclidean geometry does not provide "direct" definitions for the concepts of point, line and plane. In other words, Euclidean geometry does not tell us "what is a point, a line, a plane", but assumes some propositions, the *axioms* or *postulates*, which are reasonable properties that satisfy our intuition, are accepted as true, and do not need proofs. The postulates define relations between objects called "points", "lines" and "planes".

It is customary to denote points with capital letters, A, B, C, ..., lines with lower-case letters *a*, *b*, *c*, ..., planes with lowercase Greek letters, α, β, γ, ... When we say "two points", "three points", we usually mean distinct points; the same goes for lines and planes.

Some axioms of Euclidean geometry are the following:

- given two distinct points A, B, there is one and only one line, called the line AB, passing through the two points;
- given three distinct points A, B, C not belonging to the same line, there is one and only one plane that contains them, which is called the plane ABC;
- the line passing through two distinct points of a plane is all contained in the plane.

For example, from the axioms stated above we deduce the properties:

1. if the point A does not belong to the line *a*, then there is one and only one plane that contains both the point and the line, the plane *a*A;
2. if two distinct lines *a*, *b* have only one point in common, then there is one and only one plane that contains both lines, the plane *ab*.

Historical background. Ptolemy I, who reigned over Egypt in the third century B.C., founded in Alexandria the Museum, a school of absolute excellence. He called eminent scholars as teachers and, among them, the author of the most successful mathematics book ever written: the author was Euclid of Alexandria and the book *Elements.* Euclid's ideas and teachings have not known the wear and tear of time. The Elements have been, since the beginning, a constant reference for scholars of every civilization and a scholastic text that has remained intact in the original version, translated into all languages, and considered for many centuries not scratched as a prototype of rational science. The proofs of theorems in the Elements became a model for all rigorous demonstrations in mathematics, the excellence that a proof should aspire to.

4.3 The Axiomatic Method

For over two millennia Euclidean geometry has crossed, albeit between processes of growth and re-foundations, entire civilizations, attracts scholars, is a foundation for applied sciences and an important part of modern education systems. Euclid's geometry is based on the axioms, or postulates, on which theorems are developed: this process is known as the *axiomatic method*. Rejecting or altering only one Euclid's axiom involves an overall change of the theory. This is the case of the *axiom of the unique parallel*, which states that through any point not belonging to a given line *one and only one* line can be drawn parallel to the given line. Every alteration of this axiom gives rise to the foundation of as many geometries such that the objects as points, lines and planes have models far from the expected and familiar shapes.

From the end of the seventeenth century, Euclid's work even managed to spark a heated debate among scholars, not all mathematicians. The question was: can the axiom *of the unique parallel* be deduced from the other Euclidean axioms? In other words, is the axiom of the unique parallel *independent* of the other axioms? The first, in chronological order, to tackle the question was a brilliant Jesuit, Giovanni Girolamo Saccheri (1667–1733), professor of philosophy and mathematics. He tried to prove by contradiction the dependence of the axiom of the unique parallel on the other axioms. But the proof contained an error and Saccheri's attempt was wrecked. After this effort, the problem ran out of rest and dozens of scholars tried to demonstrate the axiom of parallelism. Among them: Gauss, his Hungarian friend Farkas Bólyai, his son János Bólyai (1802–1860), Nikolai Lobačewskij (1793–1856), Bernhard Riemann (1826–1866), Eugenio Beltrami (1835–1900). These studies and those of other enthusiasts led to the definition of geometries and spaces called non-Euclidean.

4.3.1 Further Axioms of Euclidean Geometry

Let us consider a line r and a point P of it. If we remove the point P from the line we obtain on the line two disjoint sets of points which, joined to P, give back the line r. Each of the two disjoint sets is called an *open half-line* and the point P the *origin* of each half-line. The point P joined to one of the two open half-line is a geometric figure called the *half-line of origin* P. Each of the two half-lines of origin P, lying on the line r, is said the half-line *opposite* to the other.

Two distinct lines r and s of space are called *coplanar* if there is a plane that contains them. The following axiom is postulated.

If two lines r and s are coplanar one of the following alternatives occurs:

(a) the lines have one and only one point in common, $r \cap s = P$;
(b) the lines are coincident, $r = s$;
(c) the lines have no point in common, $r \cap s = \emptyset$.

Fig. 4.1 Half-planes

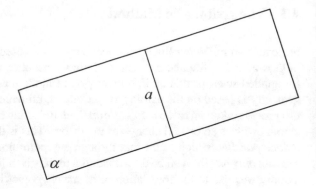

In case (b) or (c) the lines r and s are said to be *parallel*, indicated $r//s$. In particular, in case (c) the lines are said to be properly *parallel*. In case (a) the lines are said to be *incident* in the common point.

Given a plane α and a line a in the plane, if the line is removed from the plane, two disjoint parts of the plane are identified, which joined to a, give back the plane α. Each of the two parts of the plane, in union with the line a, is named *half-plane* and the line a is called the *origin* of each half-plane (Fig. 4.1).

The following axioms related to the half-planes of α, having common origin the line a, are admitted:

- a point, not belonging to the line a, belongs to only one of the two half-planes;
- two points of the same half-plane are the extremes of a segment entirely contained in the half-plane;
- two points not in the line a and belonging to different half-planes are extremes of a segment that meets the line a in a single point.

If a plane α is removed from the space, two disjoint parts in the space are identified; the union of each part with the plane α is called an *half-space* and the plane α is named the *origin* of each half-space.

4.4 The Refoundation of Geometry

We have seen that Euclid's geometry lays on a system of axioms, propositions assumed to be true. This approach is valid and is still shared and used today in the construction of mathematics. In 1898, David Hilbert published a book, *The Foundations of Geometry* (Hilbert 1968), which immediately achieved success and was translated into many languages. Geometry had made considerable progresses, especially in the nineteenth century, and Hilbert gave it a formal set-up. In fact, although Euclid's work is a deductive structure, some approximation in the enunciation of some concepts could be found in it. Hilbert too considered the point, the line and the plane as undefined objects, and formulated a set of propositions, known as Hilbert's axioms.

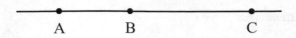

Fig. 4.2 B is between A and C

Let us state two among the Hilbert's axioms, give an intuitive description of them and deduce a theorem.

Axiom 4.1 Fixed any two points A and C, there exists at least one point B on the line AC, such that B is between A and C.

The notion of "betweenness" (Fig. 4.2) is related to the ordering of points on a line. Axiom 4.1 expresses in geometric terms the concept of *density* (Sect. 2.5.1): between two points of a line we find another point of the line, therefore, by repeating the operation, between two points of the line there are infinite points of the line.

Axiom 4.2 Let A, B, C be three non-aligned points and let *a* be a line of the plane ABC that does not pass through any of the points A, B, C. If the line *a* passes through a point of the segment AB, then it passes through a point of the segment AC, or through a point of the segment BC.

Intuitively, Axiom 4.2 states that if a line enters the interior of a triangle, then it also exits (Fig. 4.3).

Theorem 4.1 *If A, B, C are three points on a line, then one of them lies between the other two.*

Proof Suppose that A is not between B and C and that C is not between A and B. We want to prove that B is between A and C. We connect a point D, which is not on the line AC, with B; for Axiom 4.1 we take a point G on the line BD such that D is between G and B. We apply Axiom 4.2 to the triangle BCG and to the line AD: the lines AD and CG intersect at a point E between C and G. Similarly, the lines CD and AG intersect at a point F between A and G. We now apply Axiom 4.2 to the triangle AEG and to the line CF: we obtain that D is between A and E, and again applying Axiom 4.2 to triangle AEC and line BG, we consequently infer that B is between A and C, that is what we claimed (Fig. 4.4).□

Fig. 4.3 About Axiom 4.2

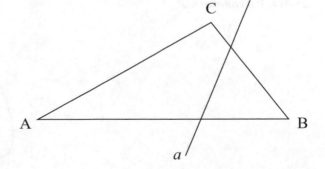

Fig. 4.4 About Theorem 4.1

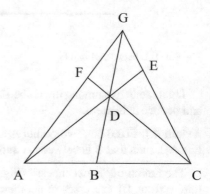

4.5 Geometric Figures

A geometric figure F is, by definition, any non-empty set of points in a plane or in the space.

4.5.1 Convex and Concave Figures

A figure F is said to be a *convex figure* if the line segment joining two points however fixed in F is all contained in F (Fig. 4.5). A non convex figure is called a *concave figure* (Fig. 4.6).

Examples of convex figures are a line segment, a line, a half-line, the plane, a half-plane, the whole space, the circle, any triangle, the cube understood as a solid body, part of the space (Fig. 4.7).

Concave figures are the circumference, an arc of circumference, the spherical surface, the surface of the cube, a banana (Fig. 4.8).

There are convex quadrilaterals and concave quadrilaterals (Fig. 4.9).

Fig. 4.5 Convex figure F,
the segment joining any two
points A, B is contained in F

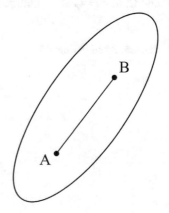

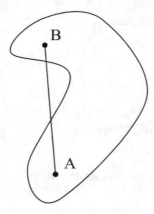

Fig. 4.6 Concave figure *F*

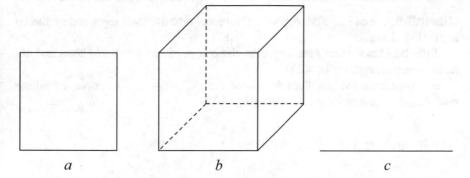

a   *b*   *c*

Fig. 4.7 **a** square, **b** full cube, **c** segment are convex figures

Fig. 4.8 Concave figures

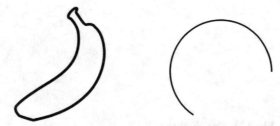

4.5.2 Angles

Two half-lines, *r* and *s*, having common origin P, and hence being coplanar, determine two regions of the plane called *angles* both having *vertex* P and *sides r* and *s*. Both angles are denoted $\widehat{rs}$, or $\widehat{rPs}$, and are such that:

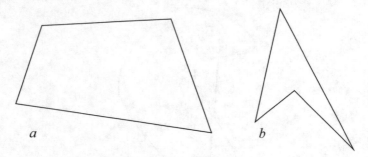

Fig. 4.9 convex quadrilateral (**a**), concave quadrilateral (**b**)

- each region contains the half-lines r and s;
- the two regions have in common only the half-lines r and s;
- the union of the two regions is equal to the plane that contains the two half-lines.

If the half-lines r and s are not opposite, they determine a convex angle and a concave angle (Fig. 4.10).

If the half-lines r and s are opposite, the angles $\widehat{rs}$ are two half-planes and are called *straight angles* (Fig. 4.11).

Two superimposed half-lines determine the *null angle* and an angle, called the *round angle*, equal to the entire plane (Fig. 4.12).

Fig. 4.10 Angles $\widehat{rs}$

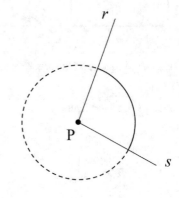

Fig. 4.11 Straight angles

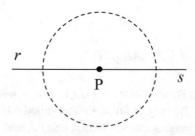

Fig. 4.12 Round angle $\hat{rs}$

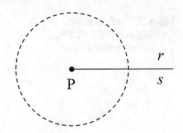

Congruence relation

The *congruence relation* is defined between angles: two angles are *congruent* if there is a movement that leads one of them to coincide with the other.

Remark 4.1 Although improperly, about two congruent angles it is sometimes said that they are "equal". Similarly, we improperly speak about figures that are equal to each other, instead of being "congruent".

Two coplanar and non-parallel lines, r and s, form two pairs of congruent angles. These angles are opposite to each other at the common vertex and are called (*vertically*) *opposite angles* (Fig. 4.13).

Skew lines

Two non-coplanar lines are said to be skew lines (Figs. 4.14 and 4.15).

Fig. 4.13 Coplanar non-parallel lines r and s

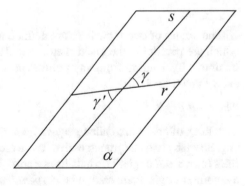

Fig. 4.14 Skew lines

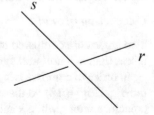

Fig. 4.15 *a//c, a* and *b* coplanar and incident, *d//b, d* and *a* skew, *c* and *d* skew

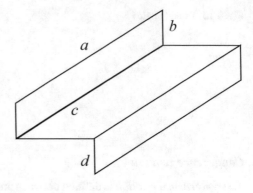

Fig. 4.16 The angles of the coplanar lines *r'* and *s'* define the angles of the skew lines *r* and *s*

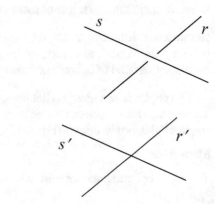

The *angles* of two skew lines are defined as the angles of any two coplanar lines, which are parallel to the given skew lines. The angles of two given skew lines may be drawn by sending from any point P in the space the parallel lines to the given skew lines (Fig. 4.16).

The right angle

Two lines of space are called *perpendicular*, or *orthogonal*, if they determine four angles two by two congruent; each of these angles is called a *right angle*. Two parallel lines form a *null* angle. Two half-lines *r* and *s* lying on perpendicular lines and having a common origin P are said to be *perpendicular*; the half-lines *r* and *s* determine a convex angle, which is called a *right angle*, and a concave angle (Fig. 4.17).

Comparison of angles

Two angles are compared to each other in magnitude. We accept that the null angle is less than any non-null angle.

In order to compare two non-null angles, the angle $\gamma = \widehat{rs}$ and the angle $\delta = \widehat{tu}$, move γ (Fig. 4.18) so that the vertices of γ and δ coincide and a side of γ, say *r*, coincides with *t* while *s* and *u* fall on the same half-plane of origin *r*. Then one of the following alternatives occurs:

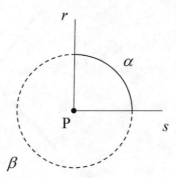

Fig. 4.17 Right angle α, region β concave

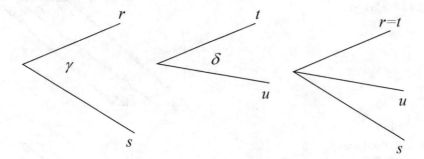

Fig. 4.18 γ is greater than δ

- the angle γ is congruent to δ;
- the angle γ is a proper subset of δ, then γ is said to be less than δ;
- the angle γ properly includes δ, then γ is said to be greater than δ.

4.5.3 *Relations Between Lines and Planes*

If *r* and α are a line and a plane of the space, respectively, then one of the following circumstances occurs:

(a) the line *r* and the plane α have only one point in common (Fig. 4.19);
(b) the line *r* lies on the plane α (Fig. 4.20);
(c) the line *r* and the plane α have no point in common (Fig. 4.21).

In case (a) the line and the plane are said to be *incident* at the common point; in cases (b) and (c) the line and the plane are said to be *parallel* to each other (*properly parallel*, in the case (c)).

Fig. 4.19 *r* and α have only
one point in common

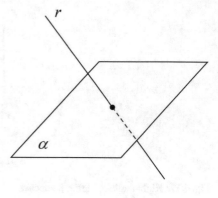

Fig. 4.20 *r* lies on α

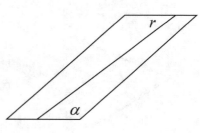

Fig. 4.21 *r* and α are
properly parallel

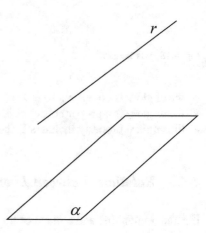

Theorem 4.2 *The lines of the space perpendicular to a line at one of its points lie in the same plane and each line of the plane passing through that point is a line perpendicular to the given line.*

The plane α containing the lines passing through a point P of the line *r*, and *perpendicular* to *r*, is called *the perpendicular* (or *orthogonal*) *plane* to *r* at P (Fig. 4.22).

By the definition of angle of two skew lines, if a line *r* and a plane α are perpendicular, each line of the plane is perpendicular to the line *r* (Fig. 4.23).

Fig. 4.22 The line *r* is perpendicular to the lines *a* and *b* passing through the point P of *r*. The line *c* in the plane *ab* and passing through P is orthogonal to the line *r*

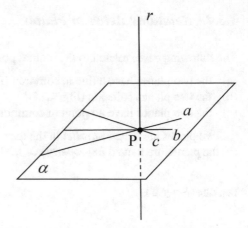

Fig. 4.23 The line *r* is perpendicular to the lines *a* and *b* passing through the point P of *r*. The line *d* in the plane *ab* is perpendicular to the line *r*

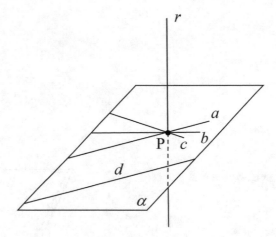

Theorem 4.3 *Given in the space a point and a line, a single plane exists which passes through the point and is perpendicular to the line.*

Theorem 4.4 *Given in the space a point and a plane, a single line exists which passes through the point and is perpendicular to the plane.*

The common point to a line and a plane that are perpendicular to each other is called the *foot of the perpendicular line* to the plane.

The *distance* of a point A and a plane α is defined as the length of the segment AB, being B the foot of the perpendicular line to the plane passing through A.

If a line and a plane are parallel, the distance of the line from the plane is defined as the distance of any point of the line from the plane.

4.5.4 Relations Between Planes

The following cases related to the mutual position of two planes α and β occur:

(a) the two planes have a line in common (Fig. 4.24);
(b) the two planes coincide (Fig. 4.25);
(c) the two planes have no point in common (Fig. 4.26).

   Two planes are called *parallel* in the cases (b) or (c), in symbols, $\alpha//\beta$; ; in case
(b) the planes are named *improperly parallel*, in case (c) *properly parallel*.

Fig. 4.24 $\alpha \cap \beta = r$

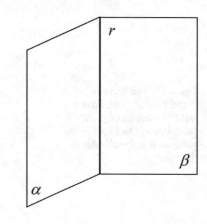

Fig. 4.25 $\alpha = \beta$

Fig. 4.26 $\alpha//\beta$ and $\alpha \neq \beta$

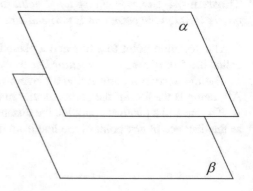

4.5.5 *Projections*

Let us consider the non-parallel lines r and d in the plane α and let P be a point of the plane. The *projection* of P onto r, with respect to the direction of the line d, is defined as the point P' of r belonging to the parallel line to d passing through P (Fig. 4.27).

In particular, the projection of P onto r with respect to the direction of a line d perpendicular to the line r is called the *orthogonal projection* of P onto the line r.

The orthogonal projection of the semi-circumference onto its diameter is the same diameter (Fig. 4.28).

The orthogonal projection of the circumference onto the line a is the segment AB of the line a; the orthogonal projection of the circumference onto the line b is the segment DE of the line b; the orthogonal projection of the arc PQ onto the segment PQ coincides with the segment PQ (Fig. 4.29).

Let a plane α *and* a line d be given with d and α non parallel. The *projection of a point* P of the space onto α, with respect to the direction of the line d, is defined as the point P' of α belonging to the line passing through P and parallel to d (Fig. 4.30). In particular, if the direction of d is perpendicular to the plane α, the point P' is called the *orthogonal projection* of the point P onto the plane.

The projection onto the plane α of a figure F of the space with respect to the direction of the line d non parallel to α is, by definition, the set F' of the projections of the points of F onto α with respect to the direction of d (Fig. 4.31).

Fig. 4.27 P' projection of P onto r w.r. to direction d

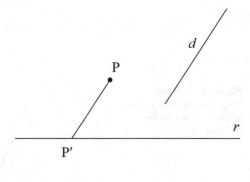

Fig. 4.28 The orthogonal projection of the semi-circumference onto its diameter

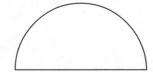

Fig. 4.29 AB is the orthogonal projection of the circumference onto the line *a*; DE is the orthogonal projection of the circumference onto the line *b*

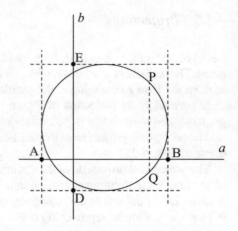

Fig. 4.30 P′ is the projection of P onto α w. r. to the direction *d*

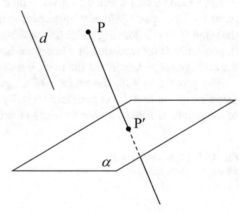

Fig. 4.31 F′ is the projection onto the plane α of figure F w. r. to the direction of *d*

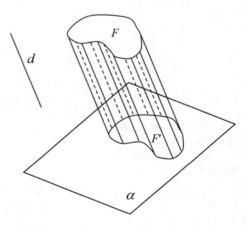

Fig. 4.32 Angle of the line r
and the plane α

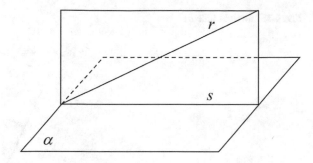

4.5.6 The Angle of a Line and a Plane

Let us consider a line r and a plane α in the space. If the line and the plane are not
perpendicular, the orthogonal projection of the line r onto the plane is a line s: the
acute angle $\widehat{rs}$ is called the *angle of the line r and the plane* α (Fig. 4.32). If the line
and the plane are perpendicular the angle of the line and any line in the plane is right:
then the line and the plane are perpendicular and their angle is the right angle.

4.5.7 Dihedrals

Two half-planes α and β having the line r as common origin determine two regions in
the space, which contain α and β and have the half-planes α and β as intersection; the
union of these two regions is the whole space and each of the two regions is called a
dihedral, or *dihedral angle*, and the half-planes α and β are called the *faces* of each
dihedral. Let us stress that both faces α and β are included in each dihedral. The
symbol $\widehat{\alpha\beta}$ denotes each of the two dihedrals with faces α and β. The line r common
to α and β is called the *edge* of each dihedral with faces α and β. If both half-planes
α and β with common edge r are both contained in the same plane γ, then α and β
are called the *extensions* of each other.

The following cases occur:

(a) the half-planes α and β are the extensions of each other. Therefore, they belong
to a plane γ and the dihedrals $\widehat{\alpha\beta}$ are the half-spaces of origin γ (Fig. 4.33). In
this case $\widehat{\alpha\beta}$ is called a *straight* dihedral.

(b) the half-planes α and β are not extensions of each other. The space is then
divided into two dihedrals which are not half-spaces. One of the two dihedrals
is concave, the other convex (Fig. 4.34).

The segment joining any two points belonging to a convex dihedral is included in
the dihedral.

The concave dihedral contains the extensions of the half-planes α and β (Fig. 4.35).

Fig. 4.33 Straight dihedral

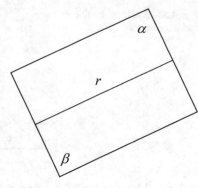

Fig. 4.34 s edge of the
dihedral $\widehat{\alpha\beta}$

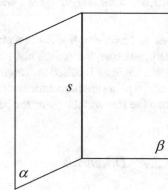

Fig. 4.35 Convex and
concave dihedrals
determined by the
intersecting planes α and β

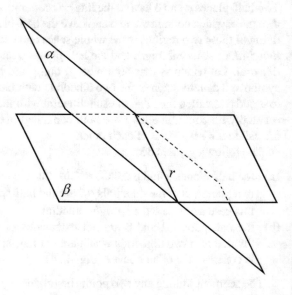

Fig. 4.36 Plane δ
perpendicular to $r = \alpha \cap \beta$

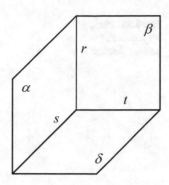

Let us consider a convex dihedral $\widehat{\alpha\beta}$ and a plane δ perpendicular to the edge r of the dihedral $\widehat{\alpha\beta}$ (Fig. 4.36). The angle of the half-lines determined by δ intersecting α and β is called the *normal section* of the dihedral. Two normal sections of a dihedral are congruent. The normal section of a straight dihedral is a straight angle. The dihedral whose normal section is a right angle is called *right dihedral*.

4.5.8   *Perpendicular Planes*

Definition 4.1 Two planes that intersect so that the four dihedrals obtained have normal two-by-two congruent sections are said to be *perpendicular* or *orthogonal* to each other.

The following properties hold:

(a)   if a line is perpendicular to a given plane, any plane containing the line is perpendicular to the given plane (Fig. 4.37);

Fig. 4.37 Plane α
perpendicular to r and to β

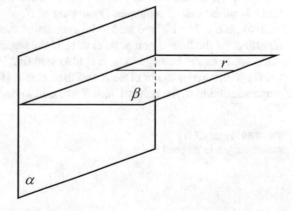

Fig. 4.38 Plane β
perpendicular to α and
passing through *r*

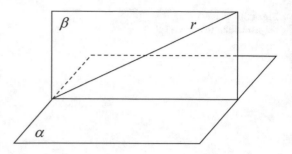

(b) if a plane is perpendicular to a line in another plane, the two planes are
      perpendicular to each other (Fig. 4.37);
(c) given a line and a plane that are not perpendicular to each other, there is one
      and only one plane containing the line and perpendicular to the given plane
      (Fig. 4.38).

4.5.9 Symmetries

Two points P and P′ are said to be *symmetric* with respect to the point C if C is the
midpoint of the segment joining P and P′; it is also said that P′ is the symmetric of
P with respect to the point C.

A figure *F* is said to be *symmetric* with respect to the point C if for each point P
of *F* the symmetric P′ of P with respect to C belongs to *F*. The segment PP′ is called
a *chord* of *F* (Fig. 4.39).

A figure *F* of the plane is said to be *symmetric with respect to the line r in the
direction* (of the line) *d*, non-parallel to *r*, if for any point P in *F*, the symmetric, in
the direction *d*, of P with respect to the line *r* belongs to *F*; the line *r* is called the *axis
of symmetry* of *F* in the direction *d* (Fig. 4.39). In particular, if *d* is perpendicular to
r (Fig. 4.40), the figure *F* is said to be *symmetric* with respect to the line *r* and the
line *r* is called *axis of orthogonal symmetry* of *F*.

Two points P and P′ are said to be *symmetric with respect to a plane* α and the
direction (of the line) *d* non parallel to α, if the segment PP′ is parallel to *d* and the
midpoint C of PP′ belongs to α; it is also said that P′ (or P) is the symmetric of P
(or P′) with respect to the plane α and direction *d* (Fig. 4.41). In particular, if *d* is
perpendicular to α, the points P and P′ are said to be *symmetric* with respect to α.

Fig. 4.39 Figure *F* is
symmetric w.r. to the point C

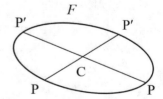

Fig. 4.40 Figure F
symmetric w.r. to the line r

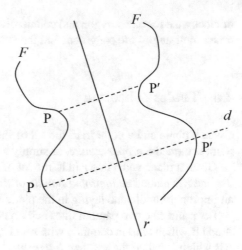

Fig. 4.41 P and P′ are
symmetric w. r. to the plane
α and the direction d

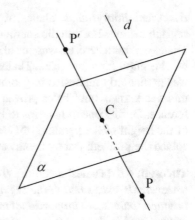

A figure F in the space is said to be *symmetric with respect to the plane* α and the direction d non parallel to α, if for each point P of F the symmetric of P with respect to the plane α and the direction d belongs to F; the plane α is called the *plane of symmetry* of F w. r. to direction d. In particular, if the line d is perpendicular to α, the figure F is said to be *symmetric* w. r. to the plane α and the plane α is called *plane of orthogonal symmetry* of F.

4.5.10 Similar Polygons

Two polygons of the same number of sides are said to be *similar* if, taken the sides of each polygon in a convenient order, e. g., both clockwise, or both counterclockwise, or clockwise for one polygon and counterclockwise for the other, then, in the chosen order, their angles are congruent and the sides are proportional.

4.6   Thales' Theorem

Given a plane and a point in it, the set of the lines of the plane passing through the point is named a *proper bundle*, or simply a *bundle*, of lines.

Given a plane and a line r in it, the set of the lines in the plane that are parallel to the line r is named an *improper bundle* of lines. Any line non parallel to the lines of an improper bundle and laying in the plane is called a *transversal* line.

Let r and s be two parallel lines and t a non-parallel line to the line a. The points A and B which t has in common with r and s, respectively, are extremes of a segment AB which is called the *segment intercepted* by t on the lines r and s.

Historical Background. Thales of Miletus (about 624–548 B.C.). We are in the seventh century B.C. on the southwestern coast of Asia Minor, in the Miletus *polis*. Thales was born here. He was one of the seven sages of the ancient Greece. Pythagoras was his pupil for a few years. Thales dealt with general subjects of mathematics and was particularly interested in geometry. Three non-aligned points define a triangle, this was known, but Thales proved that they also define one and only one circumference. During one of his trips to Egypt he ran into a problem: to find the measure of the height of the pyramid of Cheops, a problem that seemed unsolvable. Thales solved the problem just by means of his theorem (Guedj 1998).

Theorem 4.5 [Thales' theorem]. *If an improper bundle of lines is intersected by two transversals lines t and t', the segments intercepted by t on the pairs of lines in the improper bundle are proportional to the segments intercepted by t' on the same pairs of lines and in the same order. If the intercepted segments are non-null, then there exists a positive number k such that*

$$AB = kA'B', BC = kB'C', CD = kC'D'$$

or

$$\frac{AB}{A'B'} = \frac{BC}{B'C'} = \frac{CD}{C'D'} = k$$

See (Fig. 4.42).

The following propositions result from Thales' Theorem.

Proposition 4.1 *Each parallel to one side of a triangle divides the other two sides into proportional parts* (Fig. 4.43).

Proposition 4.2 *If a line cuts two sides of a triangle so that the pairs of segments identified on the two sides are proportional, then the line is parallel to the third side.*

Proposition 4.2 inverts the previous one.

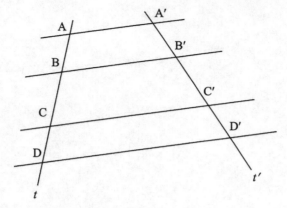

Fig. 4.42 The lines AA′, BB′, CC′, DD′ belong to an improper bundle; t, t' are transversals; AB and CD are segments intercepted by t on the pair of lines AA′ and BB′ and on the pair of lines CC′, DD′; segments A′B′ and C′D′ are intercepted by t' on the same pairs of lines of the improper bundle

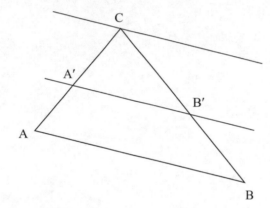

Fig. 4.43 CA′ : A′A = CB′ : B′B

References

Beutelspacher, A., Rosenbaum, U.: Projective geometry: from foundations to applications. Cambridge University Press (1998)
Guedj, D.: Le Théorème du Perroquet. Éditions du Seuil, Paris (1998)
Hilbert, D.: Grundlagen der Geometrie. G.B. Teubner, Stuttgart (1968)

Bibliography

Hartshorne, R.: Geometry: Euclid and Beyond. Springer, New York (2000)
Lobačevskij, N.I.: Nuovi Principi Della Geometria. Boringhieri, Torino (1965)

Chapter 5
Functions

5.1 Introduction

The area A of a square depends on the side l of the square, according to the formula $A = l^2$. At a precise site on the earth surface the atmospheric pressure varies depending on different inputs. When we intend to go to the theater, we buy the ticket that assigns us a seat. A relation is so stated between two sets: a set of spectators and a set of armchairs, such that each spectator is assigned one and only one seat. If the show is sold out, all seats are filled; otherwise, there are free places, not assigned to anyone.

The couples (side, square), (site, pressure), (seat, spectator) are couples of special relations called *functions* which associate one and only one element in the second component to the first. Let use arrows to indicate such relations:

$$\text{side } \rightarrow \text{ square}$$

$$\text{site } \rightarrow \text{ pressure}$$

$$\text{seat } \rightarrow \text{ spectator}$$

To state more precisely what is meant by function we remind the concept of relation (Sect. 3.2) between two sets A and B as a subset of the cartesian product $A \times B$.

Definition 5.1 Let A and B be non-empty sets. A relation f between A and B is called a *function* from A to B if for every element $a \in A$ there exists one and only one element $b \in B$ such that (a, b) is a couple in f, i. e., $(a, b) \in f \subseteq A \times B$. Along with $(a, b) \in f$ the notations $f : A \rightarrow B$ and $b = f(a)$ (which reads "b equals f of a") are used.

© The Author(s), under exclusive license to Springer Nature Switzerland AG 2023
A. G. S. Ventre, *Calculus and Linear Algebra*,
https://doi.org/10.1007/978-3-031-20549-1_5

Think of the function f as a law (or a rule, an instruction, or a procedure) which, given in the set A any element a, allows to identify or build one and only one element b of the set B. The set A is called the *domain* of f, denoted $\text{Dom}(f)$, and f is said to be *defined* in the set A. The element $a \in A$ is called the *independent variable*, or the *argument* of the function f.

Synonyms of the word "function" are *application* or *transformation*. It is also said that the function f transforms or *carries* A into B, or *associates* the element $b \in B$ with the element $a \in A$. If $a \in A$, the statement "f associates the element $b \in B$ with $a \in A$" is equivalent to the formula $b = f(a)$.

The element $b = f(a)$ is called the *value* of f at the element a, or the *corresponding* element with a by means of f, or the *image* or the *transformed* of a under (or through) f. Also, the independent variable a represents an *input* from the domain of f and b, the dependent variable, the corresponding *output* $f(a)$.

The *range* of f is the subset of B consisting of all the elements $b \in B$ such that $b = f(a)$, for every $a \in A$. The range of f is denoted $f(A)$. Of course, $f(A) \subseteq B$.

The above expressions are ways to describe how a function works.

Some relations defined in Chap.3 are functions.

The relation defined in (Sect. 3.2, Example 3.1), which associates the maximum temperature with each European capital in the set A, is a function from A to the set B of the temperatures because the relation associates exactly one number (which measures a unique temperature) to any capital in the set A.

The binary relation defined in the Example 3.2 that with every natural number n associates the next $n + 1$ is a function $f : \mathbf{N} \to \mathbf{N}$, such that $f(n) = n + 1$.

The relation R_2 defined in the example 3.3 is a function of the set of real numbers $\mathbf{R}$ to $\mathbf{R}$. In fact, for each fixed income i the amount of consumption c is determined by the equality $c = 60 + 0.8i$. If we call f this function, we write $f(r) = c$, or $f(r) = 60 + 0.8r$. The binary relation R_3 defined in the Example 3.4 is a function from the set of the lines of the plane to the set of lines of the plane passing through P.

Let $f : A \to B$ be a function. The cartesian product of the domain and the range of f is called the *graph* of the function f. Therefore, the graph of f is the set of the couples:

$$A \times f(A) = \{(a, f(a)), \text{ for every } a \in A\}.$$

If f is a function from A to B, one of the following circumstances may occur:

- there are elements of B corresponding in f with more than one element of A;
- there are elements of B which do not correspond in f with any element of A.

Both circumstances are verified in the following example. Given the sets $A = \{a, b, c, d, e\}$ and $B = \{1, 2, 3, 4, 5, 6\}$, let us define the function $f : A \to B$ as follows: $f(a) = 1, f(b) = 2, f(c) = 2, f(d) = 4, f(e) = 5$ (Fig. 5.1).

The element 2 of B is the image through f of the elements b and c; the element 3 is not the image of any element of A.

A function $g: A \to B$ such that each element of B is the image of at least one element of A is called a *surjective function* or a function from A *onto* B. The range

Fig. 5.1 Elements of B associated through f with the elements of A; each element of A has a unique corresponding element in the set B

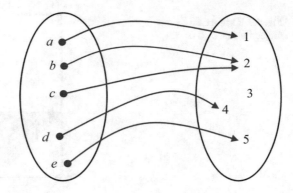

Fig. 5.2 $g : A \rightarrow B$

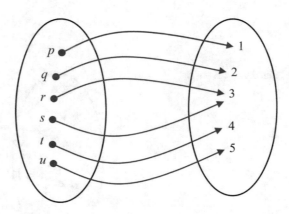

of g coincides with B, $B = g(A)$. For example, let us consider the sets $A = \{p, q, r, s, t, u\}$ and $B = \{1, 2, 3, 4, 5\}$. The function $g : A \rightarrow B$ defined by: $g(p) = 1$, $g(q) = 2$, $g(r) = 3$, $g(s) = 3$, $g(t) = 4$, $g(u) = 5$ (Fig. 5.2) is surjective.

Example 5.1 Let A be the set of the points of the space and B the set of the points of a plane α. Consider the function $k : A \rightarrow B$, which associates the point a' of α, that is the orthogonal projection (Sect. 4.5.5) of the point a in the space on the plane α (Fig. 5.3). The function k is surjective because each point of the space has the image belonging to the plane α.

A function $h : A \rightarrow B$ such that each element of B is the image of at most one element of A is called an *injective function* from A to B.

For example, let us consider the sets $A = \{a, b, c, d, e\}$ and $B = \{1, 2, 3, 4, 5, 6, 7\}$. Let $h : A \rightarrow B$ be the function defined by: $h(a) = 1$, $h(b) = 2$, $h(c) = 4$, $h(d) = 6$, $h(e) = 7$ (Fig. 5.4). The function h is an *injective* function.

Example 5.2 The function $d : N \rightarrow N$, which associates the double $2n$ with the natural number n, is an injective function.

Fig. 5.3 Orthogonal
projections on $a' \in \alpha$

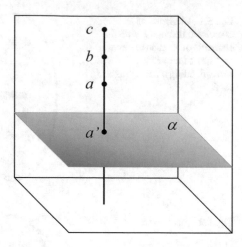

Fig. 5.4 $h : A \rightarrow B$

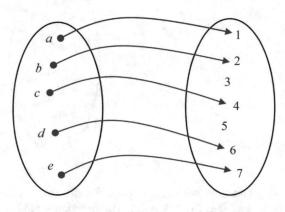

A function which is either injective and surjective, i. e., a function $f : A \rightarrow B$ such that each element b of B is the image of one and only one element a of A, $b = f(a)$, is called a *one-to-one* (or 1–1) *function*, or a 1–1 *correspondence*, between A and B, or a *bijection*, or a *bijective function* from A *onto* B. Of course, $B = f(A)$ (Fig. 5.5).

The images of two distinct elements through a one-to-one function are distinct; i. e., if f is one-to-one, if a and b are elements of the domain of f and if $a \neq b$, then $f(a) \neq f(b)$. But also, vice versa, if f is a one-to-one function from A onto B and if y and z are distinct elements of B, then there exist a and b in A such that $y = f(a)$, $z = f(b)$ and $a \neq b$.

Example 5.3 Let A be the segment with endpoints a and b and B a semi-circumference of diameter A. We construct a function $f : A \rightarrow B$ this way: the point y of the semi-circumference which belongs to the perpendicular from x to A corresponds with the point x of the segment A. The function f is a one-to-one function from A onto B (Fig. 5.6).

Fig. 5.5 $f(a) = 1, f(b) = 2, f(c) = 3, f(d) = 4$

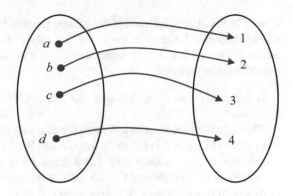

Fig. 5.6 One-to-one function $f : A \rightarrow B$

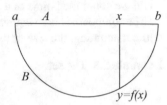

If $f : A \rightarrow B$ is a one-to-one function, it makes sense to consider the function that associates the element x of A such that $y = f(x)$ with the element y of B, in fact each element y of B is the corresponding element through f with one and only one element of A. This new function from B to A is called the *inverse function* of f and is denoted f^{-1}; so $f^{-1}(y) = x$. We immediately recognize that $f^{-1} : B \rightarrow A$ is a one-to-one function.

Given a one-to-one function, the inverse function is set out. For this reason a one-to-one function is also called an *invertible function*.

Remark 5.1 If $f : A \rightarrow B$ is a function, be careful not to confuse the symbols f and $f(x)$. However, sometimes we write $f(x)$ to indicate a function when we want to emphasize that the function is defined in a certain set whose generic element, i. e., the independent variable, is x.

5.2 Equipotent Sets. Infinite Sets, Finite Sets

Two sets A and B such that a one-to-one function $f : A \rightarrow B$ can be defined are called *equipotent*. Thus, the set of days in the week and the set of numbers {1, 2, 3, 4, 5, 6, 7} are equipotent.

As mentioned in Example 5.3, the set of the points of the semi-circumference and the set of the points of its diameter are equipotent. Let us dwell on some other examples of equipotency between sets.

Example 5.4 The function that associates the number $2n$ with the natural number n is an invertible function of the set of natural numbers onto the set of even natural numbers (see Example 5.2). Then the set of natural numbers and the set of even numbers are equipotent.

It is not difficult to realize that the interval [0, 1] and the interval [0, 2] are equipotent.

The last examples show us sets equipotent to a proper part of them. A set equipotent to a proper subset of it is called an *infinite set*. While a *finite set* is defined as a non-infinite set. It may be surprising that a *finite set* is defined as a non-infinite set, i. e., infinity is defined first and consequently finiteness. Yet this is a rational way to introduce the two concepts, which is independent of the operation of counting. How could we count the points of a segment?

The equipotency is an equivalence relation between sets that enable us to estimate, without counting, the size or the magnitude of a set.

Example 5.5 The set

$$S = \left\{ 1, \frac{1}{2}, \frac{1}{3}, \ldots, \frac{1}{n}, \ldots \right\}$$

$n \in \mathbf{N}$, is equipotent to the set of natural numbers. The interval [0, 1] and S are not equipotent.

Any set equipotent to $\mathbf{N}$ is said to be a *countable set*. For example, the set S and the set of odd natural numbers are countable. It is not difficult to prove that also the set of rational numbers is countable.

5.3   Hotel Hilbert

The deep-rooted truth "the whole is greater than any its part" seems to be contradicted by the relation of equipotency that allows that an infinite set may be equivalent to a proper subset of itself. This to say that surprises and paradoxes are to be expected in the domain of infinity.

In this regard Hilbert imagines that he is the receptionist of a prodigious hotel, a hotel with infinitely many rooms. One day a new customer shows up who, knowing about the prodigy, having asked for a room, is a bit perplexed when Hilbert replies that the rooms are all occupied. Hilbert, however, reassures the customer that he will find a room for him. He then asks all the guests, already settled, to move to the next room: who occupies the room number 1 moves to room 2, who occupies room 2 moves to room 3 and so on,... there are infinitely many rooms. The new guest will occupy the room number 1 (Fig. 5.7a).

But the wonders do not end here. The following evening the rooms are still all occupied when a bus arrives with an infinite number of new customers. Once again

Fig. 5.7 The paradox of infinity. Adding one unit to infinity (a); adding an infinite number of units to infinity (b)

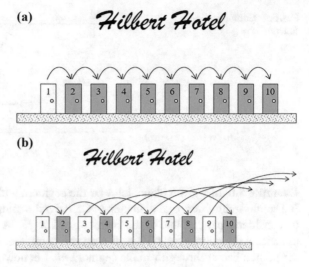

Hilbert has the solution: he asks the guest occupying room 1 to move to room 2, the guest occupying room 2 to move to room 4, … and the guest occupying room n to move to room $2n$ (Fig. 5.7b). All the odd-numbered rooms have been vacated for the new arrivals (Singh 1997).

5.4 Composite Functions

Let us consider the nonempty sets A, B and C and the functions $f : A \rightarrow B$ and $g : B \rightarrow C$. If x is an element of A and $f(x)$ its image under f, it makes sense to construct the element $g(f(x))$, the image under g of the element $f(x)$ of B. The *composite function* of f and g is the function h from A to C defined by the equality $h(x) = g(f(x))$, for every $x \in A$.

The order is important: first we construct $f(x)$ and then $g(f(x))$. The composite function of f and g is also denoted $g \circ f$,, therefore $(g \circ f)(x) = g(f(x))$ (Fig. 5.8).

Fig. 5.8 h composite function of f and g

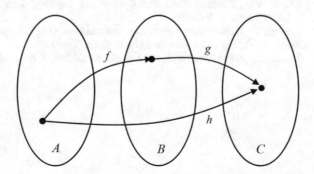

Fig. 5.9 Composite
function $h = g \circ f$

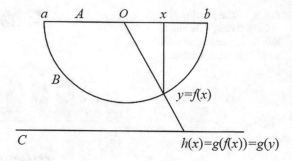

Example 5.6 In a given plane, let A be the segment with endpoints a and b and let B a semi-circumference of diameter A. Called o the midpoint of the segment ab, let us consider a parallel line to the segment A. Let $f : A \to B$ be the function that associates the point $y = f(x) \in B$ with the point $x \in A$ that is common to A and the perpendicular through x to the segment ab. Let now $g : B \to C$ be the function that associates $z \in C \cap$ (line Oy) with the point $y \in B$. The function $h : A \to C$ such that $h(x) = g(f(x))$, for every $x \in A$, is the composite function of f and g (Fig. 5.9). Observe that since the functions f and g are invertible, also the function h is invertible.

5.5   Restriction and Extension of a Function

Given the non-empty sets A and B and the function $f : A \to B$, let X be a subset of A. The function $g : X \to B$ such that $g(x) = f(x)$, for every $x \in X$, is called *the restriction* of f to X and f is said to be an *extension* of g to A.

Bibliography

Chinn, W.G., Steenrod, N.E.: First Concepts of Topology. Random House, New Mathematical Library, New York (1966)
Gowers, T.: Mathematics: A Very Short Introduction. Oxford University Press (2002)
Singh, S.: Fermat's Last Theorem. Simon Singh (1997)
Ventre, A.: Matematica. Fondamenti e Calcolo. Wolters Kluwer, Milano (2021)

Chapter 6
The Real Line

6.1 Introduction

The numerical representation of the Euclidean geometry, due to Descartes (1596–1650), is the basis of the *analytic geometry.*

Historical Background. René Descartes, is the father of modern philosophy. He wrote *La Géométrie* to explain the relation between the arithmetic and the geometry, even though in the intentions of the author, La Géométrie, published as an appendix to the *Discours de la méthode*, should have illustrate his philosophical method. Descartes' work gave rise to a rejoining of arithmetic to geometry after the fracture within the Pythagorean school, just creating the *analytic geometry.* This "new subject" has provided a method for describing algebraic formulas by means of geometric curves and shapes and, inversely, a description of geometric curves and shapes in terms of algebraic formulas.

6.2 The Coordinate System of the Axis

In Chap. 4 we dealt with Euclidean geometry in the line, the plane and the space. The key point in analytic geometry of the line is to establish a one-to-one function between real numbers and the points of the line, i. e., the equipotency (Sect. 5.2) between the set **R** of the real numbers and the set of points of the line.

Let us imagine a line r described by a point always moving in the same sense. There are exactly two senses for this motion, each defining a total order (Sect. 3.3.3) in the set of the points of the line, such that, for every pair of distinct points in the line r it is possible to establish which of the two points precedes the other. Each ordering of the line is called *orientation,* or *sense,* of the line: we conventionally define *positive* an orientation and *negative* the other. Usually, in figures an arrow indicates the positive orientation of the line (Fig. 6.1).

© The Author(s), under exclusive license to Springer Nature Switzerland AG 2023
A. G. S. Ventre, *Calculus and Linear Algebra*,
https://doi.org/10.1007/978-3-031-20549-1_6

Fig. 6.1 Oriented line *r*

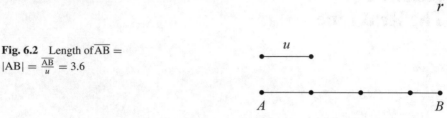

Fig. 6.2 Length of $\overline{AB}$ =
$|AB| = \frac{\overline{AB}}{u} = 3.6$

A line endowed with the positive orientation is called an *oriented line* or *axis*. The *support* of the oriented line *r* is the line itself without any orientation.

6.2.1   The Measure of a Segment

Let two distinct points A and B be given on the line *r* with A preceding B in an orientation taken as positive on *r*. The set of points of *r* that follow A and precede B along with the points A and B, define a subset of *r* said *segment* and the points A and B are called *extremes* or *endpoints* of the segment. If the points A and B coincide, the segment reduces to a point A = B, called a *null segment*. The segment having endpoints A and B is denoted $\overline{AB}$.

Let *u* be a non-null segment and $\overline{AB}$ a segment in a line. The *length* or *measure of the segment* $\overline{AB}$ with respect to the segment *u* taken as the unit of measure, is defined as the non-negative real number, denoted $|AB|_u$, or simply $|AB|$, that is the ratio of the homogeneous magnitudes $\overline{AB}$ and *u* (Sect. 2.5) (Fig. 6.2). Therefore,

$$\text{length of}\,\overline{AB} = \frac{\overline{AB}}{u} = |AB|$$

The length $\overline{AB}$ of the segment $\overline{AB}$ is a positive number if and only if A and B are distinct; the null segment has length zero.

6.2.2   The Coordinate System of an Axis

Let us fix a point O of the oriented line *r*: the point O divides the line into two half-lines with common origin O. Now let fix a point U in the axis *r*, distinct from O and following O in the positive orientation fixed on *r*. We call *unit point* of *r* axis the point U and take the segment $u = \overline{OU}$ as the unit of measure of the lengths of the segments: of course, the length of the segment $\overline{OU}$ is 1 (Fig. 6.3).

Fig. 6.3 Oriented line with origin O and unit point U

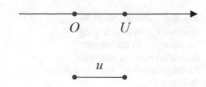

The half-line of r which has origin O and contains U, is called the *positive half-line* denoted r^+; the other half-line with origin O is called the *negative half-line* denoted r^- (Fig. 6.4).

A one-to-one function (Sect. 5.1) that associates a real number with any point of the axis r is defined this way: if the point P of r belongs to r^+, then the number $x_P = |OP|$ is associated with P; if P belongs to r^-, then let the number $x_P = -|OP|$ is associated with P (Fig. 6.5).

In particular, $x_P = 0$ if and only if P coincides with the origin O. The number x_P is called the *abscissa* or the *coordinate* of the point P, with respect to the origin O and the unit of measure $u = \overline{OU}$. The index P is omitted if there is no possibility of misunderstanding and the only symbol x denotes the abscissa of P. The point P is the *geometric representation* of the number x in the r axis.

The points O and U define a *coordinate system* of the r axis. The coordinate system is sufficient to define a one-to-one correspondence between the set of the points of the axis and the set of the real numbers **R**.

The oriented line r endowed with a coordinate system is called the *real line*. The expression *real line* or *real axis*, means that the set of real numbers and the set of the points of the line can be identified.

The symbol P(x) denotes that the abscissa of P is x in the given coordinate system of r axis. For any two points A(a) and B(b) of the r axis, the *length*, or *relative length*, or *relative measure* of the oriented segment AB is denoted (AB) and defined by the equalities:

$$\text{length of AB} = (AB) = b - a$$

whatever the positions of A and B are in r (Fig. 6.6). The oriented segment AB has the orientation opposite to that of BA. Therefore,

Fig. 6.4 Half-lines r^- and r^+ with origin O

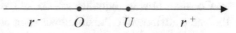

Fig. 6.5 Points and abscissas

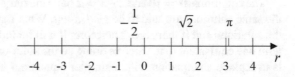

Fig. 6.6 The length of the oriented segment AB lying on the axis r is the number $b–a$, whatever the positions of the points A(a) and B(b) are

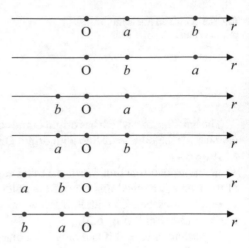

$$\text{length of AB} = (AB) = b - a = -(a - b) = -(BA) = -\text{length of BA}$$

6.3   Equalities and Identities. Equivalent Equations

The expressions made of letters and numbers, i.e., the algebraic expressions, that contain the equality sign $=$ are called *equalities*. The equalities are made of a *left-hand side* and a *right-hand side*, separated by the equality sign. The letters have the function of unspecified numbers, in fact are named *indeterminates* or *variables* or *unknowns*. The equalities including literal expressions are also named *equations*.

Equations are a tool for solving problems that require to search for one or more unknown quantities.

Let us consider a very simple problem: find the number which, added to 4, gives the sum 7. In symbols, denoted x the unknown number, the problem translates into the equation: $4 + x = 7$, where x is the number that transforms equality $4 + x = 7$ in an equality of numbers. Any numerical value attributed to the unknown x that fulfills the equality, a value which in our example is 3, is called a *solution* of the equation.

Consider now the equality $x^2 - 2x + 1 = (x-1)^2$. This is an equality whatever is the value attributed to the unknown x; an equality of this type, that does not impose any condition on the unknown x, is called *identity*.

The equations $2x = 10$ and $x - 3 = 2$ have the only solution 5. Equations that have the same solutions are said to be *equivalent*. What matters most in the equations are the solutions and if there are. Therefore, it will be indifferent to replace an equation with one equivalent to it, since in doing so the solutions remain unchanged. Our aim is to replace an equation with a simpler equivalent up to an expression of the type $x = a$, with a real number, that gives us the value sought for the unknown x. Solving an equation means to find the set of the solutions of the equation.

The following theorems, provide the rules for simplifying the form of an equation.

Theorem 6.1 *Adding the same number to the two sides of a given equation changes this into an equation equivalent to the given one.*

Theorem 6.2 *Multiplying the two sides of a given equation by a non-zero number changes it into an equation equivalent to the given one.*

An equation without solutions is said to be incompatible, otherwise it is called compatible or solvable. Let us now solve some equations where Theorems 6.1 and 6.2 are useful.

6.3.1 Examples

The equation $2y - 1 = x - 1$. has two unknowns x and y. Any couple of numerical values attributed to the unknowns x and y that achieves the equality $2y - 1 = x - 1$ is called a *solution* of the equation. For example, the couple $(x, y) = (0, 0)$ achieves the equality $2y - 1 = x - 1$ since $2 \times 0 - 1 = 0 - 1$, then the couple $(0, 0)$ is a solution of the equation $2y - 1 = x - 1$. But also $(x, y) = (2, 1)$ is a solution of the equation since, setting $x = 2$ and $y = 1$ in the equation, the numerical equality $2 \times 1 - 1 = 2 - 1$ is obtained.

1. The equation $x + 1 = x$ has no solution. The equation $0x + 1 = 0$ has no solution. So equations $x + 1 = x$ and $0x + 1 = 0$ are both incompatible.
2. The equation $x^2 + 1 = 0$ is incompatible because x^2 is nonnegative.
3. To check that a solution of $x^2 - 2x = -1$ is $x = 1$ it suffices to replace x with 1 and observe that the left-hand side equals -1.
4. The equation $2x - 3 = 0$ can be solved applying Theorems 6.1 and 6.2: indeed adding to both sides 3 we have $2x - 3 + 3 = 3$, i.e., $2x = 3$, and dividing both sides by 2, we have $\frac{2}{2}x = \frac{3}{2}$, i.e., $x = \frac{3}{2}$. The number $x = \frac{3}{2}$ is the solution of the equation $2x - 3 = 0$.
5. The equation $0x + 0y + 1 = 0$ has no solution, i.e., the equation is incompatible.
6. The equation $0x = 0$ is an identity because any value attributed to the unknown x is a solution of the equation. The equation $0x + 0y = 0$ is an identity because any couple of values given to the unknowns x and y is a solution of the equation.
7. Any solution of the equation $2x + y = 0$ is a couple (x, y) of numbers that satisfy the equation; for example, $(0, 1)$ is a solution of the equation because $2 \times 0 + 1 = 1$; $(-3, 7)$ is a solution of the equation because $2 \times (-3) + 7 = 1$; $(2, -3)$ is a solution of the equation because $2 \times 2 - 3 = 1$.
8. The equation $2x - y + 3z = 1$ three unknowns x, y and z and has infinitely many triples of solutions, each obtained by assigning values to two of the unknowns and then finding the value of the third; for example, replace in the equation the unknowns y and z with -1 and -2, respectively, to obtain $2x - (-1) + 3(-2) = 1$ by Theorems 6.1 and 6.2 we get $x = 3$, so a solution of the given equation is the triple $(3, -1, -2)$; indeed: $2 \times 3 + 1 + 3(-2) = 1$.

9. The equation $x^2 - 2x + 1 = (x - 1)^2$ reduces to the form $0 = 0$, after performing the square $(x - 1)^2$. The equation belongs to the class of *identical equations*, or *identities*, i. e., the equations satisfied by any real value attributed to the variables. Also, the equation $(x + y)^2 = x^2 + 2xy + y^2$ is an identical equation in two unknowns.

6.3.2   *Forming an Equation from Given Information*

A *pizzeria* charges on Tuesday one euro less for *pizza margherita* than on Saturdays. With the same amount you can buy 5 pizzas on Tuesday and 4 on Saturday. How much does pizza cost on Saturdays?

Let's try to translate the question into an equation. If we indicate with x the cost of pizza on Saturday, then on Tuesday pizza costs $x - 1$; then $5(x - 1)$ is the cost of 5 pizzas on Tuesdays and $4x$ the cost of 4 pizzas on Saturdays. We equal the two amounts of money:

$$5(x - 1) = 4x \tag{6.1}$$

Equation (6.1) expresses our problem in symbolic form. What is the value of x? How much does pizza cost on Saturday? To find the value of the unknown x, i. e., to solve the equation, let us perform the left-hand side multiplication:

$$5x - 5 = 4x$$

To know the cost x we need to "isolate" x, going through the following steps that apply the Theorems 6.1 and 6.2,

$$5x - 5 - 4x = 4x - 4x$$
$$5x - 5 - 4x = 0$$
$$x - 5 = 0$$
$$x = 5$$

The pizza on Saturday costs 5 euros. We have therefore solved Eq. (6.1) because we have found the solution $x = 5$, which is unique.

6.4 Order in R

The real line allows a view of the ordering of the set **R** of the real numbers. Let the real line be oriented from left to right, if x is to the left of y, then the number x is smaller than the number y, or equivalently, the number y is greater than the number x, and we write $x < y$ or $y > x$, respectively. The expression $x < y$, and the other $y > x$, are defined *inequalities*, also called *inequations*. Expressions like $x \leq y$, $x \geq y$ are also inequalities and include the case that the numbers x and y coincide.

The following statements are properties allowing to deal with inequalities.

1. If $x < y$ and $y < z$, then $x < z$.
2. If $x < y$ and z is any real number, then $x + z < y + z$.
3. If $x < y$ and $w < z$, then $x + w < y + z$.
4. If $x < y$ and $z > 0$, then $xz < yz$.
5. If $x < y$ and $z < 0$, then $xz > yz$; in particular, if $x < y$ and $z = -1$, then $-x > -y$.
6. If x and y are both positive or both negative and $x < y$, then $\frac{1}{x} > \frac{1}{y}$
7. If $0 < x < y$, then $0 < \frac{x}{y} < 1$.

The properties 1 to 6 remain satisfied if the symbols $<$ and $>$ are replaced by $\leq$ and $\geq$, respectively.

Also *inequalities* and *inequations* are expressions that may include numbers and letters which numerical values can be attributed to. These letters are called *variables* or *unknowns*.

Given an inequality, any numerical value assigned to the unknown x that fulfils the inequality is called a *solution* of the inequality. Solving an inequality means to find the set of the solutions of the inequality. If $x \leq 1$, the variable x takes any numerical value smaller than 1, or the value 1.

In Fig. 6.7 we have a representation on the real line r of the solution set of the inequality $x < 1$; in Fig. 6.8 a representation of the set of solutions of the inequality $x \geq 0$. (The symbol $\bigcirc$ on the line indicates a point which does not belong to the set, the symbol $\bullet$ indicates a point belonging to the set.) Fig. 6.9 represents the set of the points less than 1 and greater then or equal to 0.

$$1 \qquad\qquad r$$

Fig. 6.7 The set of the points of the line r smaller than 1

$$0 \qquad\qquad r$$

Fig. 6.8 The set of the points of the line r greater than or equal to 0

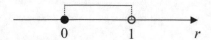

Fig. 6.9 The set of the points of the real line r smaller than 1 and greater than or equal to 0

6.4.1   Evaluating an Inequality to Making a Decision

Like the equations, the inequalities are also useful tools for translating problems into mathematical models.

Winston and Julia decide to enroll in a dance school. A renowned school asks for an annual registration fee of 315 euros for each person, plus 3 euros for each entrance. Admission to a suburban dance school instead costs 6 euros and no registration fee is required. The two aspiring dancers wonder which school is more convenient. Off the cuff they struggle to give an answer. If the two intend to attend the renowned school once a week, each will pay the registration fee plus 52 admissions in one year, that is $315 + 3 \times 52 = 471$ euros. Instead, attending the suburban school, always once the week, each would pay 52 admissions, that is $6 \times 52 = 312$. But if you attended the dance school three times a week, i. e., $3 \times 52 = 156$ times a year, the renowned school would cost $315 + 3 \times 156 = 683$, while the suburban dance school would cost $6 \times 156 = 936$. Then the renowned school would be cheaper. The annual cost therefore depends on the number of times the two will go to school. By attending x times in a year they will spend $315 + 3 \times$ at the renowned school while $6 \times$ euros at the suburban school. Therefore, the suburban school will be cheaper if the amount of money $6 \times$ is smaller than $315 + 3x$,

$$6x < 315 + 3x \qquad (6.2)$$

So we have to solve this inequality to know in which cases the suburban school is more convenient than the other. To proceed with the resolution, we use the laws allowing to deal with inequalities (Sect. 6.4). From (6.2) it follows $6x - 3x < 315$, hence $3x < 315$ and the solutions of the inequality are the numbers x such that

$$x < \frac{315}{3} = 105$$

So, if Julia and Winston intend to take classes no more than 104 times a year, or on average twice a week, the suburban school is cheaper than the renowned one. Otherwise, it is worth enrolling in the renowned school.

6.5   Intervals, Neighborhoods, Absolute Value

The equipotency and the consequent identification of the set **R** and the oriented line (Sect. 6.1) entails to state a correspondence between the geometrical and analytical nomenclatures. This seems appropriate for developing the method for describing algebraic formulas by means of geometric objects and, inversely, describing geometric objects in terms of algebraic formulas.

So it happens that the same structure can be seen either in a numerical context and in a geometrical one. For example, the terms "point" and "number" are interchangeable, a subset A of **R** and a subset of points in the line are called a *numerical set* or a *linear set*. Another example we are going to deepen is the term "oriented segment" (of geometric origin) and interval (of numerical origin), that we use as synonyms.

Given the real numbers a and b, with $a < b$, let us specify the notion of *interval*. The *closed interval* with endpoints a and b, denoted $[a, b]$, is the set defined by

$$[a, b] = \{x \in \mathbf{R} : a \leq x \leq b\}$$

The *open interval* with endpoints a and b, denoted (a, b), is defined by

$$(a, b) = \{x \in \mathbf{R} : a < x < b\}$$

The *left-open* and *right-closed* interval with endpoints a and b is the set defined by

$$(a, b] = \{x \in \mathbf{R} : a < x \leq b\}$$

The *left-closed* and *right-open* interval with endpoints a and b is the numerical set defined by

$$[a, b) = \{x \in \mathbf{R} : a \leq x < b\}$$

The interval $[a, a]$ which consists of the only point a is called a *degenerate interval*.

Let us consider the interval with endpoints a and b, $a < b$. The *width* or *length of an interval* with endpoints a and b is, by definition, the number $b–a$. Degenerate intervals have width equal to zero. The *center* of an interval with endpoints a and b, $a < b$, is, by definition, the real number c such that $c–a = b–c$, i. e., c is the *midpoint* of the interval with endpoints a and b; therefore $2c = a + b$ and $c = \frac{a+b}{2}$ (Fig. 6.10).

Definition 6.1 Any open interval (a, b) containing the point $p \in \mathbf{R}$ is called a *neighborhood* of p.

Fig. 6.10 The center $c = \frac{a+b}{2}$ of the interval $[a, b]$

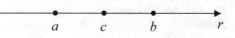

There exists a positive real number h such that the open interval $(p - h, p + h)$ is contained in the neighborhood of c; the neighborhood $(p - h, p + h)$ of p is called the neighborhood of p of *radius*, or *half-width*, h (Fig. 6.11), or *neighborhood centered on p of radius h.*

The *absolute value* of the real number x, denoted $|x|$, is the real number defined as follows:

$$|x| = x, \text{ if } x \geq 0$$

$$|x| = -x, \text{ if } x < 0$$

In other words, the absolute value of a number x is a function f that associates the number x with x, if $x \geq 0$, and associates the number $-x$ with x if $x < 0$.

The absolute value of a real number is non-negative.

For example, $|5| = 5, |-5| = -(-5) = 5$.

Moreover, the following properties hold, for every x and y in **R**:

$$|x| = 0 \text{ if and only if } x = 0,$$

$$|x - y| = |y - x|, \text{ for every } x \text{ and } y \text{ in } \mathbf{R},$$

$$-|x| \leq x \leq |x|$$

$$|xy| = |x||y|$$

$$|x + y| \leq |x| + |y|$$

Given any pair of points x, y in the real line, the nonnegative number $|x - y|$ is called the *distance* of the points x and y. The distance $|x - y|$ is nonnegative, for example:

$$|3 - 7| = |7 - 3| = 4$$
$$|3 + 7| = |3| + |7| = 10$$
$$|-7 + 3| < |-7| + |3| = 10$$
$$|8(-4)| = |-32| = 32$$
$$|8||-4| = 8 \times 4 = 32$$

Fig. 6.11 Neighborhood of p of radius h

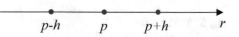

Fig. 6.12 The open interval $(-a, a)$

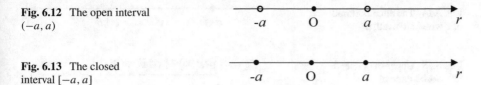

Fig. 6.13 The closed interval $[-a, a]$

If $a > 0$ the inequality $|x| < a$ is, by definition, equivalent to both the inequalities:

$$x < a, \text{ if } x \geq 0$$
$$-x < a, \text{ if } x < 0$$

which yield: $-a < x < a$.

Evidently the real numbers that are greater than $-a$ and less than a are the elements x of the open interval $(-a, a)$. The equalities between sets hold:

$$(-a, a) = \{x \in \mathbf{R} : |x| < a\} = \{x \in \mathbf{R} : -a < x < a\}$$

The open interval $(-a, a)$ is the neighborhood of 0 of radius a (Fig. 6.12). Similarly, the equalities of the following sets hold (Fig. 6.13):

$$[-a, a] = \{x \in \mathbf{R} : |x| \leq a\} = \{x \in \mathbf{R} : -a \leq x \leq a\}.$$

Let us now consider the inequation in the unknown x:

$$|x - c| < a \tag{6.3}$$

with c and a real numbers and $a > 0$. By definition of absolute value, the set of real numbers x verifying (6.3) coincides with the set of the real numbers x such that:

$$-a < x - c < a \tag{6.4}$$

and, adding c to each side of (6.4),

$$c - a < x < c + a \tag{6.5}$$

Therefore, the real numbers which satisfy (6.3), and hence (6.4), are the numbers x which fulfil (6.5). Whence the equalities of the sets:

$$(c - a, c + a) = \{x \in \mathbf{R} : |x - c| < a\} = \{x \in \mathbf{R} : -a < x - c < a\}$$

$$= \{x \in \mathbf{R} : c - a < x < c + a\} \tag{6.6}$$

Fig. 6.14 The neighborhood
of c with half-width a

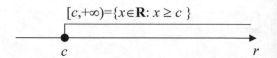

Fig. 6.15 Upper-unbounded
left-closed interval

The open interval (6.6) is the neighborhood of the point c having half-width a
(Fig. 6.14).

The half-line has to be considered an interval. The numerical set $\{x \in \mathbf{R} : x \geq c\}$
is defined the *left-closed right-unbounded interval* of origin c. It is represented on
the real line r by the right half-line of origin c, which contains the numbers greater
than or equal to c (Fig. 6.15). This interval is also denoted by $[c, +\infty)$, where $+\infty$
is the symbol of "positive infinity".

The numerical set $\{x \in \mathbf{R} : x > c\}$ is called a *left-open right-unbounded interval*.
It consists of the real numbers $x > c$. Similarly, among the left-unbounded and right-
bounded intervals we distinguish the right-open intervals $\{x \in \mathbf{R} : x < c\}$ and the
right-closed intervals $\{x \in \mathbf{R} : x \leq c\}$.

To summarize, when using the symbols $+\infty$, *plus infinity*, and $-\infty$, *minus infinity*,
the unbounded intervals are denoted and named as follows:

$[c, +\infty) = \{x \in \mathbf{R} : x \geq c\}$ left-closed right-unbounded interval.
$(c, +\infty) = \{x \in \mathbf{R} : x > c\}$ left-open right-unbounded interval.
$(-\infty, c] = \{x \in \mathbf{R} : x \leq c\}$ right-closed left-unbounded interval.
$(-\infty, c) = \{x \in \mathbf{R} : x < c\}$ right-open left-unbounded interval.

For every real number c, the interval $(c, +\infty)$ is called a *neighborhood* of $+\infty$,
the interval $(-\infty, c)$ a *neighborhood* of $-\infty$.

6.5.1 Exercises

It is worth remembering the calculation rules for solving equations (Sect. 6.3) and
inequalities (Sect. 6.4).

1. Find the solutions of the equation $|3x - 4| = 5$. Applying the definition of
    absolute value we distinguish the two cases:

    (a) $3x - 4 \geq 0$,

        then $|3x - 4| = 3x - 4$, and the equation $|3x - 4| = 5$ reduces to $3x - 4 = 5$
        which is solved by $x = 3$.

(b) $3x - 4 < 0$,

then $|3x - 4| = -(3x - 4) = -3x + 4$, so the equation $|3x - 4| = 5$ reduces to $-3x + 4 = 5$ that has the solution $x = -\frac{1}{3}$.

Therefore, the equation $|3x - 4| = 5$ has two solutions: $x = 3$ and $x = -\frac{1}{3}$.

2. Solve the inequation $|2x - 3| < 5$. We distinguish the two cases:

(a) $2x - 3 \geq 0$,

then $|2x - 3| = 2x - 3$, and the inequality $|2x - 3| < 5$ reduces to $2x - 3 < 5$ which is solved by $x < 4$;

(b) $2x - 3 < 0$,

then $|2x - 3| = -(2x - 3)$. The inequality $|2x - 3| < 5$ takes the form $-2x + 3 < 5$, whose solutions are the values $x > -1$.

Therefore, the solutions of the inequality $|2x - 3| < 5$ are the points x that belong to the union of the sets $\{x \in \mathbf{R} : x < 4\}$ and $\{x \in \mathbf{R} : x > -1\}$.

6.6 The Extended Set of Real Numbers

In relation to unbounded intervals we have introduced the symbols $+\infty$ and $-\infty$. Let us warn that $+\infty$ and $-\infty$ are not real numbers. We assume $-\infty -\infty < x < +\infty$, for all $x \in \mathbf{R}$.

The set $\mathbf{R}^* = \mathbf{R} \cup \{-\infty, +\infty\}$ is called the *extended set of real numbers*, or *the extended real line*. We also consider the intervals of $\mathbf{R}^*$ containing $+\infty$ and $-\infty$; so we have:

$$[-\infty, a) = \{-\infty\} \cup (-\infty, a) \text{ and } (a, +\infty] = (a, +\infty) \cup \{+\infty\}, \text{ for every } a \in \mathbf{R}.$$

In (Sect. 6.5) the neighborhood of a point $c \in \mathbf{R}$ has been defined as an open interval that contains the point c. Let A be a non empty subset of $\mathbf{R}$. A real number x is called an *interior point* of A if there exists a neighborhood of x contained in A. The set of interior points of A is called *the interior* of A and is denoted $\overset{\circ}{A}$. For example, if $A = [a, b)$, then $\overset{\circ}{A} = (a, b)$.

If $A = \{1, 2, 3\}$, then $\overset{\circ}{A}$ is the empty set, $\overset{\circ}{A} = \emptyset$, because every neighborhood of 1 is not contained in A and the same can be said for the points 2 and 3; so the set $A = \{1, 2, 3\}$ does not contain interior points.

The numerical set A is said to be *open* if it is equal to its interior, $A = \overset{\circ}{A}$, i. e., each point of A is an interior point of A. For example, the open interval (a, b) and $\mathbf{R}$ are open sets.

A point x of $\mathbf{R}$ is called an *accumulation point* of the set A if each neighborhood of x contains a point of A distinct from x.

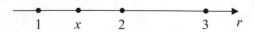

Fig. 6.16   There exists a neighborhood of x that does not contain any point of $A = \{1, 2, 3\}$

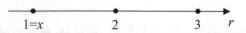

Fig. 6.17   There exists a neighborhood of x that does not contain any point of $A = \{1, 2, 3\}$ distinct from x

6.6.1   *Examples*

a.   The accumulation points of $A = (a, b)$ are the points a, b and all the points of A. Indeed: every neighborhood of the point a contains a point of A distinct from a; every neighborhood of the point b contains a point of A distinct from b; for every point x of the interval (a, b), every neighborhood of x contains a point of A distinct from x. Therefore, the set of accumulation points of the open interval (a, b) is the closed interval $[a, b]$.

b.   The set $A = \{1, 2, 3\}$ has no accumulation points; indeed, for every real number x there exists a neighborhood of x that does not contain any point of A distinct from x (Figs. 6.16, 6.17).

The elements $+\infty$ and $-\infty$, which belong to $\mathbf{R}^*$ are accumulation points of $\mathbf{R}$ that do not belong to $\mathbf{R}$. For example, the set of accumulation points of the open interval $(1, +\infty)$ is the interval $[1, +\infty]$ subset of $\mathbf{R}^*$.

It is easy to deduce that if x is an accumulation point of A, every neighborhood of x contains infinite points of A.

The set of accumulation points of A, denoted A', is called the *derived set* of A. For example, the derived set of (a, b) is $[a, b]$, $(a, b)' = [a, b]$; the derived set of the set $\{1, 2, 3\}$ is the empty set.

The set A is said to be *closed* if it contains its own derived set. For example, the sets $[a, b]$, $\mathbf{R}$ and $\{1, 2, 3\}$ are closed sets. The set $\{1, 2, 3\}$ is closed because it contains its own derived set, that is the empty set (Sect. 1.2 (P)). A point of A which is not an accumulation point of A is said to be an *isolated point*. For example, the points of the set $\{1, 2, 3\}$ are isolated points. In the set $B = \{1, 2\} \cup [3, 4]$, the points 1 and 2 are the isolated points of B, while the points of the interval $[3, 4]$ are the accumulation points of B.

A *left neighborhood* of the point c is defined as the *left-open right-closed* interval with endpoints b and c, i. e., $(b, c] = \{x \in \mathbf{R} : b < x \leq c\}$, and $b < c$. A *right neighborhood* of the point c is defined as the *left-closed right-open* interval with endpoints c and d, i. e., $[c, d) = \{x \in \mathbf{R} : c \leq x < d\}$, and $c < d$.

6.7 Upper Bounds and Lower Bounds

Let us introduce the concept of *boundedness* for numerical sets.

Definition 6.2 A numerical set A is said to be *upper bounded* if there exists a real number c such that $x \leq c$, for every element x of A. Any such number c is called an *upper bound* or a *majorant* of A (Fig. 6.18).

Example The set $A = \{-1, -2, -3, \ldots, -n, \ldots\}$, whose element are the negative integers, is upper bounded because there exists a real number, for example zero, which is greater than every element of A. Also the open interval $(-\infty, 2)$ is an upper bounded subset of **R** because the number 2, and every number greater than 2, is greater than any element of the interval $(-\infty, 2)$. The interval $(-\infty, 2]$ is still an upper bounded numerical set, since the number 2 is greater than or equal to each element of the interval $(-\infty, 2]$.

Definition 6.3 A numerical set A is said to be *lower bounded* if there is a real number d such that $d \leq x$, for every element x of A. Any such number d is called a *lower bound* or a *minorant* of A (Fig. 6.19).

Example The set $E = \{2, 4, 6, \ldots, 2n, \ldots\}$ of all the even natural numbers, is lower bounded because there exists a real number, for example zero, which is smaller than each element of E. The interval $[2, +\infty)$, is a lower bounded subset of **R**, because the number 2, and any number smaller than 2, is smaller than or equal to any element of the interval $[2, +\infty)$.

Definition 6.4 A numerical set that is upper bounded and lower bounded is called a *bounded set*.

Example The set $S = \left\{1, \frac{1}{2}, \frac{1}{3}, \ldots, \frac{1}{n}, \ldots\right\}$, whose elements are the fractions having numerator 1 and denominator a natural number, is lower bounded and upper bounded. In fact, zero is smaller than each element of S and 1 is greater than or equal to each element of S.

The open interval $(0, 1)$ is a bounded set.
The set $A = \{1, 2\} \cup [3, 4]$ is a bounded set.
Any lower bounded numerical set and any upper bounded numerical set are included in a half-line; any bounded numerical set is included in a segment.

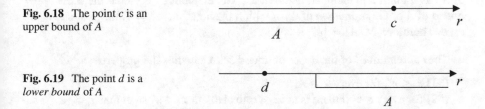

Fig. 6.18 The point c is an upper bound of A

Fig. 6.19 The point d is a lower bound of A

Let A be an upper bounded numerical set. Then one of the two properties occurs:

Property M1. An upper bound for A exists in A.

Property M2. An upper bound for A does not exist in A.

If Property M1 holds, let m'' be an upper bound for A belonging to A. This means that, for every $a \in A$, $a \leq m''$. The number m'' is unique: indeed let $n'' \in A$ be an upper bound for A, then for every $a \in A$, $a \leq m''$ and being $n'' \in A$, $n'' \leq m''$; interchanging m'' with n'', one has $m'' \leq n''$; hence $m'' = n''$. The upper bound m'' is called the *maximum element* of A, or the *maximum* of A, denoted $m'' = \max A$. Therefore, the set A is said to have a *maximum*.

Example In the set $A = \{-1, -2, -3, ..., -n, ...\}$ of the negative integers the number -1 is an upper bound for A that belongs to A; then A has a maximum. In the interval $[-\infty, 2)$ there is no upper bound for A; the number 2 is an upper bound for the interval $[-\infty, 2)$, but it does not belong to $[-\infty, 2)$; any number greater than 2 does not belong to $[-\infty, 2)$, any number smaller than 2 is not an upper bound for $[-\infty, 2)$.

Let A be a lower bounded numerical set. One of the two properties occurs:

Property m1. A lower bound for A exists in A.

Property m2. A lower bound for A does not exist in A.

If Property m1 holds, let m' be a lower bound for A belonging to A. Similarly to the property M1, m' is unique and it is named the *minimum element* of A or the *minimum* of A, denoted $m' = \min A$. Therefore, the set A is said to have a *minimum*.

The operations of addition and multiplication are defined in the set $\mathbf{R}$ of real numbers: this means that the set $\mathbf{R}$ is *closed* with respect to the addition and the multiplication (Sect. 2.5.2). Furthermore, the usual order relation $\leq$ is defined in $\mathbf{R}$. The set $\mathbf{R}$ endowed, or *structured*, with the addition and multiplication and the ordering $\leq$, is called the *real field*. In other words, if the symbol $+$ denote addition, and $\cdot$ the multiplication, then the 4-tuple (Sect. 3.2) of symbols $(\mathbf{R}, +, \cdot, \leq)$ defines the real field.

Let us accept the following propositions which express the property of *completeness* of the real field.

C1. If A is an upper bounded numerical set, then the set of the upper bounds for A has a minimum.

C2. If A is a lower bounded numerical set, then the set of the lower bounds for A has a maximum.

Definition 6.5 If A is an upper bounded subset of $\mathbf{R}$, the set $M(A)$ of the upper bounds of A has a minimum element, denoted e''. The element e'' is called the *least upper bound* of A or the *supremum* of A for which the notations $e'' = 1.\mathrm{u.b.}A = \sup A$ are used. Therefore, $M(A) = [e'', +\infty)$.

The supremum e'' of the upper bounded set A satisfies the properties:

(S1) $x \leq e''$, for every $x \in A$,

(S2) for every $\varepsilon > 0$ there is at least one element $x_\varepsilon \in A$ such that $x_\varepsilon > e'' - \varepsilon$.

Property (S1) means that e'' is an upper bound of A; property (S2) means that every number less than e'' is not an upper bound.

Definition 6.6 If A is a lower bounded subset of **R**, the set m(A) of the lower bounds for A has a maximum element, denoted e'. The element e' is called the *greatest lower bound* of A or the *infimum* of A for which the notations $e'[=$ g.l.b. $A =$ inf A are used. Therefore, $m(A) = (-\infty, e']]$.

The infimum e' of the lower bounded set A satisfies the properties:

(I1) $x \geq e'$, for every $x \in A$,
(I2) for every $\varepsilon > 0$, there is at least one element $x_\varepsilon \in A$ such that $x_\varepsilon < e' + \varepsilon$.
Property (I1) means that e' is a lower bound of A; property (I2) means that every number greater than e' is not a lower bound.

Properties (S1) and (S2) and properties (I1) and (I2) are *characteristic properties* of supremum and infimum in the sense that they are equivalent to Definitions 6.5 and 6.6 and can be assumed as definitions of supremum and infimum, respectively.

Let explicitly mention the further properties:

- $e' \leq e''$ and
- $e' = e''$ if and only if A consists of a unique element,
- among the intervals including A, the interval $[e', e'']$ has minimum length.

Example The set

$$S = \left\{ 1, \frac{1}{2}, \frac{1}{3}, \ldots, \frac{1}{n}, \ldots \right\}$$

is upper bounded and any real number greater than or equal to 1 is an upper bound of S. The number 1 belongs to S and is an upper bound of S: therefore, $maxS = 1$. The set S is lower bounded and any real number smaller than 0 or equal to 0 is a lower bound for S. The set of the lower bounds for S has a maximum that is zero. Therefore, $supS = 1$ and $infS = 0$.

Example The set **Q** of rational numbers is also ordered and structured as a field with addition and multiplication operations between rational numbers. The field **Q**, however, is not complete. In fact, consider the numerical set X contained in **Q** and defined by $X = [0, \sqrt{2}) \cap \mathbf{Q}$. Since the set of rational numbers greater than $\sqrt{2}$ does not have a minimum, **Q** is not complete.

6.8 Commensurability and Real Numbers

In (Sect. 2.5) we dealt with commensurability of segments with respect to a unit of measure: the topic is intimately linked with the geometric origin of the real numbers and can be a start point to describe a procedure for constructing real numbers. Take the non-null segment **u** as unit of measure for segments (see Sect. 2.5 and 6.2.1). A non-null segment **s** and the unit **u** are said to be *commensurable* if there exists a couple (n, m), with m natural and n nonnegative integer such that

$$S = \frac{n}{m}\mathbf{u} \tag{6.7}$$

The equality (6.7) is rewritten this way

$$s = n\left(\frac{1}{m}\mathbf{u}\right)$$

This means that the segment **s** is the sum of n segments each equal to $\frac{1}{m}\mathbf{u}$. The rational number $\frac{n}{m}$ is named the *measure* of **s** with respect to **u**. The measure of the null segment is zero.

For some segment **s** it may occur that a couple (n, m) satisfying (6.7) does not exist; then the segments **s** and **u** are defined *incommensurable*. For example, this is the case of the side and the diagonal of the square (Sect. 2.7.7).

Let us describe the steps of a procedure to associate a measure with **s**, given **u**, being or not **s** and **u** commensurable.

1st *step*. Let n be the largest nonnegative integer such that $n\mathbf{u} \leq \mathbf{s}$. Therefore, by Eudoxus-Archimedes' postulate (Sect. 2.5),

$$n\mathbf{u} \leq \mathbf{s} < (n+1)\mathbf{u}$$

The interval $[n, n+1]$ is a constraint to the measure of the segment **s**. If $n\mathbf{u} = \mathbf{s}$, then the segments **u** and **s** are commensurable and n is the measure of **s** with respect to **u**.

2nd *step*.If $n\mathbf{u} < \mathbf{s}$, let d_1 be the largest decimal digit such that

$$n.d_1\mathbf{u} \leq \mathbf{s}$$

Therefore,

$$n.d_1\mathbf{u} \leq \mathbf{s} < \left(n.d_1 + \frac{1}{10}\right)\mathbf{u}$$

The interval $\left[n \cdot d_1, n \cdot d_1 + \frac{1}{10}\right]$ is a more strict constraint to the measure of the segment **s**. If $n \cdot d_1\mathbf{u} = \mathbf{s}$, then the segments **u** and **s** are commensurable and the rational number $n.d_1$ is the measure of **s** with respect to **u**.

3rd *step*. $n \cdot d_1 \mathbf{u} < \mathbf{s}$. Let d_2 be the largest decimal digit such that

$$n.d_1 d_2 \mathbf{u} \leq \mathbf{s}$$

Therefore,

$$n.d_1 d_2 \mathbf{u} \leq \mathbf{s} < \left(n.d_1 d_2 + \frac{1}{10^2} \right) \mathbf{u}$$

The interval $\left[n.d_1 d_2, n.d_1 d_2 + \frac{1}{10^2} \right]$ is a more strict constraint to the measure of the segment $\mathbf{s}$. If $n.d_1 d_2 \mathbf{u} = \mathbf{s}$ then the segments $\mathbf{u}$ and $\mathbf{s}$ are commensurable and the rational number $n.d_1 d_2$ is the measure of $\mathbf{s}$ w. r. to $\mathbf{u}$.

4th *step*. $n.c_1 c_2 \mathbf{u} < \mathbf{s}$. The procedure continues.

Definition 6.7 Any sequence $I_1, I_2, I_3, \ldots, I_h, \ldots$ of intervals with rational end-points, each of which is contained in the preceding one and such that for every number $p > 0$ there exists a natural number n such that the length of I_n is less than p, is called a *sequence of nested intervals*.

For example, the sequence of intervals

$$[n, n+1], \left[n.d_1, n.d_1 + \frac{1}{10} \right], \left[n.d_1 d_2, n.d_1 d_2 + \frac{1}{10^2} \right], \ldots$$

is a sequence of nested intervals.

Let us accept the following proposition without proof.

Postulate. Given a sequence of nested intervals, there exists one and only one point that belongs to each interval of the sequence. This point is called a *real number*: if it is not a rational number it is called an *irrational number*.

Coming back to the procedure for constructing the measure of $\mathbf{s}$ w. r. to $\mathbf{u}$, if the unique point determined by the sequence of nested intervals is rational, then $\mathbf{s}$ and $\mathbf{u}$ are *commensurable*, if the point is irrational, then $\mathbf{s}$ and $\mathbf{u}$ are *incommensurable*.

6.9 Separate Sets and Contiguous Sets

Let the sets A and B be subsets of real numbers. If each element a of A is smaller than any element b of B, then the sets A and B are said *separate sets*.

Definition 6.8 Two separate sets are said to be *contiguous* if, for every (in particular, no matter how small) real number $d > 0$, an element a of A and an element b of B exist such that.

$$|b - a| < d$$

Example 6.10 The open interval $H = (-1, 0)$ and the set $S = \left\{1, \frac{1}{2}, \frac{1}{3}, \ldots, \frac{1}{n}, \ldots\right\}$ are separate since each element of H is smaller than any element of S. They are also contiguous since, for every $d > 0$, there exists an element h of H and an element s of S, such that $|s - h| < d$. The intervals.

$$[-1, 1], \left[-\frac{1}{2}, \frac{1}{2}\right], \left[-\frac{1}{3}, \frac{1}{3}\right], \ldots, \left[-\frac{1}{n}, \frac{1}{n}\right], \ldots$$

form a sequence of nested intervals. Then there is one and only one real number that belongs to all intervals of the sequence. This number is zero.

Example 6.11 Let us consider the irrational number $\sqrt{2} = 1.4142\ldots$ and the numerical sets.

$$A(\sqrt{2}) = \{1; 1.4; 1.41; 1.414; 1.4142; \ldots\}$$

$$B(\sqrt{2}) = \{2; 1.5; 1.42; 1.415; 1.4143; \ldots\}$$

The sequence of intervals

$$[1, 2], [1.4, 1.5], [1.41, 1.42], [1.414, 1.415], [1.4142, 1.4143], \ldots$$

is a sequence of nested intervals. Then a unique real number belonging to all intervals of the sequence exists. This point is the real number $\sqrt{2}$. The sets $A(\sqrt{2})$ and $B(\sqrt{2})$ are separate and contiguous.

If two separate numerical sets A and B are contiguous, then there exists exactly one real number r, such that $a \leq r \leq b$, for every a in A and b in B. In fact, a sequence of nested intervals $[a_n, b_n]$, $a_n \in A$, $b_n \in B$, $n \in \mathbf{N}$, can be built. Then there exists a unique real number that belongs to all intervals of the sequence. This point is called the *element of separation* of the contiguous sets A and B.

Example 6.12 If A is an upper bounded numerical set, then the $e'' = \sup A$ is the element of separation of A and the set $M(A)$ of the upper bounds of A.

Bibliography

Anton, H.: Calculus. Wiley, New York (1980)
Chinn, W.G., Steenrod, N.E.: First Concepts of Topology. Random House, New Mathematical Library, New York (1966)
Courant, R., Robbins, H.: What is Mathematics? Oxford University Press, Oxford (1978)

Chapter 7
Real-Valued Functions of a Real Variable. The Line

7.1 The Cartesian Plane

We showed (Sect. 6.2.2) that the introduction of a coordinate system in a line leads to the identification of the set of real numbers and the set of the points of the line.

We now will show that the plane can be identified with the set $\mathbf{R}^2$ of the couples of real numbers.

Let two orthogonal oriented lines, called *x axis* and *y axis*, be fixed in the plane α. The axes intersect each other at a point O and both axes are endowed with coordinate systems of common origin O and unit points U_x and U_y, such that $OU_x = OU_y = 1$.

The triple of points (O, U_x, U_y) are sufficient to define a *coordinate system* of the plane. A plane with a coordinate system is called a *coordinate plane*, or simply a *plane xy*.

Let us show that there is a 1–1 correspondence between the points of a coordinate plane and the couples of real numbers.

Indeed, if P is a point in the plane *xy*, let us draw two lines that pass through P and are perpendicular to *x* axis and to *y* axis. If the first line intersects the *x* axis at the point of coordinate *a* and the second line intersects the *y* axis at the point of coordinate *b*, then we associate the couple (a, b) to the point P: the number *a* is called the *first coordinate* or the *abscissa* of P and the number *b* is called the *second coordinate* or the *ordinate* of P. The point P is called *the point of coordinates* (a, b), denoted $P(a, b)$. The described procedure shows that a unique couple of real numbers is associated to each point of a coordinate plane.

Vice versa, from a couple (a, b) of numbers we draw the perpendicular lines to the *x* and *y* axes at the points of coordinates *a* and *b*, respectively; the intersection of the two lines is the unique point whose coordinates are (a, b). Therefore, a 1–1 correspondence exists between the set of the points of a plane and the set of the couples of real numbers. This allows the plane to be identified with the set $\mathbf{R}^2$ and the same symbol $\mathbf{R}^2$ be used to indicate both the set of the couples of real numbers and the set of points of the plane; furthermore, we say that the couple (x, y) is a point

© The Author(s), under exclusive license to Springer Nature Switzerland AG 2023   101
A. G. S. Ventre, *Calculus and Linear Algebra*,
https://doi.org/10.1007/978-3-031-20549-1_7

of the plane. The symbol P(x, y), or P $= (x, y)$, denotes the point P having coordinates (x, y).

7.1.1 Quadrants

The coordinate axes divide the plane into four regions, called *quadrants*: any point in the first quadrant has abscissa and ordinate positive, any point in the second has negative abscissa and positive ordinate, each point in the third has both abscissa and ordinate negative, each point in the fourth has positive abscissa and negative ordinate (Fig. 7.1). Of course, the ordinate of any point of the x axis is zero, the abscissa of any point of the y axis is zero.

If the point (x, y) is in the first quadrant, then the point ($-x$, y) is in the second and the two points are symmetrical with respect to the y axis; the point ($-x$, $-y$) is located in the third quadrant and is the symmetric of (x, y) with respect to the origin, i. e., O is the midpoint of the segment of endpoints (x, y) and ($-x$, $-y$); the point (x, $-y$) is located in the fourth quadrant and is the symmetric of (x, y) with respect to the x axis.

Congruent geometrical figures. We add some considerations to the concept of *congruence* already mentioned (Sect. 4.5.2). Saying that the sets A and B are equal means that A and B are two different names of the same set (Sect. 1.2). Hence it is improper to call equal the segments OU_x and OU_y because they have the unique point O in common. We say *congruent* two plane figures, i. e., two sets of points in the plane, if a (rigid) movement, i. e., a movement that preserves the distances, leads one set to coincide with the other. Then the segments OU_x and OU_y are congruent because one of them rotated overlaps the other.

Fig. 7.1 The coordinate system in the plane: O origin of the axes, U_x and U_y unit points of the axes, (x, y) the couple of coordinates of P. Quadrants I, II, III and IV; symmetrical points of P with respect to the axes and the origin

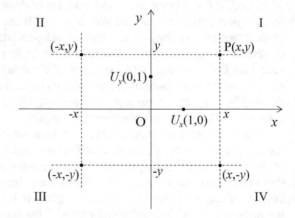

7.1.2 Distance

We extend the notion of distance introduced in (Sect. 6.5) to the pairs of the points in the plane xy. Let us define *distance* of the points $P_1(x_1, y_1)$, $P_2(x_2, y_2)$ the function d which associates the non-negative real number

$$d(P_1, P_2) = |P_1P_2| = \sqrt{(x_2-x_1)^2 + (y_2-y_1)^2}$$

to the couple of points (P_1, P_2), where $|P_1P_2|$ denotes the length of the segment $\overline{P_1P_2}$. The procedure that determines the value of d is a simple application of the Pythagorean theorem (Fig. 7.2),

and consists in

a. setting the points $P_1(x_1, y_1)$, $P_2(x_2, y_2)$ in the plane;
b. constructing the right triangle with hypotenuse $\overline{P_1P_2}$ and catheti $\overline{HP_1}$ and $\overline{HP_2}$, parallel to the coordinate axes, with $H(x_1, y_2)$;
c. calculating the lengths $|x_2 - x_1|$ and $|y_2 - y_1|$ of the catheti $\overline{HP_1}$ and $\overline{HP_2}$ (Sect. 6.2.1), respectively;
d. calculating the length of the hypotenuse:

$$d(P_1, P_2) = \sqrt{|x_2-x_1|^2 + |y_2-y_1|^2} = \sqrt{(x_2-x_1)^2 + (y_2-y_1)^2}$$

The number $d(P_1, P_2)$ is called the *value of the distance*, or simply the *distance* of the points P_1 and P_2.

For example, the distance of the points $A(6, -1)$ and $B(5, -2)$ is equal to

$$d(A, B) = \sqrt{(5 - 6)^2 + (-2 + 1)^2} = \sqrt{1 + 1} = \sqrt{2}$$

which is also the length of the segment $\overline{AB}$.

In particular, the distance of $P(x, y)$ from the origin is $|OP| = \sqrt{x^2 + y^2}$ (Fig. 7.3).

Fig. 7.2 Distance $d(P_1, P_2)$

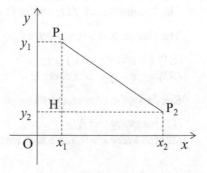

Fig. 7.3 The distance |OP|

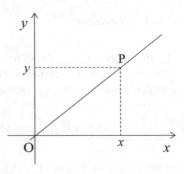

7.2   Real-Valued Functions of a Real Variable

A function whose range is a subset of **R** is called a *real-valued function*.

A function whose domain is a subset A of **R** is called a *function of a real variable*.

Therefore, a *real-valued function of a real variable* is a function $f : A \rightarrow B$, such that $A \subseteq \mathbf{R}$ and its range is $f(A) \subseteq B \subseteq \mathbf{R}$.

7.2.1   *Extrema of a Real-Valued Function*

Let $f : A \rightarrow B$ be a real-valued function defined in $A \subseteq \mathbf{R}$. Since the range $f(A)$ is a subset of **R**, the concepts and properties expressed in Chap. 6 and related with numerical sets, apply to $f(A)$, in particular what is concerned with boundedness. So instead of saying *the range of f is upper bounded*, or *the range of f is endowed with maximum*, instead of saying *the supremum of the range of f*, for sake of brevity, we prefer to talk about *upper bounded function f, f endowed with maximum, f endowed with supremum*, and so on.

Nevertheless, we want to replicate the definitions in full and adapt some symbols.

A real-valued function $f : A \rightarrow B$ is called *lower bounded, upper bounded, bounded* in A if the range $f(A)$ is lower bounded, upper bounded, bounded, respectively.

The infimum and the supremum of $f(A)$ are called the *infimum* and the *supremum* of f in A, denoted $\inf_{x \in A} f(x)$ (or *inf f*) and $\sup_{x \in A} f(x)$ (or *sup f*), respectively.

The *characteristic properties* (see Sect. 6.7) of supremum e'' of f take the form:

(S1) for every $x \in A, f(x) \leq e''$;
(S2) for every $\varepsilon > 0$, there is at least one element $x_\varepsilon \in A$ such that $f(x_\varepsilon) > e'' - \varepsilon$.

Similarly, the *characteristic properties* of infimum e' take the form:

(I1) for every $x \in A, x \geq e'$;
(I2) for every $\varepsilon > 0$, there is at least one element $x_\varepsilon \in A$ such that $f(x_\varepsilon) < e' + \varepsilon$.

If the range of f is endowed with minimum (maximum), then we say that f is endowed with *minimum* (*maximum*) in A; in other words, f is said to have a *minimum* (*maximum*) in A if there exists a point $x' \in A$ ($x'' \in A$) such that $f(x') \leq f(x)$, ($f(x) \leq f(x'')$), for every $x \in A$. The minimum and the maximum of f in A are denoted $\min_{x \in A} f(x)$ (or *min f*) and $\max_{x \in A} f(x)$ (or *max f*), respectively. If $f(x') = min\ f$, then f is said to have a *minimum point* at x'; if $f(x'') = max\ f$, then f is said to have a *maximum point* at x''.

The infimum and the supremum, and the minimum and maximum of f, are generically named *extrema* of the function.

7.2.2 The Graph of a Real-Valued Function

We know (Sect. 5.1) that the function $f : A \subseteq \mathbf{R} \to B \subseteq \mathbf{R}$ associates a unique element b of the set B to each element a belonging to the set A. We defined the *graph* of f as the set of couples $(x, f(x))$,

$$A \times f(A) = \{(x, f(x)), \text{ for every } x \in A\}$$

where A and $f(A)$ are the domain and the range of f, respectively. Any couple $(x, f(x))$ is a point P of the plane xy. We write $y = f(x)$ to indicate that the couple $(x, f(x))$ is a point (x, y) of the graph of f: this is the act that produces the birth of the "new subject" (Sect. 6.1), i. e., the analytic geometry.

The graph of the function is often identified with its geometric image: the orthogonal projection of the graph over the x axis coincides with the domain of the function and the orthogonal projection of the graph over the y axis coincides with the range.

Two different couples of elements in the graph of f have distinct abscissas. The curve in Fig. 7.5 is a set of points (x, y) which cannot be the graph of any real-valued function of a real variable, while the curve in Fig. 7.4 is a set of points $(x, f(x))$ which is the graph of a real-valued function of a real variable f.

Fig. 7.4 The curve is the graph of the function f

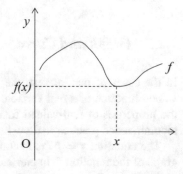

Fig. 7.5 The curve is not the graph of any function because there are two points of the curve having the same abscissa

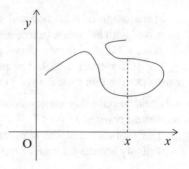

Example 7.1 Let f be the function that associates the number $2x - 3$ to the number $x, f(x) = 2x - 3$. The function f is a real-valued function of the real variable x. The domain and the range of f are equal to **R**. In order to find some couples of the graph of f, let us choose numbers for x and find $f(x)$:

$$f(x) = 2x - 3 \quad (x, f(x))$$

$$f(-1) = 2(-1) - 3 = -5 \quad (-1, -5)$$

$$f(0) = 2(0) - 3 = -3 \quad (0, -3)$$

$$f\left(\tfrac{3}{2}\right) = 2\left(\tfrac{3}{2}\right) - 3 = 0 \quad \left(\tfrac{3}{2}, 0\right)$$

$$f(1) = 2 \times 1 - 3 = -1 \quad (1, 1)$$

Therefore, $(-1, -5)$, $(0, -3)$, $\left(\tfrac{3}{2}, 0\right)$, $(1, 1)$ are four couples of the graph of f. In this way we can determine how many points we want in the graph of f.

We will show that all points of the graph of the function $f(x) = 2x - 3$ lie on a line and all the points of this line are the points of the graph of f.

7.2.3   Graph and Curve

In the cases of our interest, the geometric representation of the graph of a real-valued function of a real variable is often a curve. Such a representation highlights the properties of real-valued functions of a real variable and it is a common and recommended practice to draw the graphs of these functions, whenever possible.

The equation $y = f(x)$ is equivalent to say that the point (x, y) belongs to the graph of the function f. In particular, the expression $y = 2x - 3$ is equivalent to $f(x) = 2x - 3$.

Let us observe that drawing a graph is not always physically possible. For example, we cannot give even an idea about the graph of the real-valued function j defined in the interval $[0, 1]$:

$$j(x) = 1, \text{ if } x \text{ is a rational number}$$

$$j(x) = 0, \text{ if } x \text{ is an irrational number}$$

7.3 Lines in the Cartesian Plane

We study the analytic representation of the line in the plane xy. Such a representation can take various forms that satisfy suitable geometric and analytic features.

7.3.1 The Constant Function

The real-valued function f of the real variable x defined by:

$$f : x \rightarrow c$$

where c is a real number, is called the *constant function*, also denoted $f(x) = c$, or $y = c$. The domain of f is $\mathbf{R}$ and the range of f is the set $\{c\}$ consisting of the unique point c. Therefore, the graph of the function f is the set of points (x, c), for every point $x \in \mathbf{R}$. If $c = 0$, the graph of f is the x axis; if $c = 1$, the graph is the line parallel to the x axis passing through the points of ordinate 1 (Fig. 7.6).

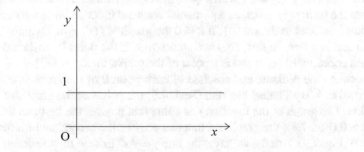

Fig. 7.6 Graph of the constant function $y = f(x) = 1$

Fig. 7.7 Graph of the
identical function $f(x) = x$

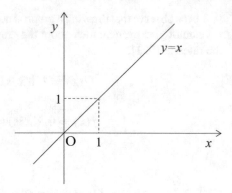

7.3.2   The Identical Function

The real-valued function f of the real variable x

$$f : x \to x$$

for every $x \in \mathbf{R}$, is called the *identical function* or *identity* (*function*) onto $\mathbf{R}$. The
domain and the range of f are equal to $\mathbf{R}$. The function f associates x to x itself and
also the notations $f(x) = x$ and $y = x$ are used. The solutions of the equation $y = x$
are the couples of the coordinates of the points $(x, f(x)) = (x, x)$. As $f(-1) = -1$,
$f(0) = 0, f(2) = 2$ the points $(-1, -1)$, $(0, 0)$, $(2, 2)$ belong to the graph of identical
function. The graph of f is the bisector of the first and third quadrant (Fig. 7.7).

7.3.3   The Function $f : x \to kx$

The function $f(x) = kx$, with $k \neq 0$, has domain and range $\mathbf{R}$. If $k = 0$, the function
has the form $f(x) = 0$, i. e., a particular constant function whose domain is $\mathbf{R}$ and the
range reduces to the set $\{0\}$. If $k \neq 0$ the graph of $f(x) = kx$ contains the origin O(0,
0) and is a line. In fact, $f(0) = 0$; moreover, if the point P, distinct from the origin,
has coordinates (x, y) and is a point of the graph, then $y = kx$, i. e., $\frac{y}{x} = k$. The ratio
between the ordinate and abscissa of each point P of the graph is constant and it is
equal to k. By Thales' theorem (Sect. 4.6) the points of the graph are the points of a
line. The graph of the function f is a line that crosses the first and third quadrant, if
$k > 0$ (Fig. 7.8); the graph is a line that crosses the second and fourth quadrant, if k
< 0 (Fig. 7.9). Briefly, we say: "the line $y = kx$" instead of "the graph of $f : x \to kx$"
or "the graph of $f(x) = kx$".

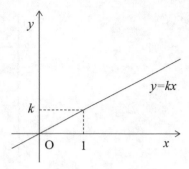

Fig. 7.8 The line $y = kx$, $k > 0$

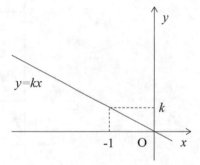

Fig. 7.9 The line $y = kx$, $k < 0$

Example 7.2 The line passing through the origin $(0, 0)$ and the point $(-1, 1)$ is the graph of the function $f(x) = -x$. Indeed, $(0, 0)$ and $(-1, 1)$ are solutions of the equation $y = -x$.

If k is positive, as k increases the graph of f is a steeper and steeper line because the ratio between the ordinate and the abscissa of the point in the line increases. The number k is called the *slope* of the line. For example, the line $y = 3 \times x$ is steeper than the line $y = 2x$ (Figs. 7.10 and 7.11).

If a segment of the line $f(x) = kx$ has the orthogonal projection of length 1 on the x axis, then it has an orthogonal projection of length k on the y axis (Fig. 7.12). The greater the absolute value of the slope k is, the closer the line is to the vertical position.

Exercise 7.1 Find the equation of the line that passes through the origin and has slope $\frac{5}{2}$.

The line passes through the origin and the point $(1, \frac{5}{2})$ and it is the graph of the function

$$f(x) = \frac{5}{2}x, \quad \left(\text{or } y = \frac{5}{2}x\right).$$

Fig. 7.10 The line $y = 2x$

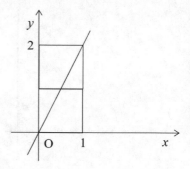

Fig. 7.11 The line $y = 3x$

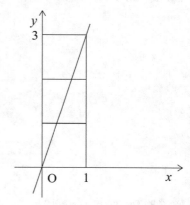

Fig. 7.12 k is the slope of the line PQ

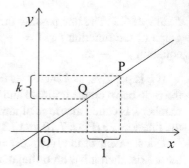

Exercise 7.2 Find the equation of the line that passes through the origin and has slope -3.

Solution. The line is the graph of $y = -3x$; it passes through the origin $O(0, 0)$ and the point $(1, -3)$.

The equality $f(x) = kx$ is an equation because it is verified if, and only if, y is equal to kx, i. e., the solutions of the equation are all the couples (x, kx), for every $x \in \mathbf{R}$, and only these couples are solutions of the equation $f(x) = kx$. The equation $y = kx$ is called the *equation of the line* having slope k and passing through the origin.

7.3.4 The Function $f : x \rightarrow kx + n$

The graph of the function $f(x) = kx + n$ is a line parallel to the line of equation $y = kx$. In fact, each point $(x, kx + n)$ of the graph of $f(x) = kx + n$ is obtained by adding n to the ordinate of the point (x, kx) of the line $y = kx$ (Fig. 7.13). Of course, the notations $f(x) = kx + n$ and $y = kx + n$ are equivalent.

If r is the line of equation $y = kx + n$, we write

$$(r) \quad y = kx + n$$

The line (r) $y = kx + n$ passes through the point $(0, n)$, intersection of r with the y axis. The number n is called the *ordinate at the origin*, or the y *intercept* of the line. The line r is identified as the line with slope k and passing through the point $(0, n)$. Two parallel lines have the same slope. The line

$$(s) \quad y = 3x - 2$$

has slope 3 and ordinate at the origin $- 2$.

Remark 7.1 Some few more words on the concept of *representation*. The equation.

$$(r) \quad y = kx + n$$

is said *to represent* the line r or to give a representation of the line r. This means that every point of the line r has coordinates (x, y) solution of the equation $y = kx + n$, and, moreover, the points of the line r are the only points of the plane whose coordinates are solutions of equation $y = kx + n$.

The equation $y = kx + n$ of the line r is called the *explicit equation* or the *explicit representation* of the line r.

Fig. 7.13 Parallel lines $y = kx$ and $y = kx + n$

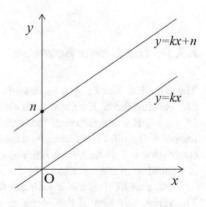

Remark 7.2 The vertical lines, i. e., the lines parallel to the y axis, are not graphs of any real function. For any vertical line, neither the slope nor the ordinate at the origin can be defined. In fact, the orthogonal projection of any vertical segment reduces to a point on the x axis; moreover, every vertical line, distinct from the y axis, does not meet the y axis.

Remark 7.3 If the point (x_0, y_0) belongs to the line.

$$\text{(r)} \quad y = kx + n$$

then the equation is satisfied by the solution (x_0, y_0), i. e.,

$$y_0 = kx_0 + n$$

and subtracting this equality from (r), the equation

$$y - y_0 = k(x - x_0)$$

with slope k and passing through the point (x_0, y_0) is obtained.

For example, the equation of the line s passing through $(-4, 8)$ and parallel to the line

$$\text{(r)} \quad y = 6x - 1$$

is

$$\text{(s)} \quad y - 8 = 6(x + 4),$$

I.e., $y = 6x + 32$.

7.3.5   *The Linear Equation*

The equation $y = kx + n$ represents a non-parallel line to the y axis. The vertical line passing through a point with abscissa h has equation $\text{x} = h$, in fact any couple $(h, y), y \in \mathbf{R}$ is a solution of the equation $\text{x} = h$. (Think at the equation $\text{x} = h$ in the form $x + 0y = h$.) For example, the vertical line passing through the point $(2, 0)$ has equation $\text{x} = 2$, in fact, each point of abscissa 2 has coordinates $(2, y)$ and this couple is a solution of the equation $\text{x} = 2$.

Both $y = kx + n$ and $x = h$ are first degree equations in the variables x and y. Therefore, any line of the plane, is represented by a first degree equation in two variables. In conclusion, the first degree equation

Fig. 7.14 The line $x - 2y + 1 = 0$

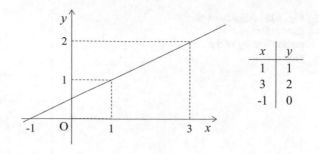

x	y
1	1
3	2
-1	0

$$ax + by + c = 0$$

with a and b not both zero, includes all the forms $y = kx + n$ and $x = h$ and it is called the *ordinary*, or *cartesian equation of the line* or the *linear equation in x and y*.

For every line r there is a linear equation with two variables x and y that represents the line and for every linear equation with two variables there is one and only one line represented by the given equation. Linear equations with proportional coefficients represent the same line.

Example 7.3 Let us find some points of the line.

$$(r) \quad x - 2y + 1 = 0$$

We proceed by giving a value to x (or y) and we find the consequent value of y (of x); for example,

$$\text{if } x = 1, \text{ then } y = 1$$

$$\text{if } y = 0, \text{ then } x = -1.$$

Let us draw the line passing through the points $(1, 1)$ and $(-1, 0)$ (Fig. 7.14).

7.3.6 The Parametric Equations of the Line

Consider the line r passing through the points $P_1(x_1, y_1)$ and $P_2(x_2, y_2)$. Whatever the point $P(x, y)$ in the line r is, the couple of oriented segments (P_1P_2, P_1P) (see. Section 6.2.2) is proportional to the couples of the lengths of the orthogonal projections $(x_2 - x_1, x - x_1)$ and $(y_2 - y_1, y - y_1)$ on the axes. In fact, by Thales' theorem the line r is a transversal that cuts the parallels to the y axis and the parallels to the x axis (Fig. 7.15). Proportionality is expressed by the equality between the segments

Fig. 7.15 Alignment of
three points and
proportionality of the
projections

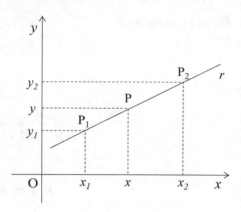

$$P_1P = tP_1P_2$$

where t is a real number and by the equalities between the lengths of the projections
on the coordinate axes:

$$x - x_1 = t(x_2 - x_1) \tag{7.1}$$

$$y - y_1 = t(y_2 - y_1) \tag{7.2}$$

where the real number t is called *parameter* or *coefficient* and $x_2 - x_1$ and $y_2 - y_1$
are the lengths of the projections of the segment P_1P_2 on x and y axes, respectively
(Fig. 7.15).

Equations (7.1) and (7.2) represent the line r passing through P_1 and P_2: indeed, if
the point P belongs to the line r, then the coordinates of P are solutions of Eqs. (7.1)
and (7.2) and vice versa, each couple of real numbers (x, y) which is the solution of
(7.1) and (7.2) is the couple of the coordinates of exactly one point of the line r.

Definition 7.1 The numbers $m = x_2 - x_1$ and $n = y_2 - y_1$ are called *direction numbers*
of the line.

Equations (7.1) and (7.2) are called *parametric equations of the line r* passing
through P_1 and P_2 and take the equivalent form

$$x = x_1 + t(x_2 - x_1) \tag{7.3}$$

$$y = y_1 + t(y_2 - y_1) \tag{7.4}$$

Example 7.4 Let $x - 2y + 1 = 0$ be the ordinary equation of the line r. Two points
of r are $(1, 1)$ and $(-1, 0)$. The parametric equations of r are, by (7.3) and (7.4):

$$x = 1 - 2t$$

$$y = 1 - t$$

with $t \in \mathbf{R}$.

If $x_2 - x_1 \neq 0$ and $y_2 - y_1 \neq 0$, (7.1) and (7.2) yield

$$t = \frac{x - x_1}{x_2 - x_1}$$

$$t = \frac{y - y_1}{y_2 - y_1}$$

which lead to the single equation $\frac{x - x_1}{x_2 - x_1} = \frac{y - y_1}{y_2 - y_1}$ called the equation of the line r in the form of the equal ratios.

For example, from the parametric equations

$$x = 1 - 2t$$

$$y = 1 - t$$

that yield $t = \frac{x-1}{-2}$. and $t = 1 - y$, we get the equation in the form of the equal ratios: $\frac{x-1}{-2} = 1 - y$.

7.4 Parallel Lines

If two lines are coplanar, then one of the alternatives occurs (see Sect. 4.3.1):

a. the lines have exactly one point in common;
b. the lines are coincident;
c. the lines have no point in common.

In the cases (b) or (c), then the lines are said *parallel* and, in particular, in case (c), the lines are said *properly parallel*. In case (a) the two lines are called *incident* at the common point.

7.4.1 Parallel Lines Represented by Parametric Equations

We state the necessary and sufficient conditions for parallelism between lines in terms of direction numbers. The parametric equations of the line r passing through the points $P_1(x_1, y_1)$ and $P_2(x_2, y_2)$ are (7.3) and (7.4); setting $m = x_2 - x_1$ and $n = y_2 - y_1$, the parametric equations take the form

$$x = x_1 + mt$$

$$y = y_1 + nt$$

with $t \in \mathbf{R}$. The numbers m and n are the direction numbers of the line r. The parametric equations of the line emphasize the direction numbers. Let now s be a line with direction numbers m', n'. By Thales' theorem we state:

Proposition 7.1 *Let (m, n) and (m', n') be the couples of the direction numbers of the lines r and s, respectively. The lines r and s are parallel if and only if the couples (m, n) and (m', n') are proportional, i.e.:*

$$m' = hm, \quad n' = hn$$

with $h \neq 0$. In particular if $n \neq 0$ and $n \neq 0$, r and s are parallel if and only if

$$m : n = m' : n'$$

Example 7.5 The line r represented by the parametric equations.

$$x = 4 + 2t$$

$$y = -1 + 3t$$

$t \in \mathbf{R}$, and the line s represented by the parametric equations

$$x = 1 + 6t$$

$$y = 6 + 9t$$

$t \in \mathbf{R}$, are parallel because their direction numbers are proportional, in fact $6 : 2 = 9 : 3$.

Example 7.6 The lines u and w,

$$(u) \quad x = 4 + 6t, \quad y = -1 - 2t$$

$$(w) \quad x = 1 + 3t, \quad y = 1 - t$$

are parallel because they have proportional direction numbers, the couple $(6, -2)$ is proportional to $(3, -1)$. The lines u and w do not coincide because the point $(4, -1)$, which belongs to the line u does not belong to w; in fact, the point $(4, -1)$ of u, does not satisfy the equations (w): indeed, there is no number t satisfying both equations: $4 = 1 + 3t$ and $-1 = 1 - t$.

7.4.2   Parallel Lines Represented by Ordinary Equations

Proposition 7.2 *A couple of direction numbers of the line.*

$$(r) \quad ax + by + c = 0$$

is $(-b, a)$.

Proof Let $P(x, y)$ and $P_0(x_0, y_0)$ be distinct points of the line r. Then the following equalities hold:

$$ax + by + c = 0$$

$$ax_0 + by_0 + c = 0$$

By subtracting we obtain

$$a(x - x_0) + b(y - y_0) = 0 \qquad\qquad (7.5)$$

Equality (7.5) is fulfilled if and only if a real number $k \neq 0$ exists such that

$$x - x_0 = -kb$$

$$y - y_0 = ka$$

As $x - x_0$ and $y - y_0$ are direction numbers of the line r, by Proposition 7.1, $(-b, a)$ and $(-kb, ka)$, $k \neq 0$, are also direction numbers of the line r. $\qquad\square$

7.4.3   Parallel Lines. Exercises

If $b \neq 0$, the ordinary equation of the line

$$(r) \quad ax + by + c = 0$$

takes the form $y = -\frac{a}{b}x - \frac{c}{b}$, for $k = -\frac{a}{b}$ and $p = -\frac{c}{b}$, we obtain the *explicit equation* of r:

$$(r) \quad y = kx + p$$

The number k, coefficient of x, is the slope of the line r. (see Sect. 7.3.4).
Let us state immediately:

Proposition 7.3 *The slope k of the line r which has direction numbers (m, n), $m \neq 0$, is equal to the ratio $\frac{n}{m}$.*

Proposition 7.4. *If the ordinary equation of the line r is*

$$(r) \quad ax + by + c = 0$$

$b \neq 0$, *then the slope of r is $k = -\frac{a}{b}$; if the line r has parametric equations*

$$x = x_1 + mt$$

$$y = y_1 + nt$$

$m \neq 0$, *then the slope of r is $k = \frac{n}{m}$.*

Proposition 7.5 *The slope of any line parallel to the line*

$$(r) \quad y = kx + p$$

is k.

For example, the lines r and s

$$(r) \quad y = 2x + 3$$

$$(s) \quad y = 2x - 1$$

are parallel.

Example 7.7

a.   The line with parametric equations

$$x = 4 + 2t$$

$$y = -1 + 3t$$

has slope $k = \frac{3}{2}$.

b. The slope of the line with ordinary Eq. $2x - y + 3 = 0$ is 2

Exercise 7.3 Find the intersection of the lines:

$$(r) \quad 2x - y + 3 = 0$$

$$(s) \quad y = -3x + 8$$

Solution The lines r and s are not parallel as their respective direction numbers, (1, 2), (–1, 3), are not proportional. Therefore, the lines are incident at a point P(x, y); let us find the coordinates of the unknown point P. The ordinate of P is $y = 2x + 3 = -3x + 8$, and the abscissa comes from the Eq. $2x + 3 = -3x + 8$, whence $5x = 5$ and $x = 1$; set $x = 1$ in the equation (r) (or (s)), to find $y = 5$. Thus P has coordinates (1, 5).

Exercise 7.4 Given the points A (–1, 2) and B (–2, 1), find the direction numbers of the line AB.

Solution The differences $(x_2 - x_1)$ and $(y_2 - y_1)$ between the abscissas and ordinates of A and B are:

$$x_2 - x_1 = -2 - (-1) = -1$$

$$y_2 - y_1 = 1 - 2 = -1$$

So direction numbers of the line AB are $m = -1$, $n = -1$ and all the non-null couples proportional to (–1, –1).

Example 7.5 Given the points A (–1, 2) and B (–2, 1), find the parametric equations of the line AB and its explicit equation.

Solution The direction numbers of the line AB are (–1, –1). Therefore, parametric equations of the line AB are

$$x = -1 - t$$

$$y = 2 - t$$

The substitution $t = -x - 1$ in the second equation yields: $y = 2 - (-x - 1)$, whence the required explicit equation: $y = x + 3$.

Fig. 7.16 Graph of $y = |x|$

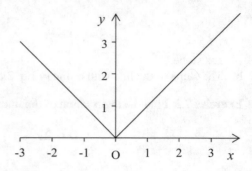

7.5 The Absolute Value Function

The absolute value function $f : x \to |x|$ associates the absolute value $|x|$ to the real number x (Sect. 5.3). The function is also denoted $y = |x|$, its domain is $\mathbf{R}$ and the range is the set of nonnegative real numbers $\mathbf{R}^+$. The graph is drawn in Fig. 7.16.

The restriction (Sect. 5.5) of the absolute value function to the set $\mathbf{R}^+$ of nonnegative real numbers is the function $g : \mathbf{R}^+ \to \mathbf{R}^+$ such that $g(x) = x$, for every $x \in \mathbf{R}^+$.

7.6 A Linear Model

Let us consider the function $c = 60 + 0.8i$, that represents a relation between income i and consumption c (Sect. 3.2, Example 3.3). This is a real-valued function of a real variable. In fact, for each fixed income expressed by the real number i the amount of consumption c is determined. Let us call f this function: $f(i) = c$, or $f(i) = 60 + 0.8i$. If in the plane we fix the axes i, for abscissas and c, for ordinates, the graph of the function $f(i) = 60 + 0.8i$ is a line, which has slope 0.8 and ordinate at the origin 60. The relation f between income and consumption is an opinion that simplifies reality. Of course, models that describe real-life situations usually fail to describe or predict all aspects of a complex situation.

7.7 Invertible Functions and Inverse Functions

We know (Sect. 5.1) that, given a one-to-one function f, if y is the image under f of x, i. e., $y = f(x)$, then x is the image under f^{-1} of y, i. e., $f^{-1}(y) = x$. Then if f is an invertible function, each of the two equalities, $y = f(x)$ and $f^{-1}(y) = x$, implies the other. In symbols (Sect. 2.7):

$$f^{-1}(y) = x \Leftrightarrow y = f(x) \qquad (7.6)$$

The domain of f^{-1} is equal to the range of f and the domain of f is equal to the range of f^{-1}. Furthermore, the inverse of the inverse of the function f is the function f itself: in fact,

$$f^{-1}(y) = x \Leftrightarrow y = \left(f^{-1}\right)^{-1}(x)$$

since the inverse exchanges x with y, and therefore $(f^{-1})^{-1}$ acts just like f:

$$\left(f^{-1}\right)^{-1}(x) = f(x) = y$$

From the equivalence of Eqs. (7.6), replacing one in the other, we obtain the following, called *cancellation identities*:

$$f\left(f^{-1}(y)\right) = y$$

$$f^{-1}(f(x)) = x$$

The cancellation identities state that the composite functions $f \circ f^{-1}$ and $f^{-1} \circ f$ are the identical functions defined in the range of f and the domain of f, respectively.

Let us now consider the graph of the inverse function of the real-valued function f of the real variable x. If the point (x, y) belongs to the graph of f, then the point (y, x) belongs to the graph of f^{-1} (Fig. 7.17).

Since the two points are symmetrical with respect to the bisector of the first and third quadrant having equation $y = x$, the graphs of f and f^{-1}, referred to the same coordinate system, and hence denoting the independent variable with x again, are symmetrical with respect to the bisector.

The graphs of f and the inverse f^{-1} are plotted in Fig. 7.18.

Exercise 7.5 Show that the function $y = f(x) = 6 - 3 \times$ is one-to-one and find the inverse.

Fig. 7.17 The points (y, x) and (x, y) are symmetrical with respect to the bisector of the first and third quadrant

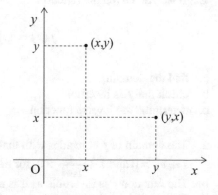

Fig. 7.18 The graphs of f
and the inverse f^{-1}

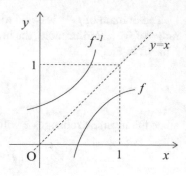

Fig. 7.19 A $f(x) = 6$
$-3x$ and the inverse
$f^{-1}(y) = x = 6 - 3y$

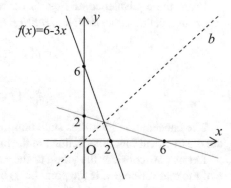

Solution Suppose $f(x_1) = f(x_2)$, then $6 - 3x_1 = 6 - 3x_2$, $-3x_1 = -3x_2$, $x_1 = x_2$. Hence, f is one-to-one. To find the inverse, solve $f(x) = 6 - 3 \times x$ for x, obtaining $x = -\frac{1}{3}(y - 6)$. Therefore, referring the graphs of f and f^{-1} to the same coordinate system, and hence denoting the independent variable with x again, the inverse of f is $f^{-1}(y) = x = 6 - 3y$. Observe that the graphs of f and f^{-1} are the lines $y = f(x) = 6 - 3 \times x$ and $x = 6 - 3y$, respectively, and are symmetrical with respect to the bisector of the first and third quadrant (Fig. 7.19).

Exercise 7.6 Given the function

$$f(x) = \ln\left(1 + \frac{1}{1 + e^{\frac{1}{x}}}\right) \tag{7.7}$$

a.   find the domain,
b.   check that f is invertible,
c.   determine the inverse function.

a.   The domain of f coincides with that of $\frac{1}{x}$. Indeed, $e^{\frac{1}{x}} > 0$, for every $x \neq 0$, then
     $1 + e^{\frac{1}{x}} > 0$ and $1 + \frac{1}{1 + e^{\frac{1}{x}}} > 0$, for every $x \neq 0$. Hence $\mathrm{Dom}(f(x)) = \mathbf{R} - \{0\}$.
b.   The function f is invertible as it is a composite function of invertible functions.
     For example, a decomposition into invertible functions is the following:

$$g = \frac{1}{x}, h = e^g, k = 1 + h, p = \frac{1}{k}, q = 1 + p, s = \ln q$$

c. We have to solve the Eq. (7.7) for x as follows:

$$\ln\left(1 + \frac{1}{1 + e^{\frac{1}{x}}}\right) = y \Rightarrow 1 + \frac{1}{1 + e^{\frac{1}{x}}} = e^y \Rightarrow \frac{1}{1 + e^{\frac{1}{x}}} = e^y - 1 \Rightarrow$$

$$1 + e^{\frac{1}{x}} = \frac{1}{e^y - 1} \Rightarrow e^{\frac{1}{x}} = \frac{1}{e^y - 1} - 1 \Rightarrow \frac{1}{x} = \ln\left(\frac{1}{e^x - 1} - 1\right)$$

Therefore, the inverse of the given function is

$$x = \frac{1}{\ln\left(\frac{1}{e^x - 1} - 1\right)}$$

Bibliography

Anton, H.: Calculus. Wiley, New York (1980)

Lax, P., Burnstein, S., Lax, A.: Calculus with Applications and Computing. Springer, New York (1976)

Spivak M.: Calculus. Cambridge University Press, Cambridge (2006)

Chapter 8
Circular Functions

8.1 Introduction

Let us now introduce the circular functions, defined by properties related with the circumference. When we dealt with the coordinate system on a line we imagined the line described by a point moving on the line. Likewise, a point in the circumference is the independent variable that defines the real-valued functions called circular functions.

8.1.1 The Equation of the Circumference

In (Sect. 7.2 et seq.) we dealt with the equation of the line. Precisely, we stated that the set of the points (x, y) of the line r are the only solutions of a linear equation in x, y.

We now want to establish the relation between the coordinates of the points of a circumference.

Let us fix the point $C(a, b)$ and the positive real number h. The *circumference* with *center* C and *radius* h is, by definition, the set of points $P(x, y)$ whose distance from C equals h. According to the formula (Sect. 7.1.1) of the distance, the circumference is the set of points $P(x, y)$ that satisfy the equality:

$$|PC| = \sqrt{(x-a)^2 + (y-b)^2} = h \tag{8.1}$$

The point P varies and maintains the distance h from C describing the circumference. The Eq. (8.1) expresses the relation between the coordinates and the points: if the point P with coordinates (x, y) belongs to the circumference, then the couple of numbers (x, y) is a solution of Eq. (8.1) and, vice versa, if the couple of numbers (x, y) is a solution of Eq. (8.1), then the point $P(x, y)$ belongs to the circumference with center C and radius h. The relation (8.1) is called the *equation of the circumference*

© The Author(s), under exclusive license to Springer Nature Switzerland AG 2023   125
A. G. S. Ventre, *Calculus and Linear Algebra*,
https://doi.org/10.1007/978-3-031-20549-1_8

Fig. 8.1 Circumference
with center C and radius h

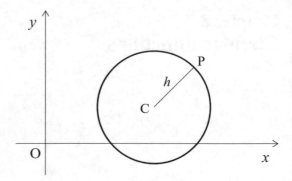

with center C and radius h; (8.1) is said *to represent* the circumference with center
C and radius h (Fig. 8.1).

Let us square both sides of equality (8.1) to obtain:

$$(x - a)^2 + (y - b)^2 = h^2 \tag{8.2}$$

It can be shown that (8.2) is an equation that has the same solutions as (8.1). This
means that if the coordinates of a point of the plane are a solution of (8.1), then
they are also a solution of (8.2) and, vice versa, if the coordinates of a point are a
solution of (8.2), then they are also a solution of (8.1). Therefore, Eqs. (8.1) and (8.2)
represent the same set of points, which is the circumference with center C and radius
h.

Remark 8.1 We moved from Eqs. (8.1)–(8.2) because we wanted to get rid of the
square root. By squaring the two sides of (8.1) we obtain Eq. (8.2) and we have
observed that (8.1) and (8.2) have the same solutions. This cannot always be said
when performing the square of the sides of an equation. For example, passing to
the squares of the two sides of the equation $x = 1$, we have $x^2 = 1$. While the first
equation has the only solution 1, the second has the solutions 1 and -1.

Example 8.1 The circumference with center $C(1, 0)$ and radius 2 has equation $(x
- 1)^2 + y^2 = 2^2$, i.e.,

$$x^2 + y^2 - 2x - 3 = 0$$

The points of the circumference are found by assigning "suitable" numerical
values to x (or y) and calculating the consequent values of y (or x). Set $x = 0$ in the
equation: then $y^2 - 3 = 0$ and $y = \pm\sqrt{3}$. Put now $y = 0$ in the equation: then $x^2
- 2x - 3 = 0$, whence $x = -1$ and $x = 3$. The points $(0, \sqrt{3})$ and $(0, -\sqrt{3})$ if $x =
0$, and the points $(-1, 0)$, $(3, 0)$ if $y = 0$ have been got. The geometric meaning of
the operations above is evident: we intersected the circumference with the y axis,
obtaining the points $(0, \sqrt{3})$, $(0, -\sqrt{3})$, and then with the x axis, obtaining the points
$(-1, 0)$, $(3, 0)$ (Fig. 8.2).

Fig. 8.2 Circumference
$(x-1)^2 + y^2 = 4$

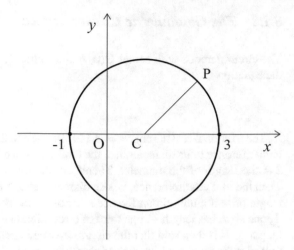

If $x = -3$ in the circumference $(x-1)^2 + y^2 = 4$, then we obtain $y^2 = 4 - 16 = -12$: there is no real number y whose square is negative. So the line $x = -3$ does not intersect the circumference at any point of the plane. This implies that the vertical line $x = -3$ is external to the circumference. The observation suggests that "suitable" values to be assigned to x and y are the points of the segments of abscissae and ordinates that are the orthogonal projections of the points of the circumference on the coordinate axes, i.e., the projections of the dashed sides of the square on the coordinate axes (Fig. 8.3, see also Fig. 4.29).

Fig. 8.3 The perpendicular lines to the coordinate axes intersecting the circumference, also intersect the traced square

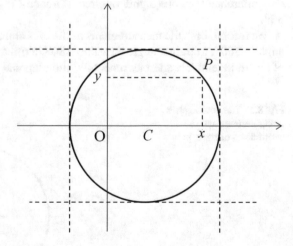

8.1.2   The Goniometric Circumference

The circumference with center O(0,0) and radius 1, called *unit circumference*, has the equation

$$x^2 + y^2 = 1$$

The irrational real number $\pi = 3.14159\ldots$ (Sect. 2.5) is the ratio of the (rectified) circumference to its diameter, i.e., the circumference has length π times the number 2 *h*, the length of the diameter. Therefore, the length of the circumference is $2\pi h$. Then the unit circumference, whose diameter is 2, has length 2π. It follows that the length of half a unit circumference is π, one quarter of a circumference has length $\frac{\pi}{2}$, one sixth has length $\frac{\pi}{3}$, one third of circumference has length $\frac{2\pi}{3}$, one twelfth has length $\frac{\pi}{6}$. It is then said that the circumference is measured in *radians*, because the unit of measurement of the arcs of circumference is the radius. For instance, the fact that the length of the unit circumference is 2π means that the unit circumference measures 2π radians.

The locution *goniometric circumference* indicates the unit circumference with a measuring system of its arcs or angles. Some points are put in evidence on the unit circumference (Fig. 8.4).

Another system for measuring the lengths of the circumference and its arcs is based on the subdivision of the circumference into 360 arcs congruent to each other, any having the measure of a *degree*; each degree is divided into 60 *seconds*. Then the whole circumference measures 360 degrees (the symbol 360° is used), the semi-circumference measures 180°, one quarter of semi-circumference 90°, one sixth of circumference 60°, one eighth of circumference 45°, one twelfth of circumference 30°.

We refer to both the measurements of the arcs and angles: the measure of a round angle is 360°, a straight angle measures 180°, a right angle 90°, each internal angle of an equilateral triangle measures 60°, the diagonal of a square forms an angle of

Fig. 8.4 The goniometric circumference and the coordinates of some points

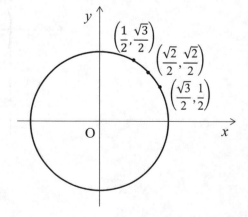

$45°$ with one side, each of the interior angles of a regular hexagon measures $120°$, the sum of the interior angles of a triangle is $180°$. This way of measuring arcs and angles, called *sexagesimal system*, was already known to the Babylonians and used by them in the study of astronomy and surveying. (see Chap. 2.)

We adopt the system of measurement of the circumference arcs in radians. In this way, the arc will be measured by a positive, negative or null real number which expresses its length with respect to the radius as the unit of measure. The null arc has its extremes coincident in one point and has measure zero. It makes sense to consider arcs of the circumference that are longer than the circumference. These are obtained by joining a number of complete circumferences to a given arc.

We continue to focus on the unit circumference. Let us define the point $A(1, 0)$ as the *origin of the arcs* of the circumference and we adopt the counterclockwise direction as the positive direction on the circumference. Consider the arc of circumference having as the first extreme A and the second extreme P; let t be the length of the arc AP: the number t individuates the position of the point P in the circumference: let denote $AP(t)$ the arc of extreme points A and P. If the number t is positive, then the arc AP is covered in the counterclockwise direction, if the number t is negative, then the arc AP is covered in the clockwise direction, if $t = 0$ then the point P coincides with point A, i. e., AP is the null arc.

Therefore, an invertible function is defined between the set of second extremes P of the oriented arcs of origin A and the real numbers.

Periodic functions

A real-valued function f of the real variable x is said to be *periodic* if there exists a positive number p such that

$$f(x + p) = f(x)$$

whenever x and $x + p$ belong to the domain of f. We say that p is a *period* of the function. The minimum period is called the *fundamental period* of f or simply the *period* of f.

8.1.3 Sine, Cosine and Tangent

We have observed that the position of P on the unit circumference is a function of the real number t. For example, if t varies from 0 to 2π, then P describes the whole circumference, starting from $A(1, 0)$ and returning to A; if t continues to vary, from 2π to 4π, then P makes another round and returns to the starting position.

Fig. 8.5 The unit
circumference; the point P =
P(*t*) belongs to the
circumference, HP = y(*t*) =
sin*t*, KP = x(*t*) = cos*t*; the
point T is common to the
line OP and the line through
A parallel to *y* axis; AT is the
tangent line to the
circumference at the point A

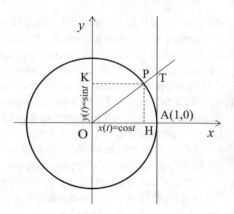

If P is the second endpoint of the arc AP having relative length *t*, then P occupies
the same position again after one complete round, i.e., the position P(*t*) coincides
with P(*t* + 2π), and the same happens if P performs an integer number *k* of complete
rounds:

$$P(t) = P(t + 2k\pi) \qquad (8.3)$$

The abscissa and the ordinate of the point P are functions of *t*, which we denote
x(*t*), y(*t*). By (8.3), y(*t*) = y(*t* + 2*k*π), x(*t*) = x(*t* + 2*k*π). The functions y(*t*) and
x(*t*) are called *sine* of *t* and *cosine* of *t*, respectively, and are denoted by: y(*t*) = sin*t*,
x(*t*) = cos*t*.

The functions sin*t* and cos*t* are real-valued functions of the real variable *t*, their
domain is **R** and their range is the interval [−1, 1]. It makes sense to speak of sine
and cosine of the arc AP: the value of the sine of the arc AP is the ordinate of the
point P, the value of the cosine of the arc AP is the abscissa of P. In particular, sin0
= 0, cos0 = 1, sin $\frac{\pi}{2}$ = 1, cos $\frac{\pi}{2}$ = 0 (Fig. 8.5).

Properties of the sine, cosine and tangent functions

The functions *sine* and *cosine* are functions of the arc AP(*t*), but also it makes sense
to define the sine of the angle $\widehat{AOP}(t)$, the cosine of the angle $\widehat{AOP}(t)$, and put
$\sin \widehat{AOP}(t) = \sin t$ and $\cos \widehat{AOP}(t) = \cos t$.

a.   By (8.3), the functions sine and cosine are periodic with period 2π, i. e., for any
real number *t*

$$\sin t = \sin(t + 2k\pi), \cos t = \cos(t + 2k\pi),$$

k relative integer.

b.   The point P(*x, y*) = P(x(*t*), y(*t*)) lies on the unit circumference if and only if x^2
+ y^2 = 1. Hence, the remarkable property that links sine and cosine: (sin*t*)^2 +
(cos*t*)^2 = 1, also written

Fig. 8.6 |AB|sinα = |CB|
and |AB|cosα = |AC|

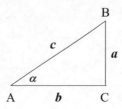

$$\sin^2 t + \cos^2 t = 1$$

equality that holds for any real number t. The equations

$$x(t) = \cos t$$

$$y(t) = \sin t,$$

with the condition $t \in [0, 2\pi]$, are the parametric equations that provide a representation of the unit circumference.

c. Given the triangle ACB right-angled at C, put $\alpha = \widehat{bc}$, we have |AB|sinα = |CB| and |AB|cosα = |AC| (Fig. 8.6).

d. The *tangent function* of the variable t, denoted tan$t$, is defined starting from sine and cosine functions, by

$$\tan t = \frac{\sin t}{\cos t}$$

The tangent function has domain in the set of real numbers such that the denominator cos$t$ is non-null, which occurs if and only if $t \neq \frac{\pi}{2} + k\pi$, k relative integer.

Referring to Fig. 8.7, let us consider some geometric properties of the tangent function. Let the line s pass through the point A and be parallel to the y axis with the same orientation; let a system of abscissas of origin A and unit of measure equal to the radius of the circumference be defined on s. The line s is the tangent line to the circumference at A. Let T be the common point to the lines s and OP. The value tan$t$ is the relative length of the segment AT, in fact OA : AT = OH : HP. The line OP intersects the circumference also at the point P'$(t + \pi)$ and then tan$t$ = tan$(t + \pi)$. Therefore, the tangent function is a periodic function with period π, i. e., for any k relative integer:

$$\tan t = \tan(t + k\pi)$$

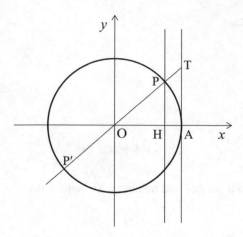

Fig. 8.7 T$\tan t$ = length of AT. The proportion OA : AT = OH : HP is equivalent to 1 : $\tan t$ = $\cos t$: $\sin t$, i. e., $\tan t = \frac{\sin t}{\cos t}$

The couple of the components of the oriented segment OP, i. e., its direction numbers, is ($\cos t$, $\sin t$). In other words, the slope (Sect. 7.3.3) of the line OP is the ratio $\frac{\sin t}{\cos t} = \tan t$.

Now name x the variable t (variables are "dummy"). The functions $\sin x$, $\cos x$, $\tan x$ are called *circular functions*.

The graph of cosine function is symmetrical with respect to the y axis because $\cos x = \cos(-x)$; the sine and tangent functions are symmetrical with respect to the origin of the coordinate system because $-\sin x = \sin(-x)$, $-\tan x = \tan(-x)$.

Some values of the circular functions are listed below.

x	$\sin x$	$\cos x$	$\tan x$
0	0	1	0
$\frac{\pi}{6}$	$\frac{1}{2}$	$\frac{\sqrt{3}}{2}$	$\frac{1}{\sqrt{3}}$
$\frac{\pi}{4}$	$\frac{\sqrt{2}}{2}$	$\frac{\sqrt{2}}{2}$	1
$\frac{\pi}{3}$	$\frac{\sqrt{3}}{2}$	$\frac{1}{2}$	$\sqrt{3}$
$\frac{\pi}{2}$	1	0	Undefined
π	0	-1	0
$\frac{3\pi}{2}$	-1	0	Undefined

8.1.4 Further Goniometric Identities

$$\cos\left(\frac{\pi}{2} - x\right) = \sin x$$

$$\sin\left(\frac{\pi}{2} - x\right) = \cos x$$

$$\cos(x + y) = \cos x \cos y - \sin x \sin y$$

$$\sin(x + y) = \sin x \cos y + \cos x \sin y$$

$$\cos(x - y) = \cos x \cos y + \sin x \sin y$$

$$\sin(x - y) = \sin x \cos y - \cos x \sin y$$

$$\sin 2x = 2 \sin x \cos x$$

$$\cos^2 x = 2 \cos^2 x - 1 = 1 - 2 \sin^2 x$$

$$1 - \cos x = 2 \sin^2 \frac{x}{2}$$

$$1 + \cos x = 2 \cos^2 \frac{x}{2}$$

$$2 \sin^2 x = 1 - \cos 2x$$

$$2 \cos^2 x = 1 + \cos 2x$$

8.1.5 The Graphs of Sinx, Cosx and Tanx

See Figs. 8.8, 8.9 and 8.10.

Fig. 8.8 Graph of the function $y = \sin x$

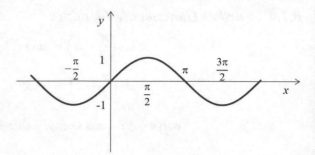

Fig. 8.9 Graph of the function $y = \cos x$

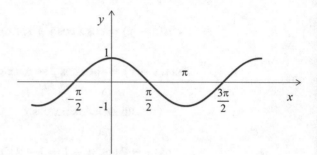

Fig. 8.10 Graph of the function $y = \tan x$

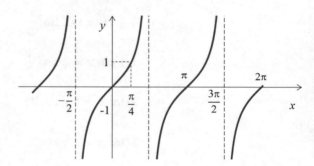

Bibliography

Anton, H.: Calculus. Wiley, New York (1980)

Lax, P., Burnstein, S., Lax, A.: Calculus with Applications and Computing. Springer, New York (1976)

Spivak M.: Calculus. Cambridge University Press, Cambridge (2006)

Chapter 9
Geometric and Numeric Vectors

9.1 *n*-Tuples of Real Numbers

We introduced (Sect. 3.1) the Cartesian product of sets, in particular the set $\mathbf{R}^2$ of the couples of real numbers.

Given two couples $(a, b), (a', b') \in \mathbf{R}^2$ and a real number h an operation of *addition* of two couples $(a, b), (a', b') \in \mathbf{R}^2$

$$(a, b) + (a', b') = (a + a', b + b')$$

and an operation of *multiplication* of the real number h by a couple $(a, b) \in \mathbf{R}^2$

$$h(a, b) = (ha, hb)$$

are defined. For example,

$$(-4, 8) + (3, -8) = (-1, 0)$$

$$-7\left(6, \frac{4}{9}\right) = \left(-42, -\frac{28}{9}\right)$$

$$6(1, -7) - 12(5, 4) = 6(1, -7) + (-12)(5, 4) = (-54, -90)$$

Similarly, the addition of two elements of $\mathbf{R}^3 = \mathbf{R} \times \mathbf{R} \times \mathbf{R}$ and the multiplication of a real number by an element of $\mathbf{R}^3$ are defined:

$$(a, b, c) + (a', b', c') = (a + a', b + b', c + c')$$

$$h(a, b, c) = (ha, hb, hc)$$

© The Author(s), under exclusive license to Springer Nature Switzerland AG 2023   135
A. G. S. Ventre, *Calculus and Linear Algebra*,
https://doi.org/10.1007/978-3-031-20549-1_9

whatever the triples (a, b, c), (a', b', c') in $\mathbf{R}^3$ and $h \in \mathbf{R}$ are. For example,

$$(-11, 0, 1) + (2, -2, 0) = (-9, -2, 1)$$

$$-5(4, 2, 1) = (-20, -10, -5)$$

An element $(a_1, a_2, \ldots, a_n)$ of $\mathbf{R}^n$, $n \geq 1$, is called an *n-tuple* of real numbers, or a *numeric vector*; the number a_1 is called the *first component* of the *n*-tuple, the number a_2 the *second component*, ..., the number a_i the *i*th *component*, for $i = 1, 2$, ..., n. The *n*-tuple $\mathbf{O} = (0, 0, \ldots, 0)$ which has all the components equal to zero is called the *null n*-tuple. Usually an *n*-tuple $(a_1, a_2, \ldots, a_n)$ of $\mathbf{R}^n$ is denoted in bold type by letters, $\boldsymbol{a} = (a_1, a_2, \ldots, a_n)$, $\boldsymbol{b} = (b_1, b_2, \ldots, b_n)$, ...

With a natural extension, for $n \geq 2$, the operations of addition of two elements of $\mathbf{R}^n$ and multiplication of a real number by an element of $\mathbf{R}^n$ are defined: the sum of two *n*-tuples $\boldsymbol{a} = (a_1, a_2, \ldots, a_n)$ and $\boldsymbol{b} = (b_1, b_2, \ldots, b_n)$ is an *n*-tuple of real numbers and the product of a real number h by an *n*-tuple of real numbers is an *n*-tuple of real numbers:

$$\boldsymbol{a} + \boldsymbol{b} = (a_1, a_2, \ldots, a_n) + (b_1, b_2, \ldots, b_n) = (a_1 + b_1, a_2 + b_2, \ldots, a_n + b_n)$$

$$h\boldsymbol{a} = h(a_1, a_2, \ldots, a_n) = (ha_1, ha_2, \ldots, ha_n)$$

The operations of addition of two *n*-tuples of real numbers and multiplication of a real number by an *n*-tuple of real numbers, satisfy the following properties, whatever the elements $\boldsymbol{a}$, $\boldsymbol{b}$ and $\boldsymbol{c}$ of $\mathbf{R}^n$ and the real numbers h and k are:

1. the addition is *commutative*, $\boldsymbol{a} + \boldsymbol{b} = \boldsymbol{b} + \boldsymbol{a}$;
2. the addition is *associative*, $(\boldsymbol{a} + \boldsymbol{b}) + \boldsymbol{c} = \boldsymbol{a} + (\boldsymbol{b} + \boldsymbol{c})$;
3. the null *n*-tuple $\mathbf{O} = (0, 0, \ldots, 0)$ is the only element of $\mathbf{R}^n$ such that $\boldsymbol{a} + \mathbf{O} = \boldsymbol{a}$ (existence of the *neutral element* $\mathbf{O}$ with respect to addition);
4. for each *n*-tuple $\boldsymbol{a}$, there exists the *opposite* $- \boldsymbol{a}$, which is the only *n*-tuple such that $\boldsymbol{a} + (-\boldsymbol{a}) = \mathbf{O}$;
5. the multiplication of an *n*-tuple by a real number is *associative*, $(hk)\boldsymbol{a} = h(k\boldsymbol{a})$;
6. for every *n*-tuple $\boldsymbol{a}$, $1\boldsymbol{a} = \boldsymbol{a}$;
7. the *distributive* properties hold:
    $$h(\boldsymbol{a} + \boldsymbol{b}) = h\boldsymbol{a} + h\boldsymbol{b}$$
    $$(h + k)\boldsymbol{a} = h\boldsymbol{a} + k\boldsymbol{a}$$

9.1.1 Linear Combinations of n-Tuples

Two *n*-tuples $(a_1, a_2, \ldots, a_n)$ and $(b_1, b_2, \ldots, b_n)$ of $\mathbf{R}^n$ are said to be *equal* if $a_i = b_i$, for any $i = 1, \ldots, n$. If $n \geq 1$, a choice of $m \geq 1$ elements $\boldsymbol{a}_1, \boldsymbol{a}_2, \ldots, \boldsymbol{a}_m$ of $\mathbf{R}^n$, not necessarily distinct, is called a *system* of *n*-tuples of $\mathbf{R}^n$, denoted $[\boldsymbol{a}_1, \boldsymbol{a}_2, \ldots, \boldsymbol{a}_m]$

(Sect. 3.2). The positive integer m is called the *order* of the system. The existence of the system without elements, called the *empty system*, is postulated.

Let us define *linear combination of the n-tuples* $a_1, a_2, \ldots, a_m$, or *linear combination of the n-tuples of the system* $[a_1, a_2, \ldots, a_m]$ any *n*-tuple of the form

$$b = h_1 a_1 + h_2 a_2 + \cdots + h_m a_m$$

where $h_1, h_2, \ldots, h_m$, are real numbers called *coefficients* of the linear combination.

The vector b which is the linear combination of the *n*-tuples of the system $[a_1, a_2, \ldots, a_m]$ is said to be *linearly dependent* on the *n*-tuples $[a_1, a_2, \ldots, a_m]$, or on the system $[a_1, a_2, \ldots, a_m]$.

Example 9.1 The following 4-tuples of R^4 : $a_1 = (-2, 0, 1, 2)$, $a_2 = (0, 3, 1, 2)$, $a_3 = (2, 3, 0, 0)$

form a system S:

$$S = [a_1, a_2, a_3] = [(-2, 0, 1, 2), (0, 3, 1, 2), (2, 3, 0, 0)]$$

Let us check that the null 4-tuple $\mathbf{O} = (0, 0, 0, 0)$ is a linear combination of the 4-tuples of the system S. Indeed, numbers h_1, h_2, h_3 that satisfy the equality

$$(0, 0, 0, 0) = h_1(-2, 0, 1, 2) + h_2(0, 3, 1, 2) + h_3(2, 3, 0, 0) \qquad (9.1)$$

are

$$h_1 = 0, h_2 = 0, h_3 = 0;$$

indeed, it is immediate to check:

$$0(-2, 0, 1, 2) + 0(0, 3, 1, 2) + 0(2, 3, 0, 0) =$$
$$(0, 0, 0, 0) + (0, 0, 0, 0) + (0, 0, 0, 0) = (0, 0, 0, 0)$$

However, there are non-zero values of h_1, h_2, h_3 that satisfy the equality (9.1). For example, setting the values $h_1 = 1, h_2 = -1, h_3 = 1$ in the right-hand side of (9.1) we get:

$$1(-2, 0, 1, 2) - 1(0, 3, 1, 2) + 1(2, 3, 0, 0) =$$
$$(-2, 0, 1, 2) + (0, -3, -1, -2) + (2, 3, 0, 0) =$$
$$(-2 + 0 + 2, 0 - 3 + 3, 1 - 1 + 0, 2 - 2 + 0) = (0, 0, 0, 0)$$

Example 9.2 Let us consider the couples $a_1 = (0, 1)$, $a_2 = (1, 0)$ of $\mathbf{R}^2$ and the system.

$$T = [a_1, a_2] = [(0, 1), (1, 0)]$$

We ask ourselves if the couple $b = (1, 1)$ of $\mathbf{R}^2$ is a linear combination of the system T. To answer the question we must verify if there are real numbers h_1, h_2 that satisfy the equality:

$$h_1(0, 1) + h_2(1, 0) = (1, 1) \tag{9.2}$$

that can be written, performing the multiplications

$$h_1(0, 1) + h_2(1, 0) = (0, h_1) + (h_2, 0) = (h_2, h_1).$$

Therefore, (9.2) is verified if $(h_2, h_1) = (1, 1)$, i. e., $h_1 = 1, h_2 = 1$. Observe that $(h_2, h_1) = (1, 1)$ is the unique couple of numbers that satisfies (9.2).

Let us state some properties of the linear combinations of n-tuples.

Proposition 9.1 *The null n-tuple O is linearly dependent on any non-empty system of n-tuples.*

In fact, for every non-empty system of n-tuples S $= [a_1, a_2, \ldots, a_m]$, *the equality holds*:

$$O = 0a_1 + 0a_2 + \ldots + 0a_m \tag{9.3}$$

Remark 9.1 The Example 9.1 shows that if a linear combination of the n-tuples of a system has all the coefficient equals to zero, then the linear combination is equal to O; furthermore, there exist linear combinations with non-null coefficients that equal zero; i.e., the condition that the coefficients of a linear combination of n-tuples are all zeros is a sufficient condition for the linear combination to be equal to zero, but the condition is not necessary in that there exist linear combinations with coefficients not all null whose value is O. For example, the null couple $O = (0, 0)$ is a linear combination of the system $[a, -a]$, indeed:

$$O = 0a + 0(-a)$$

However, we obtain a null linear combination of a and $-a$ with non-null coefficients:

$$O = 1a + 1(-a) = a - a.$$

It is also true that $O = ha + h(-a)$, for every real number h.

Remark 9.2 If a is a nonnull n-tuple of $\mathbf{R}^n$ and h a real number, the n-tuples a and ha are said to be *proportional*, or *parallel*. For example, the following triples are two by two proportional.

$$(1, 0, -4), (-1, 0, 4), (4, 0, -16), \left(\frac{1}{2}, 0, -2\right)$$

Proposition 9.2 *If* ***a*** *is a non-zero n-tuple of* ***R**^{n}, *then the null n-tuple and the n-tuples proportional to* ***a*** *depend linearly on* ***a***. *The null n-tuple and the n-tuples proportional to* ***a*** *are the only n-tuples of* ***R**^{n} *linearly dependent on* ***a***.

The system of n-tuples $S = [\boldsymbol{a}_1, \boldsymbol{a}_2, \ldots, \boldsymbol{a}_m]$ *is said to be linearly dependent (or the n-tuples* $\boldsymbol{a}_1, \boldsymbol{a}_2, \ldots, \boldsymbol{a}_m$ *are said to be linearly dependent), if there exist coefficients* $h_1, h_2, \ldots, h_m$ *not all zero such that* $h_1\boldsymbol{a}_1 + h_2\boldsymbol{a}_2 + \ldots + h_m\boldsymbol{a}_m = \boldsymbol{O}$.

The system $S = [(-2, 0, 1, 2), (0, 3, 1, 2), (2, 3, 0, 0)]$ *of 4-tuples of* ***R**^{4} *is linearly dependent (see Example* 9.1).

9.2 Scalars and Vectors

In the description or identification of geometric and physical characteristics it is sometimes sufficient to use a number. For example, the length of a river, the extension of the surface of a field, the volume of a box, the body temperature, are features described by a number, or *scalar*. Therefore, such characteristics are called *scalar quantities*.

More detailed descriptions are required when describing other phenomena. A feather on the desert sand, blown by the wind, moves from point A to point B 400 m away.

We intend to define the position of B. The only distance from A does not allow us to identify the position of B; in fact, the feather can be found in any point of the circumference with radius 400 and center A (Fig. 9.1).

Afterwards, we know that the shift occurred along a certain line, then the possible positions of B reduce to two, the common points to the circumference and the line (Fig. 9.2).

Any ambiguity is eliminated when we know the direction of the movement of the feather, i. e., which of the two half-lines of origin A the feather covers in its trajectory (Fig. 9.3).

Therefore, the shift of the feather is determined by three elements: a number, 400; a direction, that of the line; the sense of the movement on the line.

Definition 9.1 A *vector* is defined as a triple made of a nonnegative real number called *magnitude* or *modulus*, a *direction* of a line and an *orientation* or *sense* of the direction.

Fig. 9.1 Circumference
with center A

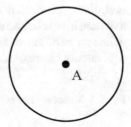

Fig. 9.2 The point B is one
of the two diametrically
opposite points

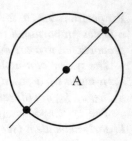

Fig. 9.3 The point B is
determined

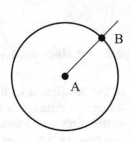

Velocity, acceleration on a curved trajectory, gravity are vectors.

9.3   Applied Vectors and Free Vectors

A *geometric applied vector*, or simply, a *vector applied* at the point A is, by definition, an oriented segment AB with *first endpoint* A and *second endpoint* B. In Fig. 9.4, an arrow at B is put to highlight the orientation of AB. The point A is also called the *origin*, or the *point of application*, of the applied vector AB.

The segment $\overline{AB}$ is called the *support* of the applied vector AB. The length |AB| of the segment $\overline{AB}$ is called the *magnitude,* or *modulus,* of the applied vector AB. Therefore, the magnitude of the applied vector AB is a nonnegative real number.

Definition 9.2 Two oriented segments of the line (plane, space) are said to be *equipollent* if they have the same direction, the same orientation and their supports are congruent. Any two null segments are equipollent by definition.

The equipollence relation in the set of oriented segments of the line (plane, space) is an equivalence relation (Sect. 3.5). In fact, equipollence is a reflexive, symmetrical and transitive relation, as it can be easily verified. By Theorem 3.5 the equipollence defines a partition of the set of oriented segments of the line (plane, space) into equivalence classes, called *equipollence classes*.

Fig. 9.4 Vector applied at A   A ●————————————▶ B

Definition 9.3 Any equipollence class of oriented segments is named a *free vector* of the line (plane, space); in particular, the null vector is the equipollence class whose elements are the null segments, i. e., the segments that reduce to a point. The null vector is denoted with **O**.

Let us denote with V_3 the set of free vectors of the space. Similarly, we define the set V_α of the free vectors in the plane α and the set V_r of the free vectors of the line *r*.

Therefore, any non-null oriented segment AB identifies a free vector, which is the set of the oriented segments equipollent to AB, and the *direction* and *orientation* of the free vector are defined by the direction and the orientation of the oriented segment AB; the *magnitude,* or *modulus,* of the free vector is, by definition, the magnitude of any oriented segment in the class of equipollence identified by AB.

Free vectors are denoted in bold type by letters, $\boldsymbol{a}, \boldsymbol{b}, \boldsymbol{c}, \ldots$, the free vector containing the oriented segment AB is denoted **AB**.

Let $\boldsymbol{a}$ be a non-null free vector. The vector $- \boldsymbol{a}$, which has the same direction and the same magnitude of $\boldsymbol{a}$ and the orientation opposite to that of $\boldsymbol{a}$, is defined as the *opposite of the vector $\boldsymbol{a}$.*

9.4 Addition of Free Vectors

A free vector has no fixed position in the plane or in the space and then may be moved under parallel displacement wherever. Some operations are defined on free vectors. We deal with the *addition* of free vectors and the *multiplication* of a real number by a free vector. In order to define the addition of two free vectors, we first refer to the addition of two vectors applied at the same point A, i. e., two oriented segments of origin A. So let AB and AC be two non-null oriented segments of the plane. We distinguish the following cases:

Case 1. The points A, B, C are aligned and the segments AB and AC have the same orientation (Fig. 9.5).

The sum of the oriented segments AB and AC is defined as the oriented segment AD which has the same direction and orientation as AB and AC, and magnitude $|AD| = |AB| + |AC|$, i.e., the magnitude of AD is equal to the sum of the magnitudes of the oriented segments AB and AC. Then we write AD = AB + AC.

Case 2. The points A, B, C are aligned and the segments AB and AC have opposite orientations (Fig. 9.6).

If AB and AC have the same magnitude, then the sum of AB and AC is the null vector. If AB and AC do not have the same magnitude, the sum of the oriented

Fig. 9.5 AD = AB + AC

A   B   C   D

Fig. 9.6 AD = AB + AC

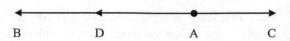

segments AB and AC is defined as the oriented segment AD which has the same direction of AB, the orientation of the segment of greater magnitude (referring to Fig. 9.6, the orientation of AB is that of the sum) and has the magnitude $|AD| = ||AB| - |AC||$, which is the absolute value of the difference of the magnitudes of AB and AC. Then we write AD = AB + AC.

Definition 9.4 Two non-null applied vectors such that the second *endpoint* of a segment coincides with the origin of the other are called *consecutive*.

Referring to Fig. 9.7, the oriented segments AB and BC are consecutive.

Case 3. If A, B and C are not aligned, the segments AB and BC form two sides of a triangle whose third side AC is called the sum of AB and BC: AC = AB + BC. This procedure for carrying out the sum is called the *triangle rule* (Fig. 9.8).

The triangle rule is easily extended to the *polygon rule* for carrying out the sum of several vectors. If AB, BC and CD are oriented segments that belong to the free vectors **a**, **b** and **c**, respectively, then the oriented segment AD, the closing side AD of the polygon ABCD (Fig. 9.9), is an oriented segment of the free vector sum **d** = (**a** + **b**) + **c**.

Case 4. If the points A, B, C are not aligned, the sum of the oriented segments AB and AC is defined as the oriented segment AD, the diagonal of the parallelogram with adjacent sides AB and AC, oriented from A to D. Then AD = AB + AC (Fig. 9.10). The procedure, known as the *parallelogram rule*, is equivalent to the triangle rule.

The *addition of two free vectors* is defined by referring to the addition of two vectors applied in the same point, each identifying a free vector. Precisely, given the

Fig. 9.7 AB and BC
consecutive

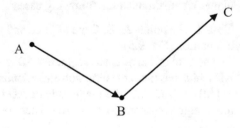

Fig. 9.8 AB + BC = AC

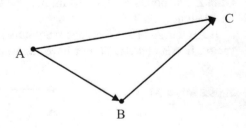

Fig. 9.9 AB + BC + CD = AD

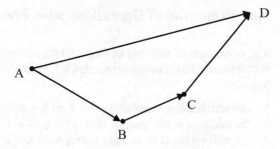

Fig. 9.10 AD = AB + AC

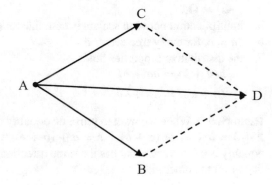

free vectors *a* and *b*, and the segments AB and AC belonging to vectors *a* and *b*, respectively, the vector sum of *a* and *b* is defined as the free vector *c* identified by the applied vector AD = AB + AC, i. e., *c* = *a* + *b*.

9.5 Multiplication of a Scalar by a Free Vector

Given a real number, or *scalar*, *h* and a free vector *a*, an operation, called *multiplication* of a scalar by a free vector which associates the vector *ha* with the pair (*h*, *a*) is thus defined:

ha is the null vector, if *h* = 0 or *a* is the null vector;

if *h* ≠ 0 and *ha* is different from the null vector, two alternatives occur:

if *h* > 0, then *ha* is the vector that has the same direction and the same orientation of *a*, and has magnitude *h*|*a*|;

if *h* < 0, then *ha* is the vector that has the same direction of *a*, the opposite orientation of *a* and magnitude |*h*||*a*|, i. e., the absolute value of *h* by the magnitude of *a*.

9.6   Properties of Operations with Free Vectors

The operations of addition of free vectors and multiplication of a scalar by a free vector satisfy the following properties, whatever the free vectors a, b, c and the scalars h, k are:

1. the addition is *commutative*: $a + b = b + a$,
2. the addition is *associative*: $(a + b) + c = a + (b + c)$,
3. the null vector O is the only vector such that $a + O = a$,
4. for every free vector a, the opposite $-a$ is the only free vector such that $a + (-a) = O,,$
5. multiplication between scalars is associative: $(hk)\, a = h(ka)$
6. $1a = a$, for every free vector a
7. the distributive properties hold:
$$h(a + b) = ha + hb$$
$$(h + k)a = ha + ka$$

Remark 9.3 When we want to prove an equality between two vectors, for example, $a + b = b + a$, or $(a + b) + c = a + (b + c)$, we must show that each side of the equality is a free vector that has the same direction, orientation and magnitude as the vector on the other side.

A visual proof (Sect. 2.7.5) of the equality that expresses the associative property is obtained from the inspection of Fig. 9.11, where the vectors a, b, c are identified by consecutive oriented segments.

Remark 9.4 As the addition of vectors is associative and commutative the sum of several vectors does not depend on the order in which they are added, nor on the number of addends in the partial sums.

Fig. 9.11 Check of the associative property of the addition of vectors

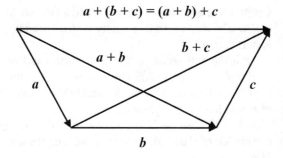

Fig. 9.12 The projection
vectors of AB. The applied
vector AB is the sum of AM
and AN

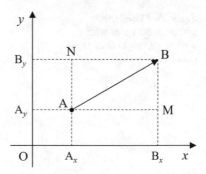

9.7 Component Vectors of a Plane Vector

Let a coordinate system on the plane α be given with origin O and unit points U_x and U_y. Let AB be an oriented segment of the plane, i. e., an applied vector with origin A. The orthogonal projections of AB on the coordinate axes are oriented segments $A_x B_x$ and $A_y B_y$, called the *projection vectors*, or the *vector components* of AB in the coordinate axes (Fig. 9.12).

Let a and b be the free vectors containing the oriented segments $A_x B_x$ and $A_y B_y$, respectively, and c the free vector containing AB. By the parallelogram rule we obtain:

$$c = a + b$$

The free vectors a and b are the *vector components* of the free vector c.

There is a 1–1 correspondence between the set of vectors, applied or free, and the set of the couples of their projections on the coordinate axes. The vectors a and b are called the *vector components* of c.

The relative lengths $(A_x B_x)$ and $(A_y B_y)$ of the oriented segments $A_x B_x$ and $A_y B_y$ are called the *scalar components* of the free vector c.

Thales' theorem implies:

Proposition 9.3
Parallel oriented segments have proportional scalar components and free vectors with the same direction have proportional vector components.

9.8 Space Coordinate System and Vectors

The results related to the plane vectors are extended to the vectors of the space. Indeed, from the coordinate system of the plane (see Sect. 7.1), having axes x, y, origin O and unit points U_x and U_y, let us introduce a new axis z of the space,

Fig. 9.13 Coordinate
system of the space with
axes x, y, z, origin O and unit
points U_x, U_y, U_z

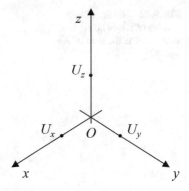

perpendicular to the plane xy, where the x, y axes lie, and passing through the point O (Fig. 9.13).

Let a system of abscissae be fixed on the axis z (see Sect. 5.1), with origin O and unit point U_z, such that the same unit of measure is adopted on the three axes (Fig. 9.13).

The points O, U_x, U_y, U_z define a *coordinate system of the space*, where the axes are the oriented lines x, y, z and O is the origin.

The position of each point P in the space is identified by a triple of real numbers (x, y, z), called *the coordinates* of the point, determined as illustrated in (Fig. 9.14), i.e.:

- x is the abscissa of the point common to the x axis and the plane through P and perpendicular to the x axis (plane that is parallel to the yz plane);
- y is the abscissa of the point common to the y axis and the plane through P and perpendicular to the y axis (plane that is parallel to the xz plane);
- z is the abscissa of the point common to the z axis and the plane through P and perpendicular to the z axis (plane that is parallel to the xy plane).

Fig. 9.14 Coordinates (x, y, z) of P.

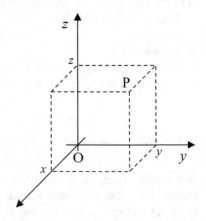

Fig. 9.15 The projections of AB

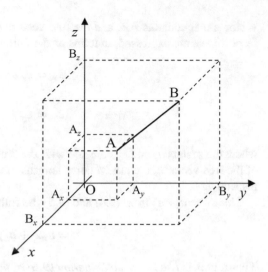

The point P having coordinates (x, y, z) is denoted $P = (x, y, z)$ or $P(x, y, z)$. So, the set of the points in the space is identified with the set $\mathbf{R}^3$ of the triples of real numbers.

The distance of two points $A(x_1, y_1, z_1)$, $B(x_2, y_2, z_2)$ in space is defined by the Pythagorean formula

$$d(A, B) = \sqrt{(x_2 - x_1)^2 + (y_2 - y_1)^2 + (z_2 - z_1)^2}$$

Let AB be an oriented segment of the space, i. e., an applied vector of the space with origin A. The orthogonal projections of AB on the coordinate axes are the oriented segments $A_x B_x$, $A_y B_y$, $A_z B_z$, the *projection vectors*, or the *component vectors* of AB on the coordinate axes. As in the case of plane, the oriented segments $A_x B_x$, $A_y B_y$, $A_z B_z$ (Fig. 9.15) identify respectively the free vectors a, b, c of the space which are the component vectors of the free vector d identified by the segment AB. Therefore, $d = a + b + c$.

9.9 Unit Vectors

Let the coordinate system having origin O and unit points U_x, U_y be assigned in the plane α. The free vectors x and y containing the oriented segments OU_x and OU_y are called the *unit vectors* of the x axis and y axis, respectively. Therefore, the x and y vectors have magnitude 1 and the same direction and orientation as the x and y axes, respectively.

Recalling the operation of multiplication of a number by a vector (Sect. 9.5), given the oriented segment AB and the orthogonal projections $A_x B_x$, $A_y B_y$, the free vector a that contains AB, and the free vectors a_x and a_y, containing A_xB_x, A_yB_y, respectively, are expressed, in terms of the unit vectors x and y of the axes, by:

$$a_x = a_x x \tag{9.4}$$

$$a_y = a_y y \tag{9.5}$$

where the scalars a_x and a_y coincide with the scalar components (A_xB_x) and (A_yB_y) of the free vector a, i. e., the relative lengths of the oriented segments A_xB_x, A_yB_y, respectively.

Since $a = a_x + a_y$ from (9.4) and (9.5) the following equality holds:

$$a = a_x x + a_y y \tag{9.6}$$

Theorem 9.1 *The linear combination* (9.6) *of the vector a with coefficients* (a_x, a_y) *is unique.*

Proof Let us suppose that (9.6) and

$$a = a'_x x + a'_y y \tag{9.7}$$

with $(a_x, a_y) \neq (a'_x, a'_y)$ both hold. Then, subtracting (9.7) from (9.6), we obtain:

$$a - a = (a_x - a'_x)x + (a_y - a'_y)y$$

By the properties of the operations with vectors (see Sect. 9.6),

$$\text{null vector} = O = (a_x - a'_x)x + (a_y - a'_y)y \tag{9.8}$$

Since $(a_x, a_y) \neq (a'_x, a'_y)$ one of the differences $a_x - a'_x$, $a_y - a'_y$ is non-null. Let $a_y - a'_y \neq 0$. Divide the two sides of (9.8) by $a_y - a'_y$ to obtain

$$y = -\frac{a_x - a'_x}{a_y - a'_y}x$$

The equality implies that the vector y is the product of a scalar $-\frac{a_x - a'_x}{a_y - a'_y}$ by the vector x. Then the vectors x and y are parallel in contradiction with the hypothesis that they are perpendicular. Therefore, the linear combination of the vector a in the form (9.6) is unique.

If we assume $a_x - a'_x \neq 0$ instead of $a_y - a'_y \neq 0$ we come to the same conclusion.   □

Similarly, let the coordinate system having origin O and unit points U_x, U_y, U_z be assigned in the space. The oriented segments OU_x, OU_y, OU_z determine the free vectors x, y, z, called the *unit vectors* of the axes x, y and z, respectively. Therefore, the unit vectors x, y and z have magnitude 1, the same direction and the same orientation of the axes x, y and z, respectively.

Furthermore, any free vector a of the space, is a linear combination of the *unit vectors x, y, z*

$$a = a_x x + a_y y + a_z z \tag{9.9}$$

and the triple (a_x, a_y, a_z) of the scalar components of a is univocally determined.

9.10 The Sphere

Given a coordinate system of the space and the point $C(a, b, c)$, let h be a positive real number. Let us define *sphere* or *spherical surface* with *center* C and *radius h*, the set of points $P(x, y, z)$ having distance h from C. The point P varies and maintains the distance h from C describing the sphere.

The point $P(x, y, z)$ belongs to the sphere if and only if

$$|PC| = h$$

From the formula of the distance in space (Sect. 9.8), the point P belongs to the sphere if and only if

$$d(P, C) = \sqrt{(x - a)^2 + (y - b)^2 + (z - c)^2} = h \tag{9.10}$$

Let us square both sides of (9.10):

$$(x - a)^2 + (y - b)^2 + (z - c)^2 = h^2 \tag{9.11}$$

The point $P(x, y, z)$ satisfies (9.10) if and only if satisfies (9.11). Therefore, the Eq. (9.11) is the equation of the sphere with center C and radius h.

For example, the equation of the sphere with center $O(0, 0, 0)$ and radius 3 is

$$x^2 + y^2 + z^2 = 9.$$

If we put

$$m = -2a, n = -2b, p = -2c, q = a^2 + b^2 + c^2 - h^2 \tag{9.12}$$

Equation (9.11) takes the form

$$x^2 + y^2 + z^2 + mx + ny + pz + q = 0 \tag{9.13}$$

From (9.12) we get

$$m^2 + n^2 + p^2 - 4q = 4h^2 > 0 \qquad (9.14)$$

It is easy to check that every equation of the type (9.13), with m, n, p, q satisfying (9.14) represents a sphere, precisely the one with center $C(a, b, c)$, where

$$a = -\frac{m}{2}, \quad b = -\frac{n}{2}, \quad c = -\frac{p}{2},$$

and radius

$$h = \frac{1}{2}\sqrt{m^2 + n^2 + p^2 - 4q}.$$

Exercise 9.1 Find the equation of the sphere that contains the points $(0, 0, 0)$, $(1, 0, 0)$, $(0, 2, 0)$, $(0, 0, 3)$.

The four points must satisfy the Eq. (9.13). Therefore, the following equalities hold:

$$q = 0$$
$$1 + m + q = 0$$
$$4 + 2n + q = 0$$
$$9 + 3p + q = 0$$

Hence, $m = -1, n = -2, p = -3, q = 0 \ 0$. Replacing these values in (9.13), we obtain the equation of the sphere, as requested:

$$x^2 + y^2 + z^2 - x - 2y - 3z = 0$$

Bibliography

Albert, A.A.: Solid Analytic Geometry. Dover Publications Inc., New York (2016)
Anton, H.: Calculus. Wiley, New York (1980)
Apostol, T.M.: Calculus. Wiley, New York (1967)

Chapter 10
Scalar Product. Lines and Planes

10.1 Introduction

We define the operation of scalar product to deepen the concepts of parallelism and orthogonality and provide for the effective formalization of the analytic properties of lines and planes, their representations and mutual positions.

10.2 Scalar Product

The angle $\widehat{ab}$ between the free vectors a and b is the smaller angle between the vectors (A, a) and (A, b) applied at a common initial point A and belonging to the free vectors a and b, respectively.

Definition 10.1 The *scalar product* of vectors a and b, denoted $a \cdot b$ or $a \times b$, is defined by:

$a \cdot b = 0$, if one of the vectors is null.
$a \cdot b = |a||b| \cos \widehat{ab}$, otherwise.
The function that associates the scalar product $a \cdot b$ to the couple of vectors (a, b) is called *scalar multiplication* of vectors.
The following properties hold:

1. $a \cdot b = b \cdot a$
2. If a and b are non-null vectors, then $a \cdot b = 0$ if and only if $\cos \widehat{ab} = 0$, i. e., a and b are perpendicular.
3. However the vectors a, b and c are fixed, it is: $a \cdot (b + c) = a \cdot b + a \cdot c$.
4. For every real number k, it is $k(a \cdot b) = ka \cdot b = a \cdot kb$.

© The Author(s), under exclusive license to Springer Nature Switzerland AG 2023   151
A. G. S. Ventre, *Calculus and Linear Algebra*,
https://doi.org/10.1007/978-3-031-20549-1_10

Fig. 10.1 Orthogonal
projection of b on a

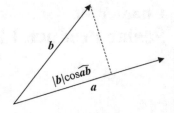

10.2.1 Orthogonal Projections of a Vector

For every non-null vector a and every vector b, let r be a parallel axis with the same
orientation of a. Let the orthogonal component of b on r be denoted by b', i.e., b' be
the relative length of the orthogonal projection of an applied vector of b on r. Then b'
$= 0$ if $b = O$ and $b' = |b| \cos \widehat{ab}$ (see Sect. 8.1.3. c)) if $b \neq O$. Therefore (Fig. 10.1),

$$a \cdot b = |a| b'$$

Similarly, if b is non-null and a' is the orthogonal component of a on the line s
having the same direction and orientation as b, we have:

$$a \cdot b = |b| a'$$

10.2.2 Scalar Product in Terms of the Components

Let a coordinate system be given in the plane α. Let a and b be vectors of the plane
α and x, y the unit vectors of the coordinate axes.

It is known (Sect. 9.9) that

$$a = a_x x + a_y y$$
$$b = b_x x + b_y y$$

where (a_x, a_y) and (b_x, b_y) are the components of a and b. From properties 3 and 4,
we obtain:

$$a \cdot b = \left(a_x x + a_y y\right) \cdot \left(b_x x + b_y y\right)$$
$$= a_x b_x x \cdot x + a_x b_y x \cdot y + a_y b_x y \cdot x + a_y b_y y \cdot y$$

As

$$x \cdot x = y \cdot y = 1 \cdot 1 \cdot \cos 0 = 1$$

and x and y are orthogonal to each other,

$$x \cdot y = y \cdot x = 0$$

the representation of scalar product in terms of the components,

$$a \cdot b = a_x b_x + a_y b_y \tag{10.1}$$

follows.

10.3 Scalar Product and Orthogonality

Let a be a non-null vector of the plane. From the definition of scalar product we have

$$a \cdot a = |a|^2$$

and by (10.1),

$$a \cdot a = a_x^2 + a_y^2$$

Equating the right-hand sides of the equations above:

$$|a| = \sqrt{a_x^2 + a_y^2} \tag{10.2}$$

If a and b are non-null vectors then, by definition of scalar multiplication,

$$\cos \widehat{ab} = \frac{a \cdot b}{|a||b|} \tag{10.3}$$

which expresses, in virtue of (10.1) and (10.2), the cosine of the angle $\widehat{ab}$ in terms of the components of the vectors a and b.

In particular, the vectors a and b are orthogonal if and only if

$$a \cdot b = a_x b_x + a_y b_y = 0$$

Analogous formulae hold if a and b are non-null vectors of the space, having components a_x, a_y, a_z and b_x, b_y, b_z, respectively. In particular, the vectors a and b in the space are orthogonal if and only if

$$a \cdot b = a_x b_x + a_y b_y + a_z b_z = 0$$

10.3.1   Angles of Lines and Vectors

Let a coordinate system with origin O and unit points U_x and U_y, be given in the plane α. Let r and s be oriented lines of the plane. In order to determine the cosine of the angle $\widehat{rs}$ we consider the couples (m, n) and (m', n') of direction numbers of r and s, respectively The vectors a and b having components (m, n) and (m', n') are parallel to r and s, respectively. Let r and s be supposed to have the same orientations of r and s, respectively. Then it turns out $\widehat{rs} = \widehat{ab}$, and by (10.3),

$$\cos \widehat{rs} = \frac{mm' + nn'}{\sqrt{m^2 + n^2}\sqrt{m'^2 + n'^2}} \tag{10.4}$$

Let us now consider the angles that an oriented line r forms with the coordinate axes. Let $(1, 0)$ and $(0, 1)$ be two couples of direction numbers of the x axis and the y axis, respectively. If (m, n) is the couple of components of a vector a parallel to r and having the same orientation of r, by (10.4), we have

$$\cos \widehat{xr} = \frac{m}{\sqrt{m^2 + n^2}} \quad \text{and} \quad \cos \widehat{yr} = \frac{n}{\sqrt{m^2 + n^2}}$$

The numbers $r_x = \cos \widehat{xr}$ and $r_y = \cos \widehat{yr} = \sin \widehat{xr}$ (Sect. 8.1.4) are called *direction cosines* of the r axis: they are the components of the unit vector r parallel to r with the same orientation of r. A couple of direction numbers of r is (m, n) and it is proportional to any couple of direction numbers of r (Sect. 7.2).

From the equality

$$\cos \widehat{yr} = \sin \widehat{xr}$$

we obtain

$$rx^2 + ry^2 = \cos^2 \widehat{xr} + \cos^2 \widehat{yr} = 1$$

Furthermore, if a is parallel to r with the same orientation of r and r is the unit vector of r, the equality holds

$$r = \frac{a}{|a|}$$

10.3.2   Orthogonal Lines in the Plane

Two lines, oriented or not, are perpendicular to each other if and only if the cosine of their angle is zero. By (10.4) we have:

Proposition 10.1 *Let r and s be lines of the plane with direction numbers* (m, n) *and* (m', n'), *respectively. Then r and s are orthogonal if and only if.*

$$mm' + nn' = 0 \tag{10.5}$$

Let the lines r and s of the plane have the equations

$$\begin{aligned} r)\, & ax + by + c = 0 \\ s)\, & a'x + b'y + c' = 0 \end{aligned} \tag{10.6}$$

It is known (Sect. 7.4.2) that $(-b, a)$ and $(-b', a')$ are couples of direction numbers of r and s, respectively.

Proposition 10.2 *The lines r and s represented by Eqs.* (10.6) *are orthogonal if and only if.*

$$aa' + bb' = 0 \tag{10.7}$$

Observe that the coefficients a and b of the equation of the line r are the components of a vector v perpendicular to the line. In fact, if $P_0 = (x_0, y_0)$ is a point of the line r, from the equation of r, we get (Sect. 7.4.2):

$$a(x - x_0) + b(y - y_0) = 0$$

i.e., the scalar product $(a, b) \times P_0P$ is null and the vector P_0P, with components $(x - x_0, y - y_0)$ which is parallel to r, is orthogonal to the vector of components (a, b); in other words, the couple $(x - x_0, y - y_0)$ is proportional to $(b, -a)$, so $(x - x_0, y - y_0)$ is the couple of components of a vector orthogonal to (a, b).

10.4 The Equation of the Plane

Theorem 10.1 *Let a plane* α *and a line r not parallel be given and* $Q(x_0, y_0, z_0)$ *be the point common to* α *and r. Moreover, let* $v = (a, b, c)$ *be a non-null vector parallel to the line r. The plane* α *and the line r are perpendicular to each other if and only if the equation.*

$$a(x - x_0) + b(y - y_0) + c(z - z_0) = 0 \tag{10.8}$$

is fulfilled, for every point $P(x, y, z)$ in the plane α.

Fig. 10.2 Vectors $v = (a, b, c)$ and $w = QP$ perpendicular to each other

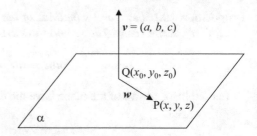

Proof By Theorem 4.2, the line r, and thus the vector v, are perpendicular to the plane α if and only if r is perpendicular to the line passing through $Q(x_0, y_0, z_0)$ and $P(x, y, z)$, whatever the point P distinct from Q and belonging to the plane α is. Hence, if w is the vector containing the oriented segment QP, then r and α are perpendicular if and only if the scalar product $v \cdot w$ is zero (Fig. 10.2): $v \cdot w = a(x - x_0) + b(y - y_0) + c(z - z_0) = 0$.

By Theorem 4.3 given the point P and the line r in the space, there is one and only one plane α containing P and perpendicular to the line r. Therefore, all the points $P(x, y, z)$ of the plane α satisfy Eq. (10.8), which is the *equation of the plane passing through* P *and perpendicular to the vector* $v = (a, b, c)$. $\square$

Example 10.1 The plane passing through the point $P(2, -5, 1)$ and perpendicular to the vector of components $(4, 3, -6)$ has the Eq. $4(x - 2) + 3(y + 5) - 6(z - 1) = 0$, that takes the form $4x + 3y - 6z + 13 = 0$.

If the vector $v = (a, b, c)$ is perpendicular to the plane α, then the vector $hv = (ha, hb, hc)$, for any non-null scalar h, is perpendicular to the plane α.

As a consequence of (10.8), each plane passing through the origin of the coordinates $(0, 0, 0)$ has the equation

$$ax + by + cz = 0$$

Any plane has an equation of first degree in x, y, z, and any equation

$$ax + by + cz + d = 0 \tag{10.9}$$

with (a, b, c) non-null, is the equation of a plane. Indeed, from Eq. (10.8) we obtain $a(x - x_0) + b(y - y_0) + c(z - z_0) = ax + by + cz - ax_0 - by_0 - cz_0 = 0$ and, putting $d = -ax_0 - by_0 - cz_0$, the Eq. (10.9) follows. Vice versa, every Eq. (10.9) is the equation of a plane.

Corollary 10.1 *The planes* α *and* β *of equations* $ax + by + cz + d = 0$ *and* $a'x + b'y + c'z + d' = 0$, *respectively, are parallel if and only if the triples* (a, b, c) *and* (a', b', c') *are proportional.*

Corollary 10.2 *The plane containing the point* (x_1, y_1, z_1) *and parallel to the plane of equation* $ax + by + cz + d = 0$ *has equation* $a(x - x_1) + b(y - y_1) + c(z - z_1) = 0$.

Fig. 10.3 Vector $(1, 1, 1)$
perpendicular to the plane $x + y + z = 1$

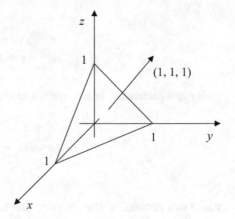

In particular, the planes identified by the z and y axes, z and x axes, x and y axes, have equations $x = 0$, $y = 0$, $z = 0$, respectively. The generic plane parallel to the xy plane has the equation $z = h$; the generic plane parallel to the xz plane has equation $y = k$; the generic plane parallel to the yz plane has equation $\mathrm{x} = l$.

Given three distinct non-aligned points, there is one and only one plane α that contains them (Sect. 4.2).

For example, if we want to determine the plane α of equation $ax + by + cz + d = 0$ which contains the unit points $(1, 0, 0)$, $(0, 1, 0)$, $(0, 0, 1)$ we must determine the coefficients a, b, c, d, imposing that the required plane passes through each of the three points. Let first observe that the given points are non-aligned. Therefore, we obtain:

- if $(1, 0, 0)$ belongs to the α plane, then $a1 + b0 + c0 + d = 0$, i. e., $a + d = 0$;
- if $(0, 1, 0)$ belongs to α, then $a0 + b1 + c0 + d = 0$, i. e., $b + d = 0$;
- if $(0, 0, 1)$ belongs to α, then $a0 + b0 + c1 + d = 0$, i. e., $c + d = 0$;

Hence, we obtain the equalities $a = -d, b = -d, c = -d$ and Eq. (10.9) becomes $-dx - dy - dz + d = 0$. Let us divide by d as the plane does not contain the origin of the coordinates. Therefore, the equation of the plane passing through the unit points is $x + y + z - 1 = 0$ (Fig. 10.3).

10.5 Perpendicular Lines and Planes

The parametric representation of the line in a coordinate system of the space still stems from Thales' theorem.

The parametric equations of the line r through distinct points $P_1(x_1, y_1, z_1)$ and $P_2(x_2, y_2, z_2)$ take the form, similar to the representation of the line in the plane (Sect. 7.3.6):

$$x - x_1 = t(x_2 - x_1)$$
$$y - y_1 = t(y_2 - y_1)$$
$$z - z_1 = t(z_2 - z_1)$$

with t real parameter, or the equivalent form:

$$x = x_1 + t(x_2 - x_1)$$
$$y = y_1 + t(y_2 - y_1)$$
$$z = z_1 + t(z_2 - z_1)$$

with t real parameter. The *direction numbers* of the line r are

$$m = x_2 - x_1$$
$$n = y_2 - y_1$$
$$p = z_2 - z_1$$

Since the points P_1 and P_2 are distinct the triple of direction numbers (m, n, p) is non-null. Any non-null triple proportional to (m, n, p) is a triple of direction numbers of r, and also the triple of the components of a space vector parallel to r.

Therefore, by Theorem 10.1 we obtain:

Proposition 10.3 (Perpendicular line and plane) *The plane α) $ax + by + cz + d = 0$ and the line r are perpendicular if and only if the vector (a, b, c) is parallel to r. In other words, the plane α and the line r are perpendicular if and only if (a, b, c) is a triple of direction numbers of r.*

Exercise 10.1 Let the point $P(3, -2, 1)$ and the plane α) $5x - 6y + 7z + 1 = 0$ be given. Find the plane α') parallel to α and passing through the point $(1, -1, 8)$.

By Corollary 10.2 the equation of α' is $5(x - 1) - 6(y + 1) + 7(z - 8) = 0$, which can be written in the form $5x - 6y + 7z - 67 = 0$.

Example 10.2 Let the point $P(-1, 2, 3)$ and the plane α) $2x - 7y + 6z - 1 = 0$ be given. Find the parametric equations of the line r, passing through P and perpendicular to α.

The required line r is parallel to the vector $(2, -7, 6)$, has direction numbers $m = 2, n = -7, p = 6$ and parametric equations

$$x - x_1 = 2t$$
$$y - y_1 = -7t$$
$$z - z_1 = 6t$$

Moreover, the line r contains the point P(-1, 2, 3) and therefore its parametric equations are:

$$x = -1 + 2t$$
$$y = 2 - 7t$$
$$z = 3 + 6t$$

Example 10.3 Let the point P(0, 7, -3) and the line r of equations

$$x = 1 + 2t$$
$$y = 5 - t$$
$$z = 6t$$

be assigned. Find the plane α passing through P and perpendicular to r.

The coefficients a, b, c of the equation of α are direction numbers 2, -1, 6 of r, respectively. Therefore, the equation of α is, by (10.8),

$$2x - (y - 7) + 6(z + 3) = 0$$

which can be put in the form

$$2x - y + 6z + 25 = 0.$$

Bibliography

Albert, A.A.: Solid analytic geometry. Dover Publications Inc., New York (2016)
Anton, H.: Calculus. Wiley, New York (1980)
Apostol, T.M.: Calculus. Wiley, New York (1967)
Spivak M.: Calculus. Cambridge University Press, Cambridge (2006)

Chapter 11
Systems of Linear Equations. Reduction

11.1 Linear Equations

We studied (Chap. 7) the equation $ax + by + c = 0$ also called the *linear equation* in the variables x and y and real coefficients a, b and c. We have also considered the geometric meaning of this equation.

A linear equation with n unknowns is an expression of the form

$$a_1 x_1 + a_2 x_2 + \dots + a_n x_n = b \tag{11.1}$$

where, for $i = 1, 2, \dots, n$, the x_i's are the *unknowns*, or *variables*, the a_i's are real numbers, called *coefficients of the unknowns* and b is the *constant term*, or simply the *constant* of the equation.

Example 11.1 We considered (Sect. 7.2.2) the linear equation $3x - 2y = -1$ in the unknowns x, y, with coefficients of the unknowns 3 and -2, and constant -1. We found couples of values to give to x and y that satisfy the equation; these couples, called *solutions* of the equation (see Example 7.1) are obtained by assigning a value to x (or y) and finding the consequent value for y (or x).

A solution of the linear Eq. (11.1) is an n-tuple $(k_1, k_2, \dots, k_n)$ of real numbers such that the equality

$$a_1 k_1 + a_2 k_2 + \dots + a_n k_n = b$$

obtained by substituting in (11.1) k_i for x_i, $i = 1, 2, \dots, n$, is verified: then the n-tuple $(k_1, k_2, \dots, k_n)$ is said to *satisfy* the Eq. (11.1), and the equation is said to be *consistent* or *compatible*. If an n-tuple satisfying (11.1), does not exist, then the equation is said to be *inconsistent* or *incompatible*. Let us consider some cases of compatibility and incompatibility.

case (i): one of the coefficients of the Eq. (11.1), say a_1, is non-null. Then we divide both sides of Eq. (11.1) by a_1:

© The Author(s), under exclusive license to Springer Nature Switzerland AG 2023   161
A. G. S. Ventre, *Calculus and Linear Algebra*,
https://doi.org/10.1007/978-3-031-20549-1_11

$$x_1 = \frac{1}{a_1}b - \frac{a_2}{a_1}x_2 - \ldots - \frac{a_n}{a_1}x_n$$

Whatever the values assigned to the unknowns $x_2, \ldots, x_n$ are, a value for x_1 is got; then the n-tuple $(x_1, x_2, \ldots, x_n)$ is a solution of (11.1) and in this way, changing the choice for $x_2, \ldots, x_n$, we obtain other possible solutions of the equation.

Example 11.2 Find some more solutions of $3x - 2y = -1$.

Rewrite the equation as

$$x = \frac{2}{3}y - \frac{1}{3}$$

For every value assigned to y a unique value is obtained for x. For example, if $y = 0$, then $x = -\frac{1}{3}$; for $y = 1$ we obtain

$$x = \frac{1}{3}$$

Therefore, the couples $\left(-\frac{1}{3}, 0\right)$, $\left(\frac{1}{3}, 1\right)$ are two among the solutions of equation $3x - 2y = -1$.

case (ii): all the coefficients in the Eq. (11.1) are zero and the constant is not zero:

$$0x_1 + 0x_2 + \ldots + 0x_n = b, \ b \neq 0$$

Then the equation has no solution, *i.* e., it is incompatible.

case (iii): all the coefficients in the Eq. (11.1) are zero and the constant is zero too:

$$0x_1 + 0x_2 + \ldots + 0x_n = 0$$

Then every n-tuple of real numbers is a solution; the equation is called an *identity* or an *identical equation* (Sect. 6.3).

11.1.1 Systems of Linear Equations

Let us consider a set of m linear equations and n unknowns $x_1, x_2, \ldots, x_n$:

$$
\begin{aligned}
a_{11}x_1 + a_{12}x_2 + \ldots + a_{1n}x_n &= b_1 \\
a_{21}x_1 + a_{22}x_2 + \ldots + a_{2n}x_n &= b_2 \\
&\ldots \\
a_{m1}x_1 + a_{m2}x_2 + \ldots + a_{mn}x_n &= b_m
\end{aligned}
\tag{11.2}
$$

The elements $a_{ij} \in \mathbf{R}$ are called *coefficients* (or *coefficients of the unknowns*) and the elements $b_i \in \mathbf{R}$ are the *right-hand sides* (or *known coefficients*) of the equations, for $i = 1, 2, \ldots, m; j = 1, 2, \ldots, n$.

The set of Eq. (11.2) is defined as a *system of linear equations* with real coefficients, real constants, and the unknowns $x_1, x_2, \ldots, x_n$.

The system is said to be *homogeneous* if all constants $b_1, b_2, \ldots, b_m$ are null. Hence, a homogeneous system takes the form

$$\begin{aligned} a_{11}x_1 + a_{12}x_2 + \ldots + a_{1n}x_n &= 0 \\ a_{21}x_1 + a_{22}x_2 + \ldots + a_{2n}x_n &= 0 \\ &\ldots \\ a_{m1}x_1 + a_{m2}x_2 + \ldots + a_{mn}x_n &= 0 \end{aligned} \qquad (11.3)$$

The system (11.3), obtained by the system (11.2) substituting zeros to all b_i's is said to be the *homogeneous linear system associated* with the system (11.2).

Definition 11.1 An n-tuple $(k_1, k_2, \ldots, k_n)$ of real numbers is said to be a *solution* of the system of linear Eq. (11.2), if it is a solution of all the equations of the system; a solution of the system is said to *satisfy* (all the equations of) the system; the system is said to *have*, or *admit*, the solution $(k_1, k_2, \ldots, k_n)$. The set of all solutions is said to be the *solutions set*, or the *general solution*, of the system.

If the system (11.2) has a solution, then the system is said to be *consistent* or *compatible*. If the system has no solution, then it is defined an *inconsistent* or *incompatible* system. A consistent system may admit a unique solution, or more than one solution.

A homogeneous system is always consistent since it admits the *null*, or *trivial*, *solution* $(0, 0, \ldots, 0)$, which is the null vector of $\mathbf{R}^n$.

Example 11.3 Let

$$\begin{aligned} 3x + 2y &= 6 \\ 3x + 2y &= 5 \end{aligned}$$

be a system of two equations and unknowns x and y. The left-hand sides of the two equations are equal while the right-hand sides are different from each other. We realize that there are no numbers k_1, k_2 that replace x, y in the two equations and make the left-hand side $3x + 2y$ of the equations equal to 6 and also equal to 5. Therefore, the system is inconsistent because it does not admit solutions.

Example 11.4 In the system of two equations and two unknowns.

$$\begin{aligned} 4x - 2y &= 5 \\ 8x - 4y &= 10 \end{aligned}$$

the second equation can be put in the form $2(4x - 2y) = 2(5)$ and, after dividing the two sides by 2, it reduces to a copy of the first equation (see Sect. 6.3). Therefore, the second equation has the same solutions as the first. We can obtain as many solutions of the equation as we want. In fact, we fix, at will, a value for x (for y) and then we find a value of y (of x). For example, if $x = 0$ we have: $4 \times 0 - 2y = 5$, i. e., $y = -\frac{5}{2}$; if $y = 0$, then $x = \frac{5}{4}$. Therefore, the two couples $\left(0, -\frac{5}{2}\right)$ and $\left(\frac{5}{4}, 0\right)$ are solutions of the given system. Hence, the system is consistent.

Example 11.5 The linear system of two equations and two unknowns

$$x - y = 2$$
$$y = 3$$

admits the solution $(x, y) = (5, 3)$. This is easily checked by replacing 5 for x and 3 for y in each equation: $5 - 3 = 2$; $3 = 3$. The solution $(5, 3)$ is unique because if 3 is the unique value that y can take, then 5 is the unique value that x can take. Therefore, the system is consistent and has only one solution.

11.2   Equivalent Systems

Definition 11.2 Two systems of linear equations with the same number of unknowns are said to be *equivalent* if they have the same solutions.

Remark 11.1 In particular, we realize that two linear equations are equivalent if one is obtained from the other by multiplying this by a non-null real number. In Example 11.4, the equations $4x - 2y = 5$ and $8x - 4y = 10$ are equivalent by Theorem 6.2 because the former is obtained from the latter by dividing this by 2.

We study the systems of linear equations in order to determine their compatibility and find the possible solutions.

A basic operation over the equations of a linear equation system is the *linear combination*.

Definition 11.3 A *linear combination of the equations* of system (11.2) is, by definition, an equation of the form:

$$c_1(a_{11}x_1 + a_{12}x_2 + \ldots + a_{1n}x_n) + c_2(a_{21}x_1 + a_{22}x_2 + \ldots + a_{2n}x_n) + \ldots$$
$$+ c_m(a_{m1}x_1 + a_{m2}x_2 + \ldots + a_{mn}x_n) = c_1b_1 + c_2b_2 + \ldots + c_mb_m$$

where the c_i's are real numbers called the *coefficients* of the linear combination. A linear combination of linear equations is a linear equation.

Example 11.6 A linear combination of the equations

$$x - y = 2$$
$$-x + 2y = 3$$

is the equation

$$c_1(x - y) + c_2(-x + 2y) = 2c_1 + 3c_2$$

which can be expressed in the equivalent form

$$(c_1 - c_2)x + (-c_1 + 2c_2)y = 2c_1 + 3c_2$$

The coefficients of the linear combination are c_1, c_2. In particular, if $c_1 = -2$ and $c_2 = 5$, the linear combination becomes

$$(-2 - 5)x + (2 + 10)y = 2(-2) + 3(5)$$

i. e., $-7x + 12y = 11$.

We state the following theorem and apply it.

Theorem 11.1 *Every solution of a system of equations is also a solution of every linear combination of the equations of the system.*

Example 11.7 The system

$$x - y = 2$$
$$x - 2y = -3$$

has the solution $(x, y) = (7, 5)$. Indeed, replacing 7 for x and 5 for y in each of the two equations, we have $7 - 5 = 2; 7 - 2 \times 5 = -3$. Let us consider the generic linear combination of the equations of the system

$$c_1(x - y) + c_2(x - 2y) = 2c_1 - 3c_2$$

and verify that this equation has the solution $(7, 5)$; in fact, if $(x, y) = (7, 5)$, we obtain

$$c_1(7 - 5) + c_2(7 - 10) = 2c_1 - 3c_2$$

11.2.1 Elementary Operations

Let us name the m equations of a system with the symbols $E_1, E_2, \ldots, E_m$. The notations simplify the indication of the linear combinations; the expression $hE_i + kE_j$ means: multiply the equation E_i by h, the equation E_j by k, and add. For example, let us consider the system of two equations

$$E_1 : \quad x - y = 2$$
$$E_2 : \quad -x + 2y = 3$$

The sum $2E_1 - 5E_2$ is an abbreviation of the Equation $2(x - y) - 5(-x + 2y) = 2 \times 2 + 3 (-5)$, i.e., $7x - 12y = -11$.

Theorem 11.2 *If in the system* (11.2) *the equation* E_i *is replaced with the sum* $E_i + kE_j$, *with* $i, j = 1, \ldots, m$, *for* k *any real number, a system equivalent to the system* (11.2) *is obtained.*

Example 11.8 Let us verify that the system

$$E_1 : \quad x - y = 2$$
$$E_2 : \quad -x + 2y = 3$$

is equivalent to the system

$$E_1 : \quad x - y = 2$$
$$E_1 + E_2 : \quad y = 5$$

obtained from the previous one by replacing the equation E2 with the sum of the two equations. It is easy to check that the second system also has the solution (7, 5). Since the second system has the unique solution (7, 5) (Example 11.4), the first system has the unique solution (7, 5) too.

The following is a generalization of Theorem 11.2.

Theorem 11.3 *If in the system* (11.2) *the equation* E_i *is replaced with the equation* $hE_i + kE_j$, *with* $h \neq 0$, *for any* $k \in \mathbf{R}$, *a system equivalent to the system* (11.2) *is obtained.*

From Remark 11.1 and Theorems 11.2 and 11.3 the following operations, which transform an equation system into an equivalent system, are defined in the set of the equations of a system:

operation 1. replacing the equation E_i with the equation hE_i, with $h \neq 0$;
operation 2. replacing the equation E_i with the equation $E_i + kE_j$, for any real k;
operation 3. replacing the equation E_i with the equation $hE_i + kE_j$, $h \neq 0$, and for any real k.

In other words, the equation E_i can be replaced by an equation which is the sum of E_i (possibly multiplied by a real number $h \neq 0$) added to an equation of the system multiplied by the number k. (In short, in the equation that replaces E_i must appear E_i, or hE_i, with $h \neq 0$.) Let us express the actions of the operations by means of the following symbols:

operation 1. $E_i \leftarrow hE_i, h \neq 0$ (replace E_i with hE_i)

operation 2. $E_i \leftarrow E_i + kE_j$ (replace E_i with $E_i + kE_j$)

operation 3. $E_i \leftarrow hE_i + kE_j, h \neq 0$ (replace E_i with $hE_i + kE_j$)

Operation 3 includes operations 1 and 2 as special cases.

The *interchange operation* of two equations E_i and E_j, denoted $E_i \leftrightarrow E_j$, transforms a system into an equivalent system: this means that the ordering of the equations in a system does not affect the general solution of the system. The operations 1, 2, 3 and the interchange operation are called *elementary operations* on a system of linear equations.

The application of a finite number of elementary operations to a system transforms it into an equivalent system.

From Theorem 11.1 it follows:

Theorem 11.4 *If the equation E_i in the system* (11.2) *is a linear combination of the remaining equations, the system is equivalent to that obtained by eliminating the equation E_i.*

For example, the system

$$
\begin{aligned}
E_1 : &\quad 2x - y = 1 \\
E_2 : &\quad -x + y = 0 \\
E_3 : &\quad x = 1
\end{aligned}
$$

is equivalent to the system of the last two equations because $E_1 = E_3 - E_2$. Therefore, the system admits the unique solution $(1, 1)$.

11.3 Reduced Systems

Definition 11.4 A system of m linear equations is said to be a *reduced system*, or a system in a *reduced form*, if in each of the first $m - 1$ equations, which is not the identical equation, an unknown appears (i.e., has non-null coefficient) which does not appear (i.e., has coefficient 0) in the equations below.

Example 11.9 The system of three equations and three unknowns

$$
\begin{aligned}
x - 2y - z &= 2 \\
4x + 2y &= 6 \\
7x &= 1
\end{aligned}
$$

is reduced because in the first equation the unknown z appears which does not appear in the other two equations, and in the second equation the unknown y appears which does not appear in the third equation.

Example 11.10 The following system of four equations with three unknowns is not reduced

$$\begin{aligned} x - y + 4z &= 0 \\ 3y - 3z &= 1 \\ y &= 2 \\ 2x + 8z &= 3 \end{aligned}$$

Remark 11.2 The incompatible equation $0 = b$, with $b \neq 0$, divided by b, takes the form $0 = 1$. Hence, the incompatible equation is unique (see Sect. 6.3).

Theorem 11.5 [*Gauss elimination*] *It is always possible to apply a finite number of elementary operations to a linear system and transform it into a reduced system.*

The theorem shows a procedure called *reduction* of the system of linear equations. We will deal with the procedure in Chap. 16.

Let us now examine some examples of reduction of systems of linear equations by means of the elementary operations in order to find the solutions or check inconsistencies.

Example 11.11 Let us consider the system

$$\begin{aligned} E_1 : \quad & 3x + 4y + z = 2 \\ E_2 : \quad & x - 2z = 3 \\ E_3 : \quad & 3z = 15 \end{aligned}$$

The operation

$$E_3 \leftarrow \frac{1}{3}E_3$$

transforms the system in an equivalent system; then the last equation is written as $z = 5$. Now we replace E_3 with the equation $z = 5$, which becomes our new E_3; we obtain the equivalent system

$$\begin{aligned} E_1 : \quad & 3x + 4y + z = 2 \\ E_2 : \quad & x - 2z = 3 \\ E_3 : \quad & z = 5 \end{aligned}$$

Then we apply the operation $E_2 \leftarrow E_2 - (-2)E_3$ (observe that the number -2 between the brackets is the coefficient of z in E_2). The second equation then becomes: $x - 2z - (-2)z = 1 - (-2)5$, that we rewrite as $x = 11$; in practice, the second equation is obtained by replacing 5 for z in the second equation:

$$E_1: \quad 3x + 4y + z = 2$$
$$E_2: \quad x = 11$$
$$E_3: \quad z = 5$$

Let us replace the acquired values of z and x and set $x = 11$ and $z = 5$ in the first equation; then $33 + 4y + 5 = 2$ and $y = 9$. The given system is therefore equivalent to the system:

$$x = 11$$
$$y = -9$$
$$z = 5$$

that reveals that the triple $(x, y, z) = (11, -9, 5)$ is a solution; indeed, it is the unique solution of the given system of equations.

Remark 11.3 A system of equations with an identical equation and the system with identical equation cancelled are equivalent. For example, the system

$$x - y = 2$$
$$0 = 0$$
$$-x + 2y = 3$$

is equivalent to the system

$$x - y = 2$$
$$-x + 2y = 3$$

It is immediate to state the equivalence of the two systems since any couple of real numbers is a solution of the identical equation.

Remark 11.4 A system of equations that contains the inconsistent equation is inconsistent; in fact, the inconsistent equation has no solution and every system that contains it does not admit any solution.

Remark 11.5 If a reduced system has more equations than unknowns, then the system contains the identical equations $0 = 0$, or the incompatible equation $0 = 1$.

Example 11.12 Consider the reduced system

$$x + 2y - z = 1$$
$$y - 3z = 0$$

of two equations and three unknowns. Let us begin to carry out from below for in the last equation the minimum number of unknowns occurs. We consider as a *parameter*, called also a *free variable*, one of the unknowns such that, if moved to the right-hand side, the system in the remaining unknowns is still reduced. For example, set $z = c$, c parameter: the system of two equations and the unknowns x and y

$$x + 2y = 1 + c$$
$$y = 3c$$

is obtained. Then we proceed by *substitution* (see Example 11.11) and replace $3c$ for y in the first equation so as to obtain

$$x + 2(3c) - c = 1$$
$$y = 3c$$

and then

$$x + 5c = 1$$
$$y = 3c$$

A solution (x, y, z) of the given system is

$$x = -5c + 1$$
$$y = 3c \qquad\qquad (11.4)$$
$$z = c$$

which depends on the parameter c. For each numerical value assigned to c, a solution of the system is obtained. For example, if $c = 0$ the solution $(x, y, z) = (1, 0, 0)$ is obtained; if $c = 1$, then $(x, y, z) = (-4, 3, 1)$, etc. The system has infinite solutions that depend on the real values that we can attribute to the parameter $z = c$.

Remark 11.6 The substitution is an elementary operation. For instance, in the case of the system

$$x + 2y - z = 1$$
$$y - 3z = 0$$

the substitution that replaces $3c$ for y in the first equation means to apply the elementary operation: $E_1 \leftarrow E_1 - 2E_2$ which transforms the given system into the following:

$$E_1 : \quad x + 2y - z - 2y + 6z = 1$$
$$E_2 : \quad y - 3z = 0$$

i.e.,

$$E_1 : \quad x + 5z = 1$$
$$E_2 : \quad y - 3z = 0$$

which is equivalent to the form (11.4).

Example 11.13 Let the following system of four linear equations and three unknowns

$$E_1 : \quad x - y + 4z = 0$$
$$E_2 : \quad 3y - 3z = 1$$
$$E_3 : \quad y = 2$$
$$E_4 : \quad 2x + 8z = 6$$

be given. The system is not reduced. Let us apply the elementary operations as follows:

$$E_3 \leftarrow E_1 + E_3 : x - y + 4z + y = 2; \quad x + 4z = 2$$
$$E_4 \leftarrow (1/2)E_4 : (1/2)(2x + 8z) = (1/2)6; \quad x + 4z = 3$$

Then the given system is equivalent to:

$$E_1 : x - y + 4z = 0$$
$$E_2 : 3y - 3z = 1$$
$$E_3 : x + 4y = 2$$
$$E_4 : x + 4z = 3$$

Let now apply: $E_4 \leftarrow E_4 - E_3$. The system is equivalent to

$$E_1 : x - y + 4z = 0$$
$$E_2 : 3y - 3z = 1$$
$$E_3 : x + 4y = 2$$
$$E_4 : 0 = 1$$

As E_4 is inconsistent, also the system is inconsistent (see Remark 11.4).

11.4 Exercises

Exercise 11.1 Let us consider the following linear system of three equations and three unknowns x, y, z,

$$2x - y + 2z = 1$$
$$x - y - 2z = 0 \qquad\qquad (11.5)$$
$$2x + y - 2z = 2$$

The system of equations is not reduced. Then, first we reduce the system. To this aim, let us eliminate z from the second equation by means the operation $E_2 \leftarrow E_1 + E_2$. We get the following system equivalent to the previous one

$$2x - y + 2z = 1$$
$$3x - 2y = 1$$
$$2x + y - 2z = 2$$

Apply the operation $E_3 \leftarrow E_1 + E_3$ to eliminate y and z from the third equation:

$$2x - y + 2z = 1$$
$$3x - 2y = 1$$
$$4x = 3$$

This system is reduced. From the third equation we obtain $x = \frac{3}{4}$; replacing x with $\frac{3}{4}$ in the second equation we obtain $9 - 8y = 4$, from which $y = \frac{5}{8}$. Plugging $x = \frac{3}{4}$ and $y = \frac{5}{8}$ into the first equation we get $z = \frac{1}{16}$. The system has a unique solution, the triple

$$\left(\frac{3}{4}, \frac{5}{8}, \frac{1}{16} \right)$$

Remark 11.7 Each equation of the system (11.5) represents a plane (see Sect. 10.4). Thus, the solution $(\frac{3}{4}, \frac{5}{8}, \frac{1}{16})$ of the system of Eqs. (11.5) is the triple of the coordinates of the common point to the three planes (11.5).

Exercise 11.2 Solve the system of three equations and four unknowns

$$x - y - z + t = 13$$
$$x - 2y - z = 13$$
$$y + t = 0$$

The system is not reduced. Let us apply $E_2 \leftarrow E_1 - E_2$ to obtain:

$$x - y - z + t = 13$$
$$y + t = 0$$
$$y + t = 0$$

The third equation is eliminated as it is equal to the second:

$$x - y - z + t = 13$$
$$y + t = 0$$

Replace t with $-y$ in the first equation:

$$x = 13 - 2t + z$$
$$y = -t$$

The set of solutions is made of the 4-tuples $(13 - 2t + z, -t, z, t)$, for every $z, t \in$ **R**. The solutions depend on two parameters z, t. Therefore, the system is consistent and has infinite solutions.

Exercise 11.3 Reduce and discuss the system of linear equations:

$$2x - 3y = 3$$
$$x - y - z + t = 0$$
$$2x - 4y + 2z - 2t = 1$$

Interchange first and third equation:

$$2x - 4y + 2z - 2t = 1$$
$$x - y - z + t = 0$$
$$2x - 3y = 3$$

The unknowns z and t do not appear in the third equation. Perform the operation $E_2 \leftarrow E_2 + \frac{1}{2} E_1$. The following system is obtained:

$$2x - 4y + 2z - 2t = 1$$
$$2x - 3y = \frac{1}{2}$$
$$2x - 3y = 3$$

The system is not yet reduced because the unknowns that appear in the second equation are present in the third. Apply the operation $E_3 \leftarrow E_3 - E_2$ to obtain the reduced system

$$2x - 4y + 2z - 2t = 1$$
$$2x - 3y = \frac{1}{2}$$
$$0x + 0y = \frac{5}{2}$$

which is incompatible due to the incompatible equation E3.

Bibliography

Aitken, A.C.: Determinants and Matrices, Oliver and Boyd LTD, Edinburgh (1965)
D'Apuzzo, L., Ventre, A.: Algebra Lineare e Geometria Analitica, Cedam, Padova (1995)
Strang, G.: Linear Algebra and Its Applications. Academic Press Inc., New York (1976)
Strang, G.: Introduction to Linear Algebra. Wellesley-Cambridge Press (2016)

Chapter 12
Vector Spaces

12.1 Introduction

We deal with *structured* sets, i.e., sets on which, or between which, functions or operations are defined. We will generalize some results shown in Chap. 9. The subject is of interest for the *linear algebra*, and it will be studied in the next chapters.

Let us start by introducing the set of *complex numbers*.

12.1.1 Complex Numbers

In the set of real numbers there is no number which satisfies the equation $x^2 = -1$. To find a remedy for this lack a new kind of number, called *the imaginary unit*, denoted i and such that $i = \sqrt{-1}$, is defined. As a result we obtain $i^2 = -1$ and the two roots of the equation $x^2 = -1$ are i and $-i$, indeed $(-i)^2 = (-i)(-i) = i^2 = -1$. The number i satisfies the usual rules of calculus. For example, $0i = 0$, $1i = i$, $\sqrt{-5} = \sqrt{(-1)5} = \sqrt{-1}\sqrt{5} = i\sqrt{5}$. Furthermore,

$$i^3 = i^2 i = -1i = -i$$
$$i^4 = i^2 i^2 = (-1)(-1) = 1$$
$$i^5 = i^4 i = 1i = i$$

Therefore, we can calculate the square root of any negative number. For example, we are able to solve all the *quadratic equations* in the unknown x

$$ax^2 + bx + c = 0 \tag{12.1}$$

where a, b, c are real numbers and $a \neq 0$. The *solutions*, also named *roots*, of Eq. (12.1) are x_1, x_2 and result from the formula

© The Author(s), under exclusive license to Springer Nature Switzerland AG 2023
A. G. S. Ventre, *Calculus and Linear Algebra*,
https://doi.org/10.1007/978-3-031-20549-1_12

$$x = \frac{-b \pm \sqrt{b^2 - 4ac}}{2a}$$

Let us find the roots of the equation $x^2 - 2x + 5 = 0$, whose coefficients replace a, b, c. As $b^2 - 4ac = 2^2 - 4 \times 5 = -16$, the roots are: $x_1 = 1 + 2i$ and $x_2 = 1-2i$.

The numbers of the form $a + bi$, being a and b real numbers and i the imaginary unit, are known as *complex numbers*. The number a is called the real part of the complex number $a + bi$, the product bi is called the *imaginary part* and the real number b is the coefficient of the imaginary part.

The set of complex numbers is denoted by **C**. Real numbers are considered particular complex numbers: those whose coefficient of the imaginary part is 0.

The roots of the equation $x^2 - 2x + 5 = 0$ are the complex numbers $1 - 2i$ and $1 + 2i$. Two complex numbers of the form $a + bi$ and $a - bi$, are said to be *conjugate numbers*. So the complex numbers $1 - 2i$ and $1 + 2i$ are conjugate to each other. The sum of two conjugate complex numbers is a real number: $a + bi + a - bi = 2a$. A quadratic equation $ax^2 + bx + c = 0$ with real coefficients and $b^2 - 4ac < 0$ has two distinct complex roots that are conjugate to each other.

12.2   Operations

Let S be a non-empty set. An application of S $\times$ S in S is called an *internal binary operation*, or simply an *operation* in S. It is denoted by a symbol as $+$ or $\cdot$ and emphasizing the domain and the range, as $+ : S \times S \to S, \cdot : S \times S \to S$.

If the $+$ symbol is used, the operation is called *addition* and the element of S corresponding of the couple (x, y) of S $\times$ S, i.e., the value in (x, y) of the application $+$, is called the *sum* of x and y and is denoted by $x + y$. If the symbol $\cdot$ is used, the operation is called *multiplication*, its value in the couple (x, y) of S $\times$ S is called the *product* of x and y and is denoted by $x \cdot y$ or simply xy.

Examples of operations in a set are:

1. the usual operations of addition and multiplication in the set **R** of the real numbers or in the set **C** of complex numbers;
2. the addition of two free vectors of the set V_α of the free vectors of the plane α or the set V_3 of the free vectors of space;
3. the addition of two n-tuples of $\mathbf{R}^n$;
4. the following applications

$$(X, Y) \in p(A) \times p(A) \to X \cup Y$$
$$(X, Y) \in p(A) \times p(A) \to X \cap Y$$

where $p(A)$ denotes the power set of the set S, i.e., the set of all subsets of S (see Sect. 3.3.2), that associate the union $X \cup Y$ and the intersection $X \cap Y$ to any couple (X, Y) of subsets of the set A, respectively.

If S and T are non-empty sets, an application of $T \times S$ in S is called an *external binary operation* of S with the elements of T as *operators*, is denoted with the symbol $\cdot$ and its value on the pair (t, x) of $T \times S$ with $t \cdot x$ or tx. The operation $\cdot$ is also called *multiplication* of an element of T by an element of S and the corresponding element tx of S is called the *product* of t by x. The multiplication of a real number by a free vector of V_α (resp. V_3) and the multiplication of a real number by an n-tuple of $\mathbf{R}^n$ are examples of external binary operations of V_α (resp. V_3) and $\mathbf{R}^n$ with operators the elements of $\mathbf{R}$. One or more sets in which, or between which, some operations, internal or external are defined, form an *algebraic structure*.

12.3 Fields-

Let S be a non-empty set and let $+$ and $\cdot$ be two operations in S; the triple $(S, +, \cdot)$ is called a *field* if the following axioms are satisfied, for any $a, b, c \in S$ (the notations $a \cdot b$ and ab are equivalent):

1. $a + b = b + a$ (commutative property of addition);
2. $(a + b) + c = a + (b + c)$ (associative property of addition);
3. there exists in S one and only one element 0 (called *zero*) such that $a + 0 = a$;
4. there exists, for every element a of S, one and only one element, which is denoted by $-a$, called the *opposite* of a, such that $a + (-a) = 0$;
5. $ab = ba$ (commutative property of multiplication);
6. $(ab)c = a(bc)$ (associative property of multiplication);
7. there exists in S one and only one element different from 0, which is denoted by 1, such that $a \cdot 1 = a$;
8. there exists in S, for each element a distinct from 0, one and only one element, which is denoted with a^{-1} or $\frac{1}{a}$, called the *inverse* of a, such that $aa^{-1} = 1$;
9. $a(b + c) = ab + ac$ (distributive property of multiplication with respect to addition).

The set S is called the *support* of the field $(S, +, \cdot)$ and the elements of S are called *scalars*. If it does not give rise to misunderstandings, the field $(S, +, \cdot)$ can be denoted by the symbol S of its support.

Examples

a. $(\mathbf{Q}, +, \cdot)$, where $\mathbf{Q}$ is the set of rational numbers and $+$ and $\cdot$ are the usual operations of addition and multiplication, respectively, is a field;
b. $(\{0,1\}, +, \cdot)$ is a field if $+$ and $\cdot$ are the internal binary operations of the set of two elements $\{0,1\}$ defined by the positions

$$0 + 1 = 1 + 0 = 1; \quad 0 + 0 = 0; \quad 1 + 1 = 0;$$

$$1 \cdot 0 = 0 \cdot 1 = 0; \quad 0 \cdot 0 = 0; \quad 1 \cdot 1 = 1;$$

c. The set of relative integers with the usual operations of addition and multiplication is not a field because axiom 8 is not satisfied;
d. $(\mathbf{R}, +, \cdot)$ is a field (Sect. 5.5);
e. $(\mathbf{C}, +, \cdot)$ is a field.

12.4   Vector Spaces

Let S be a field, V a non-empty set whose elements are called *vectors*, $+$ an internal binary operation in V, called *addition*, $\cdot$ an external binary operation of V whose operators are the scalars of S, which is called *multiplication of a scalar* of S *by a vector* of V. The notations $a \cdot x$ and ax are equivalent for every scalar a and vector x. The triple $(V, +, \cdot)$ defines an algebraic structure called *vector space* over the field S and the elements of V are called *vectors* of the vector space if the following axioms, called *vector space axioms*, are satisfied, for every $x, y, z \in V$ and whatever the scalars a and b are:

1. $x + y = y + x$ (commutative property of the addition in V);
2. $(x + y) + z = x + (y + z)$ (associative property of the addition in V);
3. there exists in V a unique vector O, the null vector, such that $x + O = x$;
4. for any vector x, there exists one and only one vector, which is denoted by $-x$, called the *opposite* of x, such that $x + (-x) = O$;
5. $a(bx) = (ab)x$ (associative property of the multiplication of a scalar by a vector);
6. $1x = x$;
7. the distributive properties of multiplication hold with respect to the addition of elements of V and with respect to the addition of scalars:

$$7_1. \, a(x + y) = ax + ay;$$
$$7_2. \, (a + b)x = ax + bx.$$

To ease the notation, we can denote the vector space $(V, +, \cdot)$ with the symbol V, unless this causes misunderstanding.

The sum $x + (-1)y$ is usually written $x - y$; furthermore we will write $-y$ instead of $(-1)y$.

Let us use 0 to identify the scalar zero and reserve the symbol O to the null vector defined in the axiom 3, called also *zero vector*.

It is possible to add up several elements of a vector space. If we want to add four elements w, x, y and z, we add the first two, then the third and finally the fourth. The axioms 1 (commutativity) and 2 (associativity) allow to perform the successive additions in any order. For example,

$$((w + x) + y) + z = w + (x + y) + z$$
$$= (w + y) + (x + z) = ((y + x) + w) + z = \ldots$$

Usually, parentheses are omitted and it suffices to write $w + x + y + z$.

For every scalar h and vector a, it is easy to check the following identities:

$$h(-a) = -ha, \quad (-h)a = -ha, \quad (-h)(-a) = ha.$$

Theorem 12.1 *The equality* $0x = O$ *holds, for every vector x of* V, *where* 0 *denotes the zero of* S *and* O *the null vector of* V.

Proof For every scalar a, by field axiom 3, we have $(a + 0)x = ax$ and by axiom 7 of vector space, $(a + 0)x = ax + 0x$. Then, equating the right-hand sides of both equalities:

$$ax = ax + 0x$$

Hence,

$$0x = O$$

$\square$

Definition 12.1 A vector space on the field **R** is called *real vector space*. A vector space on the field **C** is called a *complex vector space*.

Examples

a. The set V_3 (resp. V_α, V_r) (Sect. 9.3) of the geometric vectors of the space (resp. the plane α, the line) with the operations of addition of two vectors and multiplication of a real number by a vector is a real vector space. Let us call V_3, V_α and V_r *geometric vector spaces*.

b. The set $\mathbf{R}^n$ of the n-tuples of real numbers with the operation of addition of two n-tuples and the operation of multiplication of a real number by a n-tuples is a real vector space named *n-coordinate real space*, or *numerical vector space*. In particular, the set **R** of real numbers, with the addition and multiplication operations, can be considered either a field, or a vector space: in the latter case the multiplication is seen as an external operation with operators the scalars of the real field.

c. For every positive integer n, the set P_n of polynomials of degree less than n in the variable t with complex coefficients is a complex vector space with respect to the operations of addition of two polynomials and multiplication of a complex number by a polynomial. The null vector of vector space P_n is the polynomial identically zero.

12.5   Linear Dependence and Linear Independence

We already studied the systems of n-tuples in $\mathbf{R}^n$ (see Chap. 9). The concept of system of n-tuples is extended to that of system of vectors in a vector space. Hence, a system of vectors $[x_1, x_2, \ldots, x_n]$ in a vector space V is made of elements of V not necessarily distinct. The existence of the system without elements, called the *empty system*, is postulated.

Let us consider the non-empty system $[x_1, x_2, \ldots, x_n]$ of vectors of a vector space V on the field S.

Definition 12.2 The system of vectors $[x_1, x_2, \ldots, x_n]$ is said to be a *linearly dependent system* if there exists a system of scalars $[a_1, a_2, \ldots, a_n]$ not all null such that

$$\sum a_i x_i = a_1 x_1 + a_2 x_2 + \ldots + a_n x_n = O = \text{null vector}$$

The following propositions hold.

Proposition 12.1 *A vector system containing the null vector is linearly dependent.*

In fact, the proposition is trivially true if the system contains the null vector as the only element. If the system contains more elements and one of these is the null vector, this depends linearly on the remaining vectors and therefore the system is linearly dependent.

Proposition 12.2 *A vector system containing two equal elements is linearly dependent.*

Proposition 12.3 *A system of vectors, that includes a linearly dependent part, is linearly dependent.*

Proposition 12.4 *Each vector of a non-empty system of vectors linearly depends on the system itself. Indeed, the linear combination of the vectors of the system $\left[x_1, x_2, \ldots, x_{i-1}, x_i, x_{i+1}, \ldots, x_n\right]$ with the coefficients $h_1 = 0, h_2 = 0, \ldots, h_{i-1} = 0, h_i = 1, h_{i+1} = 0, \ldots, h_n = 0$ equals vector x_i.*

A system of vectors that is not linearly dependent is said to be a linearly independent system. The empty system is assumed to be linearly independent.

The expressions "linearly dependent (independent) system of vectors $[x_1, x_2, \ldots, x_n]$" and "linearly dependent (independent) vectors $x_1, x_2, \ldots, x_n$" are equivalent.

From Propositions 12.1, 12.2 and 12.3 we obtain:

Proposition 12.5 *A linearly independent vector system, which is not the empty system, includes distinct and non-null vectors and all its parts are linearly independent systems.*

Examples

a. The unit vectors of the axes in a coordinate system of the plane (space) form a linearly independent system of free vectors of the plane (space) (Sect. 9.8). Any three free vectors in the plane form a linearly dependent system.

b. Let us consider the vector space P_3 of polynomials of degree less than 3 with complex coefficients in the variable t and let x_1, x_2, x_3 be elements of P_3 defined by

$$x_1 = 2 - t^2$$
$$x_2 = 1 + t + t^2$$
$$x_3 = t - 1$$

The system of vectors $[x_1, x_2, x_3]$ is linearly dependent; in fact $x_1 - x_2 + x_3 = 2 - t^2 - (1 + t - t^2) + t - 1$ is equal to the null polynomial, which is the null vector $0t^2 + 0t + 0$ of the vector space P_3. In other words, there exist not all null scalars $a_1 = 1, a_2 = -1, a_3 = 1$, such that $a_1 x_1 + a_2 x_2 + a_3 x_3 = 0t^2 + 0t + 0 = O$.

c. The polynomials

$$x_1 = 1$$
$$x_2 = t$$
$$x_3 = t^2$$

form a linearly independent system of P_3.

Definition 12.3 If $x = a_1 x_1 + a_2 x_2 + \ldots + a_n x_n$, where the x_i's are vectors and the a_i's scalars, $i = 1, 2, \ldots, n$, then the vector x is called the *linear combination of the system* of vectors $[x_1, x_2, \ldots, x_n]$ with coefficients $a_1, a_2, \ldots, a_n$. It is also said that x is *linearly dependent on the system* $[x_1, x_2, \ldots, x_n]$, or the vector x *is a linear combination of the vectors* $x_1, x_2, \ldots, x_n$, if $x = a_1 x_1 + a_2 x_2 + \ldots + a_n x_n$.

It is straightforward to realize the following:

Proposition 12.6 *If the vector x linearly depends on the vectors of the system $[x_1, x_2, \ldots, x_n]$ and if each vector x_i linearly depends on the vectors of the system $[y_1, y_2, \ldots, y_m]$, then the vector x linearly depends on the vectors of the system $[y_1, y_2, \ldots, y_m]$.*

Theorem 12.2 *The system* $T = [x_1, x_2, \ldots, x_n], n \geq 2$, *of the vector space V is linearly dependent if and only if one of the vectors of T linearly depends on the others vectors of T.*

Proof Given the linearly dependent system T, there exist n scalars not all null, $h_1, h_2, \ldots, h_n$, such that $O = h_1 x_1 + h_2 x_2 + \ldots + h_n x_n$. Suppose, for instance, $h_1 \neq 0$; it turns out

$$O = x_1 + \frac{h_2}{h_1}x_2 + \ldots + \frac{h_n}{h_1}x_n$$

and then

$$x_1 = -\frac{h_2}{h_1}x_2 - \ldots - \frac{h_n}{h_1}x_n$$

Therefore, x_1 depends linearly on the vectors $x_2, \ldots, x_n$. Let us now suppose that one of the vectors of the system T, for example the vector x_1, depends linearly on the others. Then there exist $n - 1$ scalars $p_2, p_3, \ldots, p_n$ such that $x_1 = p_2 x_2 + p_3 x_3 + \ldots + p_n x_n$. Therefore,

$$O = -x_1 + p_2 x_2 + p_3 x_3 + \ldots + p_n x_n$$

proving that the system T is linearly dependent.   □

Theorem 12.3 *The non-empty system of non-zero vectors* $[x_1, x_2, \ldots, x_n]$ *is linearly dependent if and only if there exists a vector* x_k, *with* $2 \leq k \leq n$, *which is a linear combination of the system* $[x_1, \ldots, x_{k-1}]$.

Proof Suppose that $[x_1, x_2, \ldots, x_n]$ is a linearly dependent system of non-zero vectors. Evidently it is $n \geq 2$. Let k be the first integer between 2 and n such that $[x_1, \ldots, x_k]$ is a linearly dependent system. (Observe that x_1 is linearly independent because it is different from the null vector and, by Theorem 12.2, there is an integer k between 2 and n such that $[x_1, \ldots, x_k]$ is linearly dependent.) Then

$$a_1 x_1 + \ldots + a_k x_k = O$$

with any of a_i's non-zero. It cannot be $a_k = 0$ because this would imply linear dependence of the system $[x_1, \ldots, x_{k-1}]$, while, by definition of k, this system is linearly independent. Hence, from the previous equality it follows

$$x_k = -\frac{a_1}{a_k}x_1 - \ldots - \frac{a_{k-1}}{a_k}x_{k-1}$$

Vice versa, if in the system of non-null vectors $[x_1, x_2, \ldots, x_n]$ there is a vector x_k, $2 \leq k \leq n$, such that

$$x_k = b_1 x_1 + \ldots + b_{k-1} x_{k-1}$$

by Theorem 12.2, the system $[x_1, \ldots, x_k]$ is linearly dependent and also the system $[x_1, \ldots, x_n]$ that contains it (see Proposition 12.3).   □

Theorem 12.4 *A non-empty system* T *of the vector space* V *is linearly independent if and only if each vector of* V *linearly dependent on* T *is expressed in a unique way as a linear combination of the vectors of* T.

Proof Let $T = [x_1, x_2, \ldots, x_n]$, with $n \geq 1$, be a linearly independent system of vectors of V and let x be a vector linearly dependent on T. Suppose

$$x = h_1x_1 + h_2x_2 + \ldots + h_nx_n$$

and also

$$x = k_1x_1 + k_2x_2 + \ldots + k_nx_n$$

By subtraction we obtain,

$$x - x = O = (h_1 - k_1)x_1 + (h_2 - k_2)x_2 + \ldots + (h_n - k_n)x_n$$

Since T is linearly independent, necessarily the equalities hold:

$$h_1 = k_1, h_2 = k_2, \ldots, h_n = k_n$$

Therefore, the linear combination of the vectors of T equal to the vector x is unique.

Vice versa, suppose that each vector linearly dependent on T is expressed in a unique way as a linear combination of the vectors of T: this is also true for the null vector and therefore T is linearly independent. □

12.6 Finitely Generated Vector Spaces. Bases

Definition 12.4 A vector space V is said to be a *finitely generated* (or a *finite-dimensional*) *vector space* if there exists in V a finite system $X = [x_1, x_2, \ldots, x_m]$ of linearly independent vectors such that any vector of V is a linear combination of the elements of X. Such a system X is called a *basis* (or a *coordinate system*) of the vector space.

If y is a vector of the finitely generated vector space V and $X = [x_1, x_2, \ldots, x_m]$ is a basis of V, then there exist scalars $a_1, a_2, \ldots, a_m$ such that $y = a_1 x_1 + a_2 x_2 + \ldots + a_m x_m$. By Theorem 12.4 the scalars $a_1, a_2, \ldots, a_m$ are uniquely determined and the m-tuple $(a_1, a_2, \ldots, a_m)$ is called the m-tuple of the *components*, or *coordinates*, of the vector y in the basis X. The ordering of the vectors of the basis X induces the ordering of the components.

Examples

a. The linearly independent system $[1, t, t^2, t^3]$ is a basis for the space P_4 of polynomials in the variable t with complex coefficients of degree less than 4. In fact, each polynomial of P_4 has the form $a_1t^3 + a_2t^2 + a_3t + a_4$, with complex coefficients.

b.   The vector space P of polynomials in a variable t of however high degree with complex coefficients is not a finitely generated vector space. Indeed, there is no basis, that is a finite system of vectors, of the vector space P since the linearly independent systems of P contains polynomials of however high degree.

c.   The system of the unit vectors of a coordinate system of the plane (space) is a basis of the vector space of the free vectors of the plane (space).

Proposition 12.7 *For each $i = 1, \ldots, n$, let e_i denote the vector of $\mathbf{R}^n$ which has the i-th component equal to 1 and all the other components equal to zero. Then the system $[e_1, e_2, \ldots, e_n]$ is a basis of $\mathbf{R}^n$.*

Proof The linear combination of the vectors of the system $[e_1, e_2, \ldots, e_n]$ with coefficients $a_1, a_2, \ldots, a_n$ is equal to the vector a with components $a_1, a_2, \ldots, a_n$. Indeed,

$$a = a_1 e_1 + a_2 e_2 + \ldots + a_n e_n = a_1(1, 0, \ldots, 0) + a_2(0, 1, \ldots, 0) + \ldots$$
$$+ a_n(0, 0, \ldots, 1)$$

Hence, the two facts of interest: each vector a of $\mathbf{R}^n$ is a linear combination of the vectors of $[e_1, e_2, \ldots, e_n]$ and this system is linearly independent because each vector a of $\mathbf{R}^n$ is a unique linear combination of the vectors of the system, (the coefficients of the linear combination are the components of the vector).   □

The basis $[e_1, e_2, \ldots, e_n]$ is called the *canonical basis* of $\mathbf{R}^n$.

12.7   Vector Subspaces

Let $(V, +, \cdot)$ be a vector space over the field S and let $W \subseteq V$ a non-empty subset of V. Let us consider the following properties:

i.   the sum of any two vectors of W is a vector of W;
ii.   the product of any scalar of S by any vector of W is a vector of W.

Proposition 12.8 *Under the hypotheses (i) and (ii), the triple $(W, +, \cdot)$ is a vector space over the field S.*

Proof We must verify that the vector space axioms (Sect. 12.4) are satisfied by the triple $(W, +, \cdot)$. Indeed, Axioms 1, 2, 5, 6 and 7 are obviously verified. Axiom 3 follows from Proposition 12.1; in fact, for every vector y of W we have: $0y = O$ (where 0 indicates the zero of S and O the null vector of V). By hypothesis (ii), O belongs to W and acts like the null vector of W.

In order to verify Axiom 4 we must prove that if y is any vector of W, then the vector $-y$ opposite of y in V, is identified with the vector opposite of y in W. Indeed, whatever y is in W, from Axiom 6 we have $1y = y$ and then, $(-1)y = -y$ where $-y$ denotes the opposite of the vector y in V. From hypothesis (ii) the vector $-y$ belongs to W.   □

Definition 12.5 The vector space $(W, +, \cdot)$ is said to be a *subspace* of $(V, +, \cdot)$ over the field S.

12.7.1 Spanned Subspaces

Let V be a vector space over the field S. Let us consider a non-empty system of vectors $T = [x_1, ..., x_m]$ in V. We denote $\langle T \rangle = \langle x_{1_2}........., x_m \rangle$ the set of all linear combinations of the vectors of T with scalars of S. Therefore the set $\langle T \rangle$ is equal to the set of the vectors $\sum a_i x_i$, for every $a_j \in S$ and $x_i \in T$. By Proposition 12.8 the set $\langle T \rangle$, with the operations of addition $+$ and multiplication $\cdot$, i.e., the triple $(\langle T \rangle, + , \cdot)$, is a vector space over the field S, called the subspace *generated* or *spanned* by T, or the *linear span* of T, simply denoted $\langle T \rangle$.

In other words, $\langle T \rangle$ is the smallest subspace of V containing T.

If T is a basis of V, then $\langle T \rangle = V$, i.e., the vector space V is generated by T. It is postulated $<\emptyset> = 0$.

Let us observe that if T is a finite system of linearly dependent vectors of the vector space $\langle T \rangle = V$, then any element of V is not necessarily a linear combination of the elements of T. For instance, the set of the linear combinations of the free vectors in the plane xy is generated by the unit vectors of the axes x and y; the linear combination of three free vectors in the space parallel to the plane xy does not generate free vectors perpendicular to the plane xy.

A system T of vectors of V such that the subspace generated by T coincides with V is called a *system of generators* of V.

12.8 Dimension

Theorem 12.5 *Any two bases of a finite-dimensional vector space V have the same number of vectors.*

Proof Let us consider the following propositions related with a finite system X of elements in the vector space V:

P(X) ={every vector of V is a linear combination of the vectors of X}

Q(X) ={X is linearly independent}

Let $X = [x_1, x_2, ..., x_n]$ and $Y = [y_1, y_2, ... y_m]$ be two finite vector systems of the vector space V and assume that properties P(X) and Q(X) hold. Let us consider the system

$$T = [y_m, x_1, x_2, ... x_n]$$

of vectors of V. Since every vector of V is a linear combination of the system $[x_1, x_2, \ldots, x_n]$ it is also a combination of the vectors of the system T according to Proposition 12.6, because y_m is linearly dependent on X. Then we can consider the first of the integers i between 1 and n such that the system $[y_m, x_1, x_2, \ldots, x_i]$ is linearly dependent. By Theorem 12.3, the vector x_i linearly depends on the system $[y_m, x_1, x_2, \ldots, x_{i-1}]$ and, therefore, again every vector of V linearly depends on the system T$'$ obtained from T by eliminating x_i:

$$T' = \left[y_m, x_1, x_{2,\ldots\ldots} x_{i-1}, x_{i+1}, \ldots, x_n \right]$$

Let us set y_{m-1} in front of T$'$ apply the same reasoning to the system $[y_{m-1}, y_m, x_1, x_2, \ldots, x_{i-1}, x_{i+1}, \ldots, x_n]$ eliminating an x of the system X and adding on a new y of the system Y. After the reasoning has been made over again m times, we get a system of vectors with the same property P that the x's had and this system is different from X since m of the x's are replaced by y's. This implies $n \geq m$. Hence, if X and Y are bases, i. e., X and Y satisfy both properties P and Q, then we have $n \geq m$ and $m \geq n$. Therefore, $m = n$.   □

In the course of the proof we have shown the following:

Corollary 12.1 Given the systems X and Y, supposing that the properties P(X) and Q(Y) hold, it turns out $n \geq m$, i.e., the number of elements of X is greater than or equal to the number of elements of Y. In other words, if X is a system of generators of V, then the number n of the vectors of X is greater than or equal to the number m of vectors of any basis of V.

Definition 12.6 The number of the elements in any basis of a finite-dimensional vector space V is called the *dimension* of V

For example, the real vector space $\mathbf{R}^n$ and the complex vector space $\mathbf{C}^n$ have dimension n.

Definition 12.7 Any linearly independent system Y of vectors of the vector space V, not properly included in a linearly independent system of vectors of V, is called a *maximal system of linearly independent vectors* of V.

The following statements hold:

Proposition 12.9 *Any maximal system of linearly independent vectors of a finite-dimensional vector space V is a basis of V and every basis of V is a maximal system of linearly independent vectors of V.*

Proposition 12.10 *Two maximal systems of linearly independent vectors of the finite-dimensional vector space V have the same number of vectors.*

Proof If S and S$'$ are maximal systems of linearly independent vectors of V and n and n' are the numbers of elements of S and S$'$, respectively, by definition of maximal system of linearly independent vector, we have: $n \geq n'$ and $n' \geq n$. Hence $n = n'$.   □

Proposition 12.11 *The set of the bases of V and the set of the systems with n elements of linearly independent vectors in* V *with n elements coincide.*

Proposition 12.12 *Any system of* $n + 1$ *vectors of a finite-dimensional vector space* V *of dimension n is linearly dependent.*

12.9 Isomorphism

Definition 12.8 Two vector spaces V and W over the same field are said to be *isomorphic* if there is a one-to-one function $f : V \rightarrow W$ such that

$$f(ax + by) = af(x) + bf(y) \tag{12.2}$$

whatever the scalars a, b and the vectors x, y of V are. A function f, satisfying equality (12.2), is called an *isomorphism* between the vector spaces V and W. In other words, V and W are isomorphic if there is a one-to-one function f (the isomorphism) that preserves the vector space operations. Two isomorphic vector spaces differ, in essence, by the names of their elements, and therefore can be identified.

Proposition 12.13 *If two finite-dimsional vector spaces over the field* S *are isomorphic, then they have the same dimension.*

In fact, to each basis in one space corresponds, through the isomorphism, a basis in the other space.

Theorem 12.6 *Each vector space* V *of dimension n over the field* S *is isomorphic to* S^n.

Proof Let $[x_1, \ldots, x_n]$ be a basis of V. For every vector x of V we have $x = a_1 x_1 + \ldots + a_n x_n$, being the n-tuple of scalars $(a_1, \ldots, a_n)$ univocally determined by x (Sect. 12.6). Then the function $f : x \rightarrow (a_1, \ldots, a_n)$ from V to S^n is one-to-one. If y is a vector of V, then $y = b_1 x_1 + \ldots + b_n x_n$ and, for every pair of scalars a and b,

$$ax + by = (aa_1 + bb_1)x_1 + \ldots + (aa_n + bb_n)x_n$$

Then

$$f(ax + by) = (aa_1 + bb_1, \ldots, aa_n + bb_n) = af(x) + bf(y).$$

Therefore, V and S^n are isomorphic.   □
The following theorem holds.

Theorem 12.7 *Two finite-dimensional vector spaces* U *and* V, *isomorphic to the same vector space* W, *are isomorphic to each other.*

Consider now two vector spaces U and V of dimension n over the field S. By Theorem 12.6, U and V are isomorphic to S^n and, by Theorem 12.7, to each other. Thus, the following result which inverts Proposition 12.8, holds.

Proposition 12.14 *If two vector spaces over the field* S *have the same dimension, then they are isomorphic.*

12.10 Identification of Geometric and Numerical Vector Spaces

The sets V_r, V_α and V_3 of the vectors of the line r, the plane α and the space are vector spaces over the field **R**, called the *vector space of the geometric vectors of the line, the plane* and *the Euclidean space*, respectively. We have shown (Sect. 9.8) that, fixed a coordinate system of the plane (space), the couple (triple) of the components (a_x, a_y) $((a_x, a_y, a_z))$ is associated with the vector $\boldsymbol{a}$. The correspondence that leads to identify the free vector $\boldsymbol{a}$ with (a_x, a_y) (resp. (a_x, a_y, a_z)) is a one-to-one function f of the set of the free geometric vectors onto the set $\mathbf{R}^2$ (resp. $\mathbf{R}^3$) of couples (triples) of real numbers:

$$ f : \boldsymbol{a} \to (a_x, a_y) \quad (f : \boldsymbol{a} \to (a_x, a_y, a_z)) $$

The function f is an isomorphism between the vector spaces V_α and $\mathbf{R}^2$ (V_3 and $\mathbf{R}^3$). It seems natural to identify the vector space V_α of the free vectors of the plane with $\mathbf{R}^2$, and set $\boldsymbol{a} = (a_x, a_y)$; and so too identify V_3 with $\mathbf{R}^3$ and set $\boldsymbol{a} = (a_x, a_y, a_z)$. From the isomorphism of V_α with $\mathbf{R}^2$ and V_3 with $\mathbf{R}^3$ arises the possibility, and the utility, of being able to refer only to the components of the geometric vectors, simplifying notations and avoiding figure drawings.

Furthermore, the concepts of parallelism and proportionality (Remark 9.2) are extended to V_α, V_3 and $\mathbf{R}^n$. More precisely, two non-zero vectors of V_α or V_3 or $\mathbf{R}^n$ are parallel if and only if their respective components are proportional.

12.11 Scalar Product in $\mathbf{R}^n$

The operation of *scalar multiplication* of vectors of the plane or space (Sect. 10.2) extends to the vectors of $\mathbf{R}^n$.

Given the vectors $\boldsymbol{a} = (a_1, a_2, \ldots, a_n)$ and $\boldsymbol{b} = (b_1, b_2, \ldots, b_n)$ of $\mathbf{R}^n$, *the scalar multiplication* of vectors $\boldsymbol{a}$ and $\boldsymbol{b}$, denoted $\boldsymbol{a} \times \boldsymbol{b}$, or $\boldsymbol{a} \cdot \boldsymbol{b}$, is the operation defined as follows

$$ \boldsymbol{a} \times \boldsymbol{b} = \boldsymbol{a} \cdot \boldsymbol{b} = a_1 b_1 + a_2 b_2 + \ldots + a_n b_n $$

The number $a_1b_1 + a_2b_2 + \ldots + a_nb_n$ is called the *scalar product* of the vectors a and b. If $a \times b = 0$, then the vectors a and b are named *orthogonal*. For example, the vectors $(1, -1, 1)$ and $(1, 1, 0)$ are orthogonal because $(1, -1, 1) \times (1, 1, 0) = 1 - 1 + 0 = 0$.

12.12 Exercises

Exercise 12.1 Consider the following vectors of $\mathbf{R}^3$: $a = (2, -1, 8)$, $b_1 = (-1, 0, 2)$, $b_2 = (-7, 1, 0)$, $b_3 = (0, 0, 1)$. Check that vector a is a linear combination of vectors b_1, b_2, b_3.

Vector a is a linear combination of b_1, b_2, b_3 if and only if there exist real numbers h_1, h_2, h_3 such that

$$a = h_1b_1 + h_2b_2 + h_3b_3 \tag{12.3}$$

i.e., $(2, -1, 8) = h_1(-1, 0, 2) + h_2(-7, 1, 0) + h_3(0, 0, 1)$. Then Eq. (12.3) is satisfied if and only if

$$h_1 - 7h_2 = 2$$
$$h_2 = -1$$
$$2h_1 + h_3 = 8$$

Accordingly, the system of three equations and three unknowns has the unique solution $(h_1, h_2, h_3) = (5, -1, 2)$.

Exercise 12.2 Verify that the vector $a = (1, -2, 3)$ is a linear combination of the vectors $b_1 = (-1, 0, 2)$, $b_2 = (1, 1, 0)$, $b_3 = (0, 1, 0)$.

The problem is similar to the previous one. There is a single triple $(h_1, h_2, h_3) = \left(\frac{3}{2}, \frac{5}{2}, -\frac{9}{2}\right)$ for the coefficients of the linear combination.

Exercise 12.3 Show that $T = [(0, 0, -1), (0, 1, 1), (1, 1, 1)]$.

i.   is a system of generators of $\mathbf{R}^3$,
ii. is a base of $\mathbf{R}^3$.

i.   We have to prove that any vector $(a, b, c) \in \mathbf{R}^3$ is a linear combination of the vectors $(0, 0, -1)$, $(0, 1, 1)$ and $(1, 1, 1)$, i.e., there are scalars x, y, z such that:

$$(a, b, c) = x(0, 0, -1) + y(0, 1, 1) + z(1, 1, 1) = (z, y + z, -x + y + z)$$

Let us equate the components and obtain the system of linear equations

$$z = a$$
$$y + z = b$$
$$-x + y + z = c$$

equivalent to the system

$$-x + y + z = c$$
$$y + z = b$$
$$z = a$$

that is reduced and consistent. Operating by substitution (Sect. 11.3) we obtain the unique solution: $z = a$, $y = b - a$, $x = c - b$. Then the system $T = [(0, 0, -1), (0, 1, 1), (1, 1, 1)]$ generates $\mathbf{R}^3$.

ii.   The system T is a basis since it is a maximal systems of linearly independent vectors of $\mathbf{R}^3$ (see Definition 12.4).

Exercise 12.4 Let S be a field and S^n the vector space of the n-tuples of elements of S. Verify that the vector system $[e_1, e_2, \ldots, e_n]$, with $e_1 = (1, 0, 0, \ldots, 0)$, $e_2 = (0, 1, 0, \ldots, 0)$, $\ldots$, $e_n = (0, 0, 0, \ldots, 1)$, formed by the n distinct n-tuples of elements of S having exactly one component equal to the unit and all others zero, is a basis of S^n.

Exercise 12.5 In the vector space P_4 of the polynomials with real coefficients of degree less than 4 in the variable t let the system of polynomials $T = [x_1, x_2, x_3, x_4]$ be given, where

$$x_1 = 1 + t$$
$$x_2 = t^2$$
$$x_3 = t^2 - t^3$$
$$x_4 = -1 + t$$

Now we verify that the vectors of T form a basis of P_4. Let us first prove that every vector in P_4 is a linear combination of the system $T = [x_1, x_2, x_3, x_4]$. To this aim we will show that, for every polynomial x of P_4,

$$x = a_0 + a_1 t + a_2 t^2 + a_3 t^3 \tag{12.4}$$

there are real coefficients a, b, c, d (for now unknown) such that

$$x = a(1 + t) + bt^2 + c(t^2 - t^3) + d(-1 + t). \tag{12.5}$$

By equating the expressions (12.4) and (12.5) of x we have

$$a_0 + a_1 t + a_2 t^2 + a_3 t^3 = a + at + bt^2 + ct^2 - ct^3 - d + dt$$

then

$$a_0 + a_1 t + a_2 t^2 + a_3 t^3 = a - d + (a + d)t + (b + c)t^2 - ct^3$$

The equality is satisfied if and only if the coefficients of the addends with the same degrees are equal:

$$a_0 = a - d$$
$$a_1 = a + d$$
$$a_2 = b + c$$
$$a_3 = -c$$

Then we find the values of the coefficients a, b, c, d in terms of the system of polynomials T:

$$a = \frac{a_0 - a_1}{2}$$
$$b = a_2 + a_2$$
$$c = -a_3$$
$$d = \frac{3a_0 - a_1}{2}$$

This proves that every vector in P_4 is a linear combination of the system T.

To check that T is a basis of P_4, there is still to prove that T is linearly independent. If the linear combination, with coefficients m, n, p, q, is the null polynomial $O = 0 + 0t + 0t^2 + 0t^3$, i.e.,

$$m(1 + t) + nt^2 + p(t^2 - t^3) + q(-1 + t) = 0 + 0t + 0t^2 + 0t^3$$

then

$$m - q = 0, \, m + q = 0, \, n + p = 0, \, -p = 0$$

that implies $m = n = p = q = 0$. Thus, T is a basis.

Exercise 12.6 Let $T = \lfloor x_1, x_2, x_3 \rfloor$ be a system of vectors in $\mathbf{R}^3$, with $x_1 = (1, 0, 1)$, $x_2 = (-1, 1, 0)$, $x_3 = (0, 0, 1)$. Then:

i.   verify that the system T is a basis and
ii. find the components of the vector $e_1 = (1, 0, 0)$ in the basis T.

i.   By Proposition 12.10, each basis of $\mathbf{R}^3$ is a linearly independent system consisting of three elements. Thus, in order to verify that the system T is a basis of $\mathbf{R}^3$ it is sufficient to show that T is a linearly independent system. To this aim, if

$$ax_1 + bx_2 + cx_3 = (a, 0, a) + (-b, b, 0) + (0, 0, c) = (0, 0, 0)$$

with a, b, c, real coefficients, then the equalities hold: $a - b = 0, b = 0, a + c = 0$, which imply $(a, b, c) = (0, 0, 0)$. So T is linearly independent.
ii. The vector e_1 is linearly dependent on T.

Then we set

$$e_1 = (1, 0, 0) = mx_1 + nx_2 + px_3 = m(1, 0, 1) + n(-1, 1, 0) + p(0, 0, 1).$$

The scalars m, n, p are determined by equating the components:

$$(1, 0, 0) = (m - n, \, n, \, m + p)$$

Accordingly, $m = 1, n = 0, p = -1$ are the components of e_1 in the basis T.

Exercise 12.7 Prove that the vectors $(1, -2, 1)$, $(2, 1, -1)$, $(7, -4, 1)$ of $\mathbf{R}^3$ are linearly dependent.

Let us consider a linear combination of triples, by means of unknown scalar x, y, z, equal to the null vector:

$$x(1, -2, 1) + y(2, 1, -1) + z(7, -4, 1) = (0, 0, 0)$$

Then

$$(x, -2x, x) + (2y, y, -y) + (7z, -4z, z) = (0, 0, 0),$$

and adding the components, we get

$$(x + 2y + 7z, \, -2x + y - 4z, \, x - y + z) = (0, 0, 0)$$

Let us equate the components:

$$\begin{aligned} x + 2y + 7z &= 0 \\ -2x + y - 4z &= 0 \\ x - y + z &= 0 \end{aligned}$$

and moving on to reduction (Sect. 11.3) through the operations $E_2 \leftarrow 2E_1 + E_2$; $E_3 \leftarrow E_1 - E_3$ we obtain the equivalent system

$$\begin{aligned} x + 2y + 7z &= 0 \\ 5y + 10z &= 0 \\ 3y + 6z &= 0 \end{aligned}$$

Let us cancel the third equation multiple of the second

$$\begin{aligned} x + 2y + 7z &= 0 \\ 5y + 10z &= 0 \end{aligned}$$

The system is consistent and has non-null solutions. Indeed, the system takes the form

$$x + 2y = -7c$$
$$y = -2c$$

with $z = c$ real parameter. For every non-null value given to c a non-null solution is determined. For example, if $c = 1$, then $x = -9$, $y = -2$ and the solution $(-9, -2, 1)$ is obtained. The general solution is $(x, y, z) = (-3c, -2c, c)$, $c \in \mathbf{R}^3$. Therefore, the three given vectors are linearly dependent.

Bibliography

Aitken, A.C.: Determinants and Matrices. Oliver and Boyd LTD, Edinburgh (1965)
Halmos, P.R.: Finite-Dimensional Vector Spaces. Springer, Heidelberg, New York (1974)
Lang, S.: Linear Algebra. Addison-Wesley Publishing Company, Reading, Mass (1966)
Lipschutz, S.: Theory and Problems of Linear Algebra. McGraw-Hill, New York (1968)
Pontryagin, L.S.: Foundations of Combinatorial Topology. Graylock Press, Rochester N.Y (1952)
Schwartz, J.T.: Introduction to Matrices and Vectors. Dover Publications, New York (1961)
Strang, G.: Linear Algebra and Its Applications. Academic Press, New York (1976)
Strang, G.: Introduction to Linear Algebra. Wellesley-Cambridge Press (2016)

Chapter 13
Matrices

13.1 First Concepts

A rectangular array of the form

$$\begin{bmatrix} a_{11} & a_{12} & \dots & a_{1n} \\ a_{21} & a_{22} & \dots & a_{2n} \\ \dots & \dots & \dots & \dots \\ a_{m1} & a_{m2} & \dots & a_{mn} \end{bmatrix} \tag{13.1}$$

where the a_{ij}, $i = 1, \dots, m$; $j = 1, \dots, n$, are real numbers, is called a *matrix* over $\mathbf{R}$, or simply, a *matrix*. The matrix (13.1) is also denoted (a_{ij}). Any number a_{ij} is called an *entry*, or *component* or *element* or *scalar* in the *place* or *position ij*.

The number of elements of the matrix (13.1) is equal to the product $m \times n$ and the matrix is called an $m \times n$ (m by n) matrix. The couple (m, n) is called the *size* of the matrix.

The m vectors of $\mathbf{R}^n$

$$\begin{aligned} \boldsymbol{a}_1 &= (a_{11}, a_{12} \dots, a_{1n}) \\ \boldsymbol{a}_2 &= (a_{21}, a_{22} \dots, a_{2n}) \\ &\quad \dots \\ \boldsymbol{a}_m &= (a_{m1}, a_{m2} \dots, a_{mn}) \end{aligned} \tag{13.2}$$

are n-tuples called *row vectors* or *rows* of the matrix (13.1). The vertical m-tuples

$$\boldsymbol{a}^1 = \begin{bmatrix} a_{11} \\ a_{21} \\ \dots \\ a_{m1} \end{bmatrix} \quad \boldsymbol{a}^2 = \begin{bmatrix} a_{12} \\ a_{22} \\ \dots \\ a_{m2} \end{bmatrix} \quad \dots \boldsymbol{a}^n = \begin{bmatrix} a_{1n} \\ a_{2n} \\ \dots \\ a_{mn} \end{bmatrix} \tag{13.3}$$

© The Author(s), under exclusive license to Springer Nature Switzerland AG 2023
A. G. S. Ventre, *Calculus and Linear Algebra*,
https://doi.org/10.1007/978-3-031-20549-1_13

are called *column vectors* or *columns* of the matrix (13.1). Each vector (13.2) is a 1 × *n* matrix, each vector (13.3) is a *m* × 1 matrix. A matrix with one row (column) is named also a *row (column) vector*. The generic term *line* is used to denote a row or a column. An *m* × *n* matrix has *m* rows and *n* columns. A *null line*, i. e., a null row or a null column has all entries null.

The vectors (13.2) are in $\mathbf{R}^n$, the vectors (13.3) are in $\mathbf{R}^m$. A matrix with one raw and one column reduces to a number. A *null matrix* has, by definition, all zero entries.

Matrices are usually denoted by capital letters A, B, C, ... and the entries in the place *ij* by lowercase letters: a_{ij}, b_{ij}, c_{ij}, ...

Two matrices A and B are said to be *equal*, written A = B, if they have the same size and the same entry in the same *place ij*.

Example 13.1 The matrix

$$A = \begin{bmatrix} 1 & 0 & 1 & -2 \\ 3 & 1 & 0 & 4 \\ -1 & 1 & \sqrt{2} & 5 \end{bmatrix}$$

has size (3, 4), its rows are the following three 4-tuples:

$$a_1 = (1, 0, 1, -2), \quad a_2 = (3, 1, 0, 4), \quad a_3 = (-1, 1, \sqrt{2}, 5)$$

and the columns are the four triples:

$$a^1 = \begin{bmatrix} 1 \\ 3 \\ -1 \end{bmatrix} \quad a^2 = \begin{bmatrix} 0 \\ 1 \\ 1 \end{bmatrix} \quad a^3 = \begin{bmatrix} 1 \\ 0 \\ \sqrt{2} \end{bmatrix} \quad a^4 = \begin{bmatrix} -2 \\ 4 \\ 5 \end{bmatrix}$$

The entry in the place 33 (read: three three) is $\sqrt{2}$.

The operation of *addition* is defined in the set of matrices with the same size. Given two matrices A and B with the same size (*m, n*) the *sum* of A and B, denoted A + B, is the matrix of size (*m, n*) whose element in the place *ij* is equal to $a_{ij} + b_{ij}$; $i = 1, 2, ..., m$ and $j = 1, 2, ..., n$.

For example,

$$\begin{bmatrix} 1 & 0 & 1 & -2 \\ 0 & 1 & 0 & -4 \\ -1 & 0 & 9 & 5 \end{bmatrix} + \begin{bmatrix} 1 & 0 & 1 & -2 \\ 3 & 1 & 0 & 4 \\ -1 & 1 & -7 & -3 \end{bmatrix} = \begin{bmatrix} 2 & 0 & 2 & -4 \\ 3 & 2 & 0 & 0 \\ -2 & 1 & 2 & 2 \end{bmatrix}$$

Given a real number *h* and a matrix A of size (*m, n*), the *multiplication* of *h* by the matrix A is the operation that associates the matrix denoted by *h*A, whose entry in the place *ij* is equal to ha_{ij}, to the pair (*h*, A). For example,

$$-2\begin{bmatrix} 3 & -1 \\ -2 & 0 \end{bmatrix} = \begin{bmatrix} -6 & 2 \\ 4 & 0 \end{bmatrix}$$

If A, B and C are matrices of size (m, n) and O the null matrix of size (m, n), the following properties are satisfied for any real numbers h, k:

1. $A + B = B + A$
2. $(A + B) + C = A + (B + C)$
3. $A + O = A$
4. $A + (-1A) = A - A = O$
5. $(hk)A = h(kA)$
6. $1A = A$

    7_1. $h(A + B) = hA + hB$
    7_2. $(h + k)A = hA + kA.$

Accordingly, the set of $m \times n$ matrices over **R** is a *vector space* (see Sect. 12.4) over the field **R** with respect to the addition of $m \times n$ matrices and the multiplication of a real number by an $m \times n$ matrix.

13.2   Reduced Matrices

Definition 13.1 A matrix with m rows is said to be *row reduced* if in each non-null row of the first $m - 1$ there is a non-null element below which there are only zeros.

The concept of row reduced matrix similarly develops to that of *reduced system* (see Sect. 11.3); the operations that are applied to a system to transform it into a reduced system are in essence the same reduction operations applied to a matrix.

Similarly, a matrix with n columns is said to be *column reduced* if in each non-null column of the first $n - 1$ there is a non-null element followed by only zeros in the row which it belongs to.

Example 13.2 The matrix

$$\begin{bmatrix} 0 & 3 & -1 & 2 & 0 \\ 0 & 0 & 0 & 0 & 0 \\ 1 & 0 & 3 & 0 & 1 \\ 0 & 0 & 2 & 0 & 1 \end{bmatrix}$$

is row reduced.
The matrix

$$\begin{bmatrix} 2 & 0 & 3 \\ 1 & 0 & 0 \\ 0 & 0 & 1 \end{bmatrix}$$

is not row reduced for in the first row there is no non-null element below which there are only zeros.

Example 13.3 The matrix

$$\begin{bmatrix} 1 & 0 & 0 & 0 \\ 1 & 3 & 0 & 0 \\ 1 & 0 & 0 & 0 \\ 0 & 3 & -4 & 0 \end{bmatrix}$$

is column reduced. The matrix

$$\begin{bmatrix} 1 & 0 & 0 & 0 \\ 1 & 3 & 1 & 0 \\ 1 & 0 & 1 & 0 \\ 0 & 3 & 0 & 4 \end{bmatrix}$$

is not column reduced since in the second column there is no any non-null element followed by only zeros in the row which it belongs to.

Theorem 13.1 *The non-zero rows (columns) of a row (column) reduced matrix are linearly independent.*

Proof Consider a row reduced matrix and let $a_1, a_2, ..., a_k$ be the non-zero rows. Let

$$a_{1 j_1}, a_{2 j_2}, ..., a_{k j_k}$$

be non-null elements of the rows $a_1, a_2, ..., a_k$ below which there are only zeros (j_1, $j_2, ..., j_k$ are k distinct integers in the set $I_n = \{1, 2, ..., n\}$ of the column indices). Therefore,

$$a_{i_1 j_1} = a_{i_2 j_2} = \ldots = a_{i_k j_k} = 0$$

if $i_1 > 1, i_2 > 2, ..., i_k > k$ (it is understood that if $k = m$ the last inequality is omitted). Accordingly, the linear combination

$$h_1 a_1 + h_2 a_2 + \ldots + h_k a_k = h_1 (a_{11}, a_{12}, \ldots, a_{1n})$$
$$+ h_2 (a_{21}, a_{22}, \ldots, a_{2n}) + \ldots + h_k (a_{k1}, a_{k2}, \ldots, a_{kn})$$

has components

$$h_1 a_{1 j_1}, \; h_2 a_{2 j_2}, \; \ldots, \; h_k a_{k j_k}$$

in the places $1j_1, 2j_2, ..., kj_k$, respectively. Thus, if the linear combination $h_1 a_1 + h_2 a_2 + \cdots + h_k a_k$ is the null vector, then necessarily $h_1 = h_2 = \cdots = h_k = 0$. This proves that the rows $a_1, a_2, ..., a_k$ are linearly independent. □

13.3 Rank

The rows $a_1, a_2, ..., a_m$ (13.2) and the columns $a^1, a^2, ..., a^n$ (13.3) of the matrix

$$A = \begin{bmatrix} a_{11} & a_{12} & \cdots & a_{1n} \\ a_{21} & a_{22} & \cdots & a_{2n} \\ \cdots & \cdots & \cdots & \cdots \\ a_{m1} & a_{m2} & \cdots & a_{mn} \end{bmatrix}$$

are subsets of $\mathbf{R}^n$ and $\mathbf{R}^m$, respectively. Let R(A) and C(A) denote the vector subspaces spanned (Sect. 12.7.1) by the system of the rows $[a_1, a_2, ..., a_m]$ and columns $[a^1, a^2, ..., a^n]$, respectively.

Theorem 13.2 *The vector spaces R(A) and C(A) have the same dimension.*

Proof Let r and r' be the maximum number of linearly independent rows and the maximum number of linearly independent columns of A, respectively: then r is the dimension of the vector subspace R(A) of $\mathbf{R}^n$ spanned by the system of the rows $[a_1, a_2, ..., a_m]$ of A and r' is the dimension of the vector subspace C(A) of $\mathbf{R}^m$ spanned by the system of the columns $[a^1, a^2, ..., a^n]$ of A. Let $S = [a_{i_1}, a_{i_2}, ..., a_{i_r}]$ be a basis of R(A) ($i_1, i_2, ..., i_r$ are r distinct integers in the set $\{1, 2, ..., n\}$). Each row of A is a linear combination of the vectors of S; it turns out

$$\begin{aligned} a_1 &= k_{11} a_{i_1} + k_{12} a_{i_2} + \ldots + k_{1r} a_{i_r} \\ a_2 &= k_{21} a_{i_1} + k_{22} a_{i_2} + \ldots + k_{2r} a_{i_r} \\ &\cdots \\ a_m &= k_{m1} a_{i_1} + k_{m2} a_{i_2} + \ldots + k_{mr} a_{i_r} \end{aligned} \tag{13.4}$$

with the k_{ij} scalars. The vector equalities (13.4) imply, for every $j = 1, ..., n$, the scalar equalities:

$$\begin{aligned} a_{1j} &= k_{11} a_{i_1 j} + k_{12} a_{i_2 j} + \ldots + k_{1r} a_{i_r j} \\ a_{2j} &= k_{21} a_{i_1 j} + k_{22} a_{i_2 j} + \ldots + k_{2r} a_{i_r j} \\ &\cdots\cdots\cdots \\ a_{mj} &= k_{m1} a_{i_1 j} + k_{m2} a_{i_2 j} + \ldots + k_{mr} a_{i_r j} \end{aligned} \tag{13.5}$$

Setting $k_1 = (k_{11}, k_{21}, ..., k_{m1})$, $k_2 = (k_{12}, k_{22}, ..., k_{m2})$, ..., $k_r = (k_{1r}, k_{2r}, ..., k_{mr})$, by the Eq. (13.5), for every $j = 1, ..., n$, the following expression of the columns is obtained:

$$a^j = (a_{1j}, a_{2j}, \ldots, a_{mj}) = a_{i_1 j} k_1 + a_{i_2 j} k_2 + \ldots + a_{i_r j} k_r$$

Therefore, each column of A is a linear combination of the vectors of the system $T = [k_1, k_2, \ldots, k_r]$, which turns out to be a basis of C(A). From Corollary 12.1, the dimension of C(A) is $r' \le r$.

Likewise, it can be shown that r is less than or equal to r'.   □

Definition 13.2 The common value of the dimensions of the spaces R(A) and C(A) is named the *rank* of the matrix A, denoted p_A, or p, if A is implicit.

Of course, the rank of an $m \times n$ matrix cannot be greater than the smaller of the numbers m and n and it is zero if and only if the matrix is null.

Example 13.4 The matrix

$$A = \begin{bmatrix} 2 & 0 & 1 & 0 \\ 6 & 0 & 3 & 0 \\ 4 & 0 & 2 & 0 \end{bmatrix}$$

has rank 1. In fact, in A there are no two linearly independent columns since the only two non-zero columns are proportional to each other.

Example 13.5 The matrix

$$B = \begin{bmatrix} 2 & 0 & 0 \\ -1 & 0 & 1 \\ 1 & 0 & 0 \end{bmatrix}$$

has rank 2. In fact, the system of the three rows is linearly dependent since the third line is the sum of the first two, while any two rows are non-proportional. (Alternatively, the two non-null columns are non-proportional.)

13.4   Matrix Reduction Method and Rank

The name of *elementary row operation* is given to each of the following operations that can be performed on the rows of the matrix (13.1):

(R1) replacing the i-th row a_i with the product $k a_i$, $k \ne 0$: $a_i \leftarrow k a_i$;
(R2) replacing the i-th row with the sum of the same row with the j-th row multiplied by a real number h : $a_i \leftarrow a_i + h a_j$;

The operations (R1) and (R2) are special cases of the following elementary row operation:

(R12) replacing the i-th row with the sum of the i-th row multiplied by a real number $k \neq 0$ with the j-th row multiplied by a real number $h : a_i \leftarrow ka_i + ha_j$;

The same operations applied to the columns, are called *elementary column operations*.

The following proposition applies.

Proposition 13.1 *Each elementary row (column) operation, and therefore a finite number of them applied repeatedly, does not alter the number of linearly independent rows (columns) and consequently the rank of the matrix.*

The reduction method consists in transforming, by means of elementary row (column) operations, a given matrix into a row (column) reduced matrix. The method is effective to calculate the rank of a matrix (see Theorem 13.1).

Example 13.6 Let us row reduce the matrix

$$A = \begin{bmatrix} 2 & 1 & 0 \\ 1 & 2 & 1 \\ 5 & 1 & 2 \end{bmatrix}$$

Introduce zeros below the entry $a_{11} = 2$. First apply the operation R1: $a_2 \leftarrow -2a_2$ which results in the matrix

$$A_1 = \begin{bmatrix} 2 & 1 & 0 \\ -2 & -4 & -2 \\ 5 & 1 & 2 \end{bmatrix}$$

then the operation R2: $a_2 \leftarrow a_2 + a_1$

$$A_2 = \begin{bmatrix} 2 & 1 & 0 \\ 0 & -3 & -2 \\ 5 & 1 & 2 \end{bmatrix}$$

(Remaind that A_2 can be obtained from A by applying the operation R12: $a_2 \leftarrow -2a_2 + a_1$)

Now perform the operation R12: $a_3 \leftarrow 2a_3 - 5a_1$ to obtain

$$A_3 = \begin{bmatrix} 2 & 1 & 0 \\ 0 & -3 & -2 \\ 0 & -3 & 4 \end{bmatrix}$$

In the matrix A_3 the elements below the first component of the first row are null. Transform A_3 into a reduced matrix: use the substitution $a_3 \leftarrow a_3 - a_2 a_3 \leftarrow a_3 - a_2$

$$A_4 = \begin{bmatrix} 2 & 1 & 0 \\ 0 & -3 & -2 \\ 0 & 0 & 6 \end{bmatrix}$$

Matrix A_4 is row reduced and its rank is 3. Since in each transformation from a matrix to the next the rank did not change, the rank of the matrices A to A_4 is also 3.

Observe that the calculation of the rank of a row (column) reduced matrix is immediate: in fact, by Theorem 13.1, the rank of a matrix row (column) reduced is the number of non-zero rows (columns). The calculation of the rank of any matrix can be traced back to that of a reduced matrix.

13.5   Rouché-Capelli's Theorem

Let us consider the linear system of equations (Sect. 11.1)

$$\begin{array}{l} a_{11}x_1 + a_{12}x_2 + \ldots + a_{1n}x_n = b_1 \\ a_{21}x_1 + a_{22}x_2 + \ldots + a_{2n}x_n = b_2 \\ \ldots\ldots\ldots \\ a_{m1}x_m + a_{m2}x_2 + \ldots + a_{mn}x_n = b_m \end{array} \tag{13.6}$$

The $m \times n$ *matrix of the coefficients* of the system (13.6)

$$A = \begin{bmatrix} a_{11} & a_{12} & \ldots & a_{1n} \\ a_{21} & a_{22} & \ldots & a_{2n} \\ \ldots & \ldots & \ldots & \ldots \\ a_{m1} & a_{m2} & \ldots & a_{mn} \end{bmatrix}$$

is called the matrix of the coefficients or the *associate matrix* or the *incomplete matrix* of the system (13.6).

The matrix of size $(m, n + 1)$

$$A' = \begin{bmatrix} a_{11} & a_{12} & \ldots & a_{1n} & b_1 \\ a_{21} & a_{22} & \ldots & a_{2n} & b_2 \\ \ldots & \ldots & \ldots & \ldots & \ldots \\ a_{m1} & a_{m2} & \ldots & a_{mn} & b_m \end{bmatrix}$$

is called the *complete matrix* or the *augmented matrix* of the system (13.6).

Since the first n columns of the complete matrix are the columns of the associate matrix, it turns out that the rank p' of the complete matrix is not less than the rank p of the associate matrix,

$$p' \geq p$$

If $(h_1, h_2, \ldots, h_n)$ is a solution of the system (13.6), then the scalar equalities

$$\begin{aligned}
a_{11}h_1 + a_{12}h_2 + \ldots + a_{1n}h_n &= b_1 \\
a_{21}h_1 + a_{22}h_2 + \ldots + a_{2n}h_n &= b_2 \\
&\cdots\cdots\cdots \\
a_{m1}h_1 + a_{m2}h_2 + \ldots + a_{mn}h_n &= b_m
\end{aligned} \tag{13.7}$$

that are equivalent to the equality of n-tuples

$$\begin{aligned}
h_1(a_{11}, a_{21}, \ldots, a_{m1}) + h_2(a_{12}, a_{22}, \ldots, a_{m2}) + \ldots \\
+ h_n(a_{1n}, a_{2n}, \ldots, a_{mn}) = \big(b_1, b_2, \ldots, b_m\big)
\end{aligned} \tag{13.8}$$

are verified.

Vice versa, if equality (13.8) holds, then the equalities (13.7) are satisfied and $(h_1, h_2, \ldots, h_n)$ is a solution of system (13.6). In conclusion, we state:

Theorem 13.3 *A system of linear equations is compatible if and only if the column of known coefficients is a linear combination of the columns of the associate matrix.*

The theorem asserts that a system of linear equations is compatible if and only if a maximal system (Sect. 12.8) of linearly independent columns of the associate matrix A *is also a maximal system of linearly independent columns of the complete matrix* A′. *Therefore, from Proposition 12.6 and Definition 13.2, we can state*:

Theorem 13.4 [Rouché-Capelli's theorem] *A system of linear equations is compatible if and only if the rank of the associate matrix of the system is equal to the rank of the complete matrix.*

Thus, the following definition makes sense:

Definition 13.3 The *rank* of a compatible system is, by definition, the common value of the ranks of the associate matrix and the complete matrix.

13.6 Compatibility of a Reduced System

Let us consider the associate matrix and the complete matrix of a row reduced system. By Theorem 13.3 each of these matrices has the rank equal to the number of the non-null rows. Therefore, a reduced system is compatible if and only if the number of the non-null rows of the associate matrix is equal to the number of the non-null rows of the complete matrix; while a reduced system is incompatible if and only if the number of the non-null rows of the associate matrix is less than the number of the non-null rows of the complete matrix. In other words, in the associate matrix there is a null row and the row in the same place of the complete matrix has the last element non-null, which is equivalent to the existence of the incompatible equation in the system:

$$0x_1 + 0x_2 + \ldots + 0x_n = b, \text{ and } b \neq 0,$$

Therefore, the following theorem is proved.

Theorem 13.5 *A reduced system of linear equation is inconsistent if and only if it includes among its equations the incompatible equation* $0 = b$, *with* $b \neq 0$.

13.7 Square Matrices

A matrix with the same number n of rows and columns is named a *square matrix* of *order n*, or an *n*-square matrix.

Let

$$A = \begin{bmatrix} a_{11} & a_{12} & \ldots & a_{1n} \\ a_{21} & a_{22} & \ldots & a_{2n} \\ \ldots & \ldots & \ldots & \ldots \\ a_{n1} & a_{n2} & \ldots & a_{nn} \end{bmatrix}$$

be a square matrix of order n.

The vector $d = (a_{11}, a_{22}, \ldots, a_{nn})$ is called the *main diagonal* or the *diagonal* of A. The vector $d' = (a_{1n}, a_{2n-1}, \ldots, a_{n1})$ is named *secondary diagonal* of A. The square matrix of order n with all 1s on the diagonal and 0s elsewhere is called the *identical matrix* or *identity matrix* of order n, denoted Id_n

$$Id_n = \begin{bmatrix} 1 & 0 & 0 & \ldots & 0 \\ 0 & 1 & 0 & \ldots & 0 \\ \ldots & \ldots & \ldots & \ldots \\ 0 & 0 & 0 & \ldots & 1 \end{bmatrix}$$

An *upper triangular matrix*, or simply a *triangular matrix*, is a square matrix whose entries below the main diagonal are all zero:

$$\begin{bmatrix} a_{11} & a_{12} & a_{13} & \ldots & a_{1n} \\ 0 & a_{22} & a_{23} & \ldots & a_{2n} \\ \ldots & \ldots & \ldots & \ldots \\ 0 & 0 & 0 & \ldots & a_{nn} \end{bmatrix}$$

A *lower triangular matrix* is a square matrix whose entries above the main diagonal are all zero.

A *diagonal matrix* is a square matrix whose non-diagonal entries are all zero.

13.8   Exercises

1. Given the system of linear equations

$$2x - y - z = 0$$
$$x - z = 1 \qquad\qquad (13.9)$$
$$x + y = 1$$

   i.   reduce the associate matrix of the system,
  ii.   prove that the system is compatible,
 iii.   find the unique solution of the system.

Solution

i.   let us perform the elementary operation $a_3 \leftarrow -a_2 + a_1$ to the rows a_1, a_2, a_3 of the associate matrix

$$A = \begin{bmatrix} 2 & -1 & -1 \\ 1 & 0 & -1 \\ 1 & 1 & 0 \end{bmatrix}$$

to obtain:

$$A_1 = \begin{bmatrix} 2 & -1 & -1 \\ 1 & -1 & 0 \\ 1 & 1 & 0 \end{bmatrix}$$

Now apply $a_3 \leftarrow a_3 + a_2$:

$$A_2 = \begin{bmatrix} 2 & -1 & -1 \\ 1 & -1 & 0 \\ 2 & 0 & 0 \end{bmatrix}$$

ii.   The matrix A_2 is the row reduced associate matrix of the system

$$E_1 : 2x - y - z = 0$$
$$E_2 : x - y = 1$$
$$E_3 : 2x = 0$$

equivalent to (13.9). By Theorem 13.1, the rows of the matrix A_2 are linearly independent. Therefore, the rank of A_2 is 3 and, consequently, the rank of the complete matrix is also 3 and, by Theorem 13.3, the system E_1, E_2, E_3 and the equivalent system (13.9), are compatible.

iii. Let us solve by substitution the reduced system: from E_3 we get

$$x = 0$$

that we substitute for x in E_2 to have

$$y = 1$$

Finally, replacing 0 for x and 1 for y in E_1 we get $z = -1$. In conclusion, $(0, 1, -1)$ is the unique solution of the system (13.9).

2. Given the system of linear equations

$$E_1 : x - y - z = 13$$
$$E_2 : x - 2y - z = 0$$
$$E_3 : y = 13$$

    i.   prove that the system is compatible,
    ii. find the solutions of the system.

Solution

i.   As $E_2 \leftarrow E_1 - E_3$, the given system is equivalent to

$$E_1 : x - y - z = 13$$
$$E_2 : y = 13$$

The system is compatible because the associate matrix and the complete matrix have the same rank $p = 2$.

ii. Apply the operation $E_1 \leftarrow E_1 + E_2$

$$x = 26 + z$$
$$y = 13$$

The system is satisfied by any value of z, hence any triple $(26\text{-}z, 13, z)$ is a solution of the given system. For example, $(26, 13, 0)$, $(27, 13, 1)$, $(25, 13, -1)$,... are solutions.

Bibliography

Aitken, A.C.: Determinants and Matrices. Oliver and Boyd LTD, Edinburgh (1965)

D'Apuzzo, L., Ventre, A.: Algebra Lineare e Geometria Analitica. Cedam, Padova (1995)

Lang, S.: Linear Algebra. Addison-Wesley Publishing Company, Reading, Massachusetts (1966)

Lipschutz, S.: Theory and Problems of Linear Algebra. McGraw-Hill Book Company, New York (1968)

Schwartz, J.T.: Introduction to Matrices and Vectors. Dover Publications Inc., New York (1971)

Strang, G.: Linear Algebra and Applications. Academic Press Inc., New York (1976)

Chapter 14
Determinants and Systems of Linear Equations

14.1 Determinants

We will develop a procedure for the study of systems of linear equations, also useful in combination with reduction methods and known as the *method of the determinants*.

Let us consider a square matrix A and a function, called the *determinant function* of A which associates a real number to A, called the *value of the determinant* of A, denoted by one of the symbols

$$\det(A), \ |A|$$

The entries, the rows a_1, ..., a_n and the columns a^1, ..., a^n of the matrix A (Sect. 13.1), are, by definition, the entries, the rows and the columns of the determinant of A. To ease the background we will give an "inductive definition" of determinant. Sometimes, for the sake of brevity, we speak of calculation of the determinant meaning the calculation of its value.

If we need to mention the rows a_1, ..., a_n or the columns a^1, ..., a^n of the determinant we use the notations

$$\det(A) = \det(a_1, \ldots, a_n) = \det\left(a^1, \ldots, a^n\right)$$

The value of the determinant of the square matrix A is computed as described below.

1. If A is a square matrix of order 1, that is a matrix with one row and one column, which reduces to the number a_{11}, then we set:

$$\det(A) = |A| = a_{11}$$

   For example, if $A = [7]$, then $\det(A) = |A| = 7$; if $A = [-7]$, then $\det(A) = |A| = -7$.

© The Author(s), under exclusive license to Springer Nature Switzerland AG 2023
A. G. S. Ventre, *Calculus and Linear Algebra*,
https://doi.org/10.1007/978-3-031-20549-1_14

2. If A is a square matrix of order 2

$$A = \begin{bmatrix} a_{11} & a_{12} \\ a_{21} & a_{22} \end{bmatrix}$$

Then

$$\det(A) = |A| = a_{11}a_{22} - a_{12}a_{21} \tag{14.1}$$

namely, the value of the determinant of A is equal to the difference between the product of the entries of the main diagonal and the product of the entries of the secondary diagonal. For example, given the matrix

$$A = \begin{bmatrix} -6 & 1 \\ 2 & 0 \end{bmatrix}$$

the determinant of A is

$$\det(A) = |A| = \begin{vmatrix} -6 & 1 \\ 2 & 0 \end{vmatrix} = (-6)0 - (1)2 = 0 - 2 = -2$$

3. If A is a matrix of order 3,

$$A = \begin{bmatrix} a_{11} & a_{12} & a_{13} \\ a_{21} & a_{22} & a_{23} \\ a_{31} & a_{32} & a_{33} \end{bmatrix}$$

Then

$$|A| = a_{11} \begin{vmatrix} a_{22} & a_{23} \\ a_{32} & a_{33} \end{vmatrix} - a_{12} \begin{vmatrix} a_{21} & a_{23} \\ a_{31} & a_{33} \end{vmatrix} + a_{13} \begin{vmatrix} a_{21} & a_{22} \\ a_{31} & a_{32} \end{vmatrix}$$

and, by (14.1)

$$\det(A) = |A| = a_{11}(a_{22}a_{33} - a_{23}a_{32}) - a_{12}(a_{21}a_{33} - a_{23}a_{31})$$
$$+ a_{13}(a_{21}a_{32} - a_{22}a_{31}) = a_{11}a_{22}a_{33} - a_{11}a_{23}a_{32}$$
$$- a_{12}a_{21}a_{33} + a_{12}a_{23}a_{31} + a_{13}a_{21}a_{32} - a_{13}a_{22}a_{31} \tag{14.2}$$

In other words, det(A) is equal to a linear combination, whose coefficients are the entries of the first row with the alternating signs, a_{11}, $-a_{12}$, a_{13}, multiplied by the determinants of the second order, the first of which is obtained by deleting the first row and the first column of A, the second is obtained from A by deleting the first row and second column, and the third by deleting the first row and third column.

So, for example, it turns out

$$|A| = \begin{vmatrix} 1 & 0 & 3 \\ -1 & 2 & 1 \\ 0 & 1 & 1 \end{vmatrix} = 1 \begin{vmatrix} 2 & 1 \\ 1 & 1 \end{vmatrix} - 0 \begin{vmatrix} -1 & 1 \\ 0 & 1 \end{vmatrix} + 3 \begin{vmatrix} -1 & 2 \\ 0 & 1 \end{vmatrix}$$

$$= 1(2(1) - 1(1)) - 0 + 3(-1(1) - 2(0)) = 1 - 3 = -2$$

It is easy to verify that the same result (14.2) is obtained when choosing the elements of any row as coefficients of the linear combination; for example, if the coefficients are the ones of the second with alternating signs $-a_{21}, a_{22}, -a_{23}$, now first and third coefficient change sign, i. e.,

$$|A| = -a_{21} \begin{vmatrix} a_{12} & a_{13} \\ a_{32} & a_{33} \end{vmatrix} + a_{22} \begin{vmatrix} a_{11} & a_{13} \\ a_{31} & a_{33} \end{vmatrix} - a_{23} \begin{vmatrix} a_{11} & a_{12} \\ a_{31} & a_{32} \end{vmatrix}$$

But, also, the equalities hold when forming linear combinations with the elements of columns. For example,

$$\det(A) = |A| = a_{11} \begin{vmatrix} a_{22} & a_{23} \\ a_{32} & a_{33} \end{vmatrix} - a_{21} \begin{vmatrix} a_{12} & a_{13} \\ a_{32} & a_{33} \end{vmatrix} + a_{31} \begin{vmatrix} a_{12} & a_{13} \\ a_{22} & a_{23} \end{vmatrix}$$

The coefficients in each linear combination is taken with alternating signs: precisely, the coefficient is a_{ij} if the sum $i + j$ is even; the coefficient is $-a_{ij}$ if $i + j$ is odd.

It is an easy task to prove the following propositions.

Let A be a square matrix of order 2 or 3. Then

a. if all the entries of a row, or column, of A are zeros, then $\det(A) = 0$;
b. if A' is obtained from A interchanging two rows or two columns, then $\det(A) = -\det(A')$.

The *order* of the determinant $|A|$ is defined as the order of the square matrix A.

The computation of a third-order determinant is therefore a linear combination of three second-order determinants.

As the order m of a square matrix A increases, the calculation of the determinant becomes even more elaborated, but the principle is the same: the value of a determinant of order 4 is a linear combination of four determinants of the third order; the value of a determinant of order m is a linear combination of m determinants of order $m - 1$.

The way to perform the computation of a determinant is called the *expansion* of the determinant along a row, or column.

Example 14.1 Let us compute the determinant of the matrix

$$A = \begin{bmatrix} -2 & 0 & 1 & 5 \\ 0 & 3 & -1 & -4 \\ 1 & 0 & 0 & -3 \\ 2 & 3 & 1 & 0 \end{bmatrix}$$

The coefficients of the linear combination are 1 (with even place $3 + 1$), 0, 0, $-(-3) = 3$ (in fact -3 has an odd place $3 + 4$). To lighten the computation let us expand the determinant along the third row. Indeed,

$$det(A) = \begin{vmatrix} -2 & 0 & 1 & 5 \\ 0 & 3 & -1 & -4 \\ 1 & 0 & 0 & -3 \\ 2 & 3 & 1 & 0 \end{vmatrix} =$$

$$1 \begin{vmatrix} 0 & 1 & 5 \\ 3 & -1 & -4 \\ 3 & 1 & 0 \end{vmatrix} + 0 + 0 + 3 \begin{vmatrix} -2 & 0 & 1 \\ 0 & 3 & -1 \\ 2 & 3 & 1 \end{vmatrix} =$$

and expanding along the first row both determinants,

$$= -1 \begin{vmatrix} 3 & -4 \\ 3 & 0 \end{vmatrix} + 5 \begin{vmatrix} 3 & -1 \\ 3 & 1 \end{vmatrix} + 3 \left(-2 \begin{vmatrix} 3 & -1 \\ 3 & 1 \end{vmatrix} + 1 \begin{vmatrix} 0 & 3 \\ 2 & 3 \end{vmatrix} \right)$$

$$= -12 + 30 + 3(-2(3+3) - 6) = -36.$$

Then the value of the determinant is $|A| = -36$.

14.2 Properties of the Determinants

Let us state some properties of the determinants. Let A be a square matrix.

1. If A has a null row or column, then $|A| = 0$. Indeed, the development according to the row, or column, with all null elements is zero.
2. If B is the matrix obtained from A by interchanging two rows, or two columns, then

$$\det(B) = -\det(A)$$

3. If two rows, or two columns, of A are equal, then $\det(A) = 0$.
   In fact, the matrix B that is obtained from A by interchanging the two equal rows, or columns, coincides with A and therefore $|B| = |A|$; from property 2 we obtain $|B| = -|A|$. Thus $|A| = 0$.
4. If the matrix B is obtained from matrix A by multiplying all the elements of a row, or column, by a scalar k, then $|B| = k|A|$.

In fact, if all the elements of a row (or column) of A are multiplied by k, all the addends of the expansion along the row (or column) are also multiplied by k.

5. If two rows (or columns) are proportional, then $|A| = 0$.

The statement 5 follows from the properties 3 and 4. In fact, given the square matrix A having rows $(a_1, \ldots, a_n)$, $A = (a_1, \ldots, a_n)$, let the equality $a_j = k a_i$ hold. Hence,

$$\det(A) = \det(a_1, \ldots, a_i, \ldots, a_j, \ldots, a_n) = \det(a_1, \ldots, a_i, \ldots, k a_i, \ldots, a_n)$$
$$= k \det(a_1, \ldots, a_i, \ldots, a_i, \ldots, a_n) = k \bullet 0 = 0$$

A similar result holds when two columns are proportional.

6. If a row (or column) of the square matrix A is the sum of two n-tuples of $\mathbf{R}^n$, then the determinant of A is the sum of the determinants of the two matrices obtained from A by replacing the row (or column) with each of the two n-tuples:

$$\det(A) = \det(a_1, \ldots, c_i + d_i, \ldots, a_n) = \det(a_1, \ldots, c_i, \ldots, a_n)$$
$$+ \det(a_1, \ldots d_i, \ldots, a_n)$$

7. If B is a matrix obtained by the square matrix A adding a row (or column) to a row (or column) multiplied by a scalar h, then $|B| = |A|$.

Indeed, let us suppose $A = (a_1, \ldots, a_i, \ldots, a_j, \ldots, a_n)$ and $B = (a_1, \ldots, a_i + h a_j, \ldots, a_j, \ldots, a_n)$. Then, by properties 6 and 5, we have:

$$|B| = \det(a_1, \ldots, a_i + h a_j, \ldots, h a_j, \ldots, a_n)$$
$$= \det(a_1, \ldots, a_i, \ldots, a_j, \ldots, a_n)$$
$$+ \det(a_1, \ldots, h a_j, \ldots, a_j, \ldots, a_n) = |A| + 0$$

In other words, the elementary row operation R2, $a_i \leftarrow a_i + h a_j$ does not alter the value of the determinant $|A|$.

A similar result concerning columns holds.

8. If B is the matrix obtained from the square matrix A by adding a linear combination of the other rows (columns) to a given row (column), then $|B| = |A|$..

9. If a row (column) of the square matrix A is a linear combination of the remaining rows (columns), then $|A| = 0$.

14.3   Submatrices and Minors

Let A be an $m \times n$ matrix

$$A = \begin{bmatrix} a_{11} & a_{12} & \cdots & a_{1n} \\ a_{21} & a_{22} & \cdots & a_{2n} \\ \cdots & \cdots & \cdots & \cdots \\ a_{m1} & a_{m2} & \cdots & a_{mn} \end{bmatrix}$$

Consider p natural numbers $i_1 < i_2 < \cdots < i_p$ in the set $I_m = \{1, 2, \ldots, m\}$ and q natural numbers $j_1 < j_2 < \cdots < j_q$ in the set $I_n = \{1, 2, \ldots, n\}$. Consider also the rows $\boldsymbol{a}_{i_1}, \boldsymbol{a}_{i_2}, \ldots, \boldsymbol{a}_{i_p}$ and the columns $\boldsymbol{a}^{j_1}, \boldsymbol{a}^{j_2}, \ldots, \boldsymbol{a}^{j_q}$ of A. The $p \times q$ matrix:

$$A\big(i_1, i_2, \ldots, i_p; j_1, j_2, \ldots, j_q\big) = \begin{bmatrix} a_{i_1 j_1} & a_{i_1 j_2} & \cdots & a_{i_1 j_q} \\ a_{i_2 j_1} & a_{i_2 j_2} & \cdots & a_{i_2 j_q} \\ \cdots & \cdots & \cdots & \cdots \\ a_{i_p j_1} & a_{i_p j_2} & \cdots & a_{i_p j_q} \end{bmatrix}$$

is called a submatrix of A, or a *submatrix extracted* from A, relatively to, or identified by, the rows

$$\boldsymbol{a}_{i_1}, \boldsymbol{a}_{i_2}, \ldots, \boldsymbol{a}_{i_p}$$

and the columns

$$\boldsymbol{a}^{j_1}, \boldsymbol{a}^{j_2}, \ldots, \boldsymbol{a}^{j_q};$$

the elements of the submatrix are the elements common to the considered p rows and q columns of A.

If $p < m$ and $q < n$, the $(m - p) \times (n - q)$ submatrix identified by the remaining $m - p$ rows and the remaining $n - q$ columns is called the *complementary matrix* of the matrix $A(i_1, i_2, \ldots, i_p; j_1, j_2, \ldots, j_q)$.

Any determinant of a square submatrix of order p extracted from A is defined as a *minor* of order p of the matrix A. Each entry of a matrix is a minor of order 1 of the matrix.

For example, consider the 3×4 matrix

$$A = \begin{bmatrix} 1 & 1 & -1 & 0 \\ 2 & 1 & 3 & 0 \\ 4 & 5 & 1 & 2 \end{bmatrix}$$

The submatrix of A identified by the rows $\boldsymbol{a}_1, \boldsymbol{a}_2$ and the columns $\boldsymbol{a}^2, \boldsymbol{a}^3, \boldsymbol{a}^4$ is

$$A(1, 2; 2, 3, 4) = \begin{bmatrix} 1 & -1 & 0 \\ 1 & 3 & 0 \end{bmatrix}$$

The complementary matrix of $A(1, 2; 2, 3, 4)$ is $A(3; 1) = a_{31} = [4]$.
The submatrix of A identified by the rows a_1, a_3 and the columns a^1, a^4 is

$$A(1, 3; 1, 4) = \begin{bmatrix} 1 & 0 \\ 4 & 2 \end{bmatrix}$$

The determinant of matrix $A(1, 3; 1, 4)$ is a minor of A and its value is $|A(1, 3; 1, 4)| = 2$.

14.4 Cofactors

If A is a square matrix of order n, the complementary matrix of a square submatrix of order p, $p < n$, is a square submatrix of order $n - p$.

Therefore, it makes sense to define as a *complementary minor* of the minor $|A(i_1, i_2, \ldots, i_p; j_1, j_2, \ldots, j_p)|$ the determinant of the complementary matrix of $A(i_1, i_2, \ldots, i_p; j_1, j_2, \ldots, j_p)$.

The *cofactor* of the minor $|A(i_1, i_2, \ldots, i_p; j_1, j_2, \ldots, j_p)|$ of A is defined as its complementary minor if the sum $i_1 + i_2 + \cdots + i_p + j_1 + j_2 + \cdots + j_p$ is even, the opposite of the complementary minor if $i_1 + i_2 + \cdots + i_p + j_1 + j_2 + \cdots + j_p$ is odd.

In particular, the cofactor of the element a_{ij} of A is denoted by A_{ij}. Therefore, by definition,

$$A_{ij} = (-1)^{i+j} |A(1, 2, \ldots, i - 1, i + 1, \ldots, n; 1, 2, \ldots, j - 1, j + 1, \ldots, n)|$$

is the *cofactor* of a_{ij}.

Example 14.2 Let A be the square matrix

$$\begin{bmatrix} 3 & -4 & 5 & 6 \\ 7 & 8 & 9 & -3 \\ 1 & 2 & -1 & -2 \\ -3 & 4 & -5 & 0 \end{bmatrix}$$

The complementary minor of

$$|A(1, 2; 1, 2)| = \begin{vmatrix} 3 & -4 \\ 7 & 8 \end{vmatrix}$$

is the minor

$$|A(3, 4; 3, 4)| = \begin{vmatrix} -1 & -2 \\ -5 & 0 \end{vmatrix}$$

The cofactor of $a_{11} = 3$ is the product of $(-1)^{1+1}$ by the determinant of the matrix obtained by eliminating the first row and the first column (row and column to which a_{11} belongs).

$$A_{11} = (-1)^{1+1}|A(2, 3, 4; 2, 3, 4)| = \begin{vmatrix} 8 & 9 & -3 \\ 2 & -1 & -2 \\ 4 & -5 & 0 \end{vmatrix} = -194$$

Let us formulate the expansion procedure of a determinant in terms of cofactors. The determinant of a square matrix $A = (a_{ij})$ of order n is equal to the sum of the products of the elements of any row i (column j) by their respective cofactors:

$$|A| = a_{i1}A_{i1} + a_{i2}A_{i2} + \cdots + a_{in}A_{in}$$

$$|A| = a_{1j}A_{1j} + a_{2j}A_{2j} + \cdots + a_{nj}A_{nj}$$

Each of the linear combinations is called the *Laplace's expansion* of the determinant $|A|$ with respect to the row a_i and the column a^j.

14.5 Matrix Multiplication

Consider an $m \times n$ matrix A, an $n \times q$ matrix B, the i-th row a_i of A ($i = 1, \ldots, m$) and the j-th column b^j of B ($j = 1, \ldots, q$). Since a_i, b^j are vectors of $\mathbf{R}^n$ we can calculate their scalar product (Sect. 12.10) $a_i \times b^j$, for every i and j.

Definition 14.1 Let A and B be two matrices of sizes (m, n) and (n, q), respectively. The *product* AB is, by definition, the matrix of size (m, q) whose ij-entry c_{ij} is the scalar product $a_i \times b^j$:

$$AB = \begin{bmatrix} c_{11} & c_{12} & \cdots & c_{1q} \\ c_{21} & c_{22} & \cdots & c_{1q} \\ \cdots & \cdots & \cdots & \cdots \\ c_{m1} & c_{m2} & \cdots & c_{mq} \end{bmatrix} = \begin{bmatrix} a_1 \times b^1 & a_1 \times b^2 & \cdots & a_1 \times b^q \\ a_2 \times b^1 & a_2 \times b^2 & \cdots & a_2 \times b^q \\ \cdots & \cdots & \cdots & \cdots \\ a_m \times b^1 & a_m \times b^2 & \cdots & a_m \times b^q \end{bmatrix}$$

Example 14.3 Let A and B be the matrices

$$A = \begin{bmatrix} 1 & 0 \\ 2 & 1 \end{bmatrix} \quad B = \begin{bmatrix} 1 & 3 & 0 \\ -1 & 5 & 2 \end{bmatrix}$$

Matrix A is a 2×2 matrix and B is a 2×3 matrix. We obtain the product AB from the scalar products:

$$c_{11} = \boldsymbol{a}_1 \times \boldsymbol{b}^1 = (1, 0) \times (1, -1) = 1(1) + 0(-1) = 1$$
$$c_{12} = \boldsymbol{a}_1 \times \boldsymbol{b}^2 = (1, 0) \times (3, 5) = 3$$
$$c_{13} = \boldsymbol{a}_1 \times \boldsymbol{b}^3 = (1, 0) \times (0, 2) = 0$$
$$c_{21} = \boldsymbol{a}_2 \times \boldsymbol{b}^1 = (2, 1) \times (1, -1) = 1$$
$$c_{22} = \boldsymbol{a}_2 \times \boldsymbol{b}^2 = (2, 1) \times (3, 5) = 11$$
$$c_{23} = \boldsymbol{a}_2 \times \boldsymbol{b}^3 = (2, 1) \times (0, 2) = 2$$

Then

$$AB = \begin{bmatrix} 1 & 3 & 0 \\ 1 & 11 & 2 \end{bmatrix}$$

Remark that the product BA cannot be defined since the rows of B are triples and the columns of A are couples.

Example 14.4 Perform the product AB of matrices

$$A = \begin{bmatrix} 0 & 3 & 0 & 4 \\ 1 & 2 & 1 & -1 \\ 1 & 1 & 3 & 0 \end{bmatrix} \quad B = \begin{bmatrix} 0 \\ 3 \\ 1 \\ 0 \end{bmatrix}$$

The matrix A has size $(3, 4)$, the matrix B $(4, 1)$. The size of matrix AB is $(3, 1)$. The entries of AB are

$$c_{11} = \boldsymbol{a}_1 \times \boldsymbol{b}^1 = (0, 3, 0, 4) \times (0, 3, 1, 0) = 9$$
$$c_{21} = \boldsymbol{a}_2 \times \boldsymbol{b}^1 = (0, 2, 1, -1) \times (0, 3, 1, 0) = 7$$
$$c_{31} = \boldsymbol{a}_3 \times \boldsymbol{b}^1 = (1, 1, 3, 0) \times (0, 3, 1, 0) = 6$$

Therefore, the product is

$$AB = \begin{bmatrix} 9 \\ 7 \\ 6 \end{bmatrix}$$

If A and B are square matrices of order n, it makes sense to consider both the product AB and the product BA. In general, AB $\neq$ BA, i. e., the multiplication of the matrix A on the *right* by B does not give the same result as the multiplication A on the *left* by B.

Example 14.5 Let A and B be the square matrices of order 2

$$A = \begin{bmatrix} 1 & 1 \\ -1 & 0 \end{bmatrix} \quad B = \begin{bmatrix} 2 & 1 \\ 3 & 0 \end{bmatrix}$$

The product AB is different from BA; indeed

$$AB = \begin{bmatrix} 1 \cdot 2 + 1 \cdot 3 & 1 \cdot 1 + 1 \cdot 0 \\ -1 \cdot 2 + 0 \cdot 3 & -1 \cdot 1 + 0 \cdot 0 \end{bmatrix} = \begin{bmatrix} 5 & 1 \\ -2 & -1 \end{bmatrix}$$

$$BA = \begin{bmatrix} 2 \cdot 1 + 1(-1) & 2 \cdot 1 + 1 \cdot 0 \\ 3 \cdot 1 + 0 \cdot (-1) & 3 \cdot 1 + 0 \cdot 0 \end{bmatrix} = \begin{bmatrix} 1 & 2 \\ 3 & 3 \end{bmatrix}$$

This proves that square matrix multiplication is *not commutative*. Nevertheless, if A is a square matrix of order n and Id_n is the identity matrix of order n (see Sect. 13.8), then

$$A Id_n = Id_n A = A$$

Matrix multiplication satisfies the following properties:

a. The matrix multiplication is *associative*: if A, B, C are matrices such that the expression (AB)C makes sense, then also A(BC) makes sense and (AB)C = A(BC).

The following properties apply, provided that the operations are defined:

b. $A(B + C) = AB + AC$ (left distributive property);
c. $(B + C)A = BA + CA$ (right distributive property);
d. $k(AB) = (kA)B = A(kB)$, for every real number k;
e. [Binet's theorem] If A and B are square matrices of order n, then $\det(AB) = \det(A)\det(B)$.

14.6   Inverse and Transpose Matrices

Let us state the following property:

Proposition 14.1 *Let A be a square matrix of order n. If $|A| = 0$, there is no matrix B that fulfils any of the following equalities*

$$BA = Id_n, \quad AB = Id_n$$

Proposition 14.2 *If $|A| \neq 0$, the matrix A^{-1} defined by*

$$A^{-1} = \begin{bmatrix} \frac{A_{11}}{|A|} & \frac{A_{21}}{|A|} & \cdots & \frac{A_{n1}}{|A|} \\ \frac{A_{12}}{|A|} & \frac{A_{22}}{|A|} & \cdots & \frac{A_{n2}}{|A|} \\ \cdots & \cdots & \cdots & \cdots \\ \frac{A_{1n}}{|A|} & \frac{A_{2n}}{|A|} & \cdots & \frac{A_{nn}}{|A|} \end{bmatrix} \tag{14.3}$$

where A_{ij} is the cofactor of the entry a_{ij}, satisfies the equalities

$$A^{-1}A = Id_n \text{ and } AA^{-1} = Id_n \tag{14.4}$$

Definition 14.2 A square matrix A of order n is said to be an *invertible matrix* if there exists a matrix B such that

$$AB = BA = Id_n \tag{14.5}$$

Proposition 14.3 *If the matrix B satisfying (14.5) exists, it is unique. In fact, $AB_1 = B_1A = Id_n$ and $AB_2 = B_2A = Id_n$ imply, by the associative property (a),*

$$B_1 = B_1Id_n = B_1(AB_2) = (B_1A)B_2 = Id_nB_2 = B_2$$

To summarize, the inverse of the square matrix A, with $|A| \neq 0$, by Propositions 14.2 and 14.3, is unique and is the matrix (14.3), whose entry at the place ij is the ratio of the cofactor of a_{ji} to the determinant of A.

Example 14.6 The matrix

$$A = \begin{bmatrix} 1 & 0 \\ 2 & 2 \end{bmatrix}$$

has determinant $|A| = 2$. The inverse of A is

$$A^{-1} = \begin{bmatrix} \frac{A_{11}}{|A|} & \frac{A_{21}}{|A|} \\ \frac{A_{12}}{|A|} & \frac{A_{22}}{|A|} \end{bmatrix} = \begin{bmatrix} \frac{2}{2} & 0 \\ -\frac{2}{2} & \frac{1}{2} \end{bmatrix} = \begin{bmatrix} 1 & 0 \\ -1 & \frac{1}{2} \end{bmatrix}$$

Definition 14.3 Let A be an $m \times n$ matrix. The matrix B obtained from A by interchanging its rows and columns, i. e., the matrix B whose entries are defined by

$$b_{ij} = a_{ji}$$

is called the *transpose matrix* of A, or simply the *transpose* of A, denoted A^T.

Example 14.7 Consider the matrix

$$A = \begin{bmatrix} -1 & 2 & 3 & 4 \\ 0 & 1 & 0 & 1 \\ -5 & 6 & 7 & 8 \end{bmatrix}$$

The transpose matrix of A is the matrix

$$A^T = \begin{bmatrix} -1 & 0 & -5 \\ 2 & 1 & 6 \\ 3 & 0 & 7 \\ 4 & 1 & 8 \end{bmatrix}$$

Properties of the transpose

Let A be an $m \times n$ matrix. Let B be a matrix such that the following operations make sense. Then

a. $(A + B)^T = A^T + B^T$
b. $(A^T)^T = A$
c. $(AB)^T = B^T A^T$
d. $(hA)^T = hA^T$ where h any real number;
e. If A is a square matrix $\det(A) = \det(A^T)$.

14.7   Systems of Linear Equations and Matrices

The *matricial notation* allows to rewrite a system of linear equations in a more compact way. Given a system

$$\begin{aligned} a_{11}x_1 + a_{12}x_2 + \cdots + a_{1n}x_n &= b_1 \\ a_{21}x_1 + a_{22}x_2 + \cdots + a_{2n}x_n &= b_2 \\ \cdots\cdots\cdots \\ a_{m1}x_1 + a_{m2}x_2 + \cdots + a_{mn}x_n &= b_m \end{aligned} \tag{14.6}$$

it can be rewritten in the *matricial form*

$$\begin{bmatrix} a_{11} & a_{12} & \cdots & a_{1n} \\ a_{21} & a_{22} & \cdots & a_{2n} \\ \cdots & \cdots & \cdots & \cdots \\ a_{m1} & a_{m2} & \cdots & a_{mn} \end{bmatrix} \begin{bmatrix} x_1 \\ x_2 \\ \cdots \\ x_n \end{bmatrix} = \begin{bmatrix} b_1 \\ b_2 \\ \cdots \\ b_m \end{bmatrix} \tag{14.7}$$

equivalent to (14.6). Then, called A the matrix of the coefficients a_{ij}, b the vector of known coefficients b_i and x the vector of the unknowns x_j, the system (14.6) is rewritten in the form:

$$Ax = b \tag{14.8}$$

Obviously, the Eqs. (14.6)–(14.8) are equivalent to each other.

Theorem 14.1 *A compatible system* $Ax = b$ *that has more than one solution has an infinite number of solutions.*

Proof Let us show first that if c and d are distinct vectors of $\mathbf{R}^n$, then, for each pair of distinct real numbers h_1 and h_2, the vectors $c + h_1 (c - d)$ and $c + h_2 (c - d)$ are distinct. In fact, if by contradiction were $c + h_1(c - d) = c + h_2(c - d)$, then

$$h_1(c - d) = h_2(c - d), \quad \text{or} \quad (h_1 - h_2)(c - d) = O,$$

where O is the null vector. Therefore, it should be $c - d \neq O$, what contradicts the hypothesis. In conclusion, $c + h_1(c - d) \neq c + h_2(c - d)$.

Now, to complete the proof it suffices to show that if the system $Ax = b$ has more than one solution, then it has infinitely many of them. In fact, if c and d are distinct solutions of $Ax = b$, then $Ac = b$ and $Ad = b$. Therefore, for every real number h,

$$A(c + h(c - d)) = Ac + h(Ac - Ad) = b + h(b - b) = b$$

Thus, for each h, $c + h(c - d)$ is a solution of $Ax = h$. Since all such solutions are distinct, $Ax = b$ has infinitely many solutions. $\qquad\qquad \square$

14.8 Rank of a Matrix and Minors

Let

$$A = \begin{bmatrix} a_{11} & a_{12} & \cdots & a_{1n} \\ a_{21} & a_{22} & \cdots & a_{2n} \\ \cdots & \cdots & \cdots & \cdots \\ a_{m1} & a_{m2} & \cdots & a_{mn} \end{bmatrix} \tag{14.9}$$

be an $m \times n$ matrix, with $m > 1$ and $n > 1$. Moreover, let

$$A\left(i_1, i_2, \ldots, i_p; j_1, j_2, \ldots, j_p\right) = \begin{bmatrix} a_{i_1 j_1} & a_{i_1 j_2} & \cdots & a_{i_1 j_p} \\ a_{i_2 j_1} & a_{i_2 j_2} & \cdots & a_{i_2 j_p} \\ \cdots & \cdots & \cdots & \cdots \\ a_{i_p j_1} & a_{i_p j_2} & \cdots & a_{i_p j_p} \end{bmatrix}$$

be a square submatrix of A of order p *non maximal*, i.e., $p < m$ and $p < n$. Let a_i and a^j denote the i-th row and the j-th column of A.

Definition 14.4 We will call a *bordered minor* of the submatrix $A(i_1, i_2, \ldots, i_p; j_1, j_2, \ldots, j_p)$ of order p non maximal with the row a_i and the column a^j and denote it $|A(i_1, i_2, \ldots, i_p; j_1, j_2, \ldots, j_p)(i, j)|$, the following determinant of order $p + 1$

$$\begin{vmatrix} a_{i_1 j_1} & a_{i_1 j_2} & \cdots & a_{i_1 j_p} & a_{i_1 j} \\ a_{i_2 j_1} & a_{i_2 j_2} & \cdots & a_{i_2 j_p} & a_{i_2 j} \\ \cdots & \cdots & \cdots & \cdots & \cdots \\ a_{i_p j_1} & a_{i_p j_2} & \cdots & a_{i_p j_p} & a_{i_p j} \\ a_{i j_1} & a_{i j_2} & \cdots & a_{i j_p} & a_{i j} \end{vmatrix}$$

Obviously, if the order of the square submatrix $A(i_1, i_2, \ldots, i_p; j_1, j_2, \ldots, j_p)$ is $p = \min\{m, n\}$, the submatrix has no bordered minors.

Remark 14.1 While $A(i_1, i_2, \ldots, i_p; j_1, j_2, \ldots, j_p)$, square matrix of order p, is a submatrix of A, the square matrix $A(i_1, i_2, \ldots, i_p; j_1, j_2, \ldots, j_p)(i, j)$, which has order $p + 1$, may not be a submatrix of A. In fact, if $i < i_p$ or $j < j_p$, then the matrix $A(i_1, i_2, \ldots, i_p; j_1, j_2, \ldots, j_p)(i, j)$, is not a submatrix of A, while, if i and j are "intermediate", for example $A(i_1, i_2, \ldots, i, \ldots, i_p; j_1, j_2, \ldots, j, \ldots, j_p)$, with $i_1 < i_2 < \cdots < i < \cdots < i_p$ and $j_1 < j_2 < \cdots < j < \cdots < j_p$, is a square submatrix of A of order $p + 1$. Therefore, the determinants of the two matrices of order $p + 1$ are either both null or opposite to each other (Sect. 14.2, Property 2).

Theorem 14.2 *Let an $m \times n$ matrix A with $m > 1$ and $n > 1$ be given. Let $|A(i_1, i_2, \ldots, i_p; j_1, j_2, \ldots, j_p)|$ be a minor of A having order p non maximal. If*

$$\left| A(i_1, i_2, \ldots, i_p; j_1, j_2, \ldots, j_p) \right| \neq 0 \tag{14.10}$$

and

$$\left| A(i_1, i_2, \ldots, i_p; j_1, j_2, \ldots, j_p)(i, j) \right| = 0 \tag{14.11}$$

for every choice of i and j, with i distinct from $i_1, \ldots, i_p$, and j distinct from $j_1, \ldots, j_p$, then

$$S = \left[a_{i_1}, \ldots, a_{i_p} \right]$$

is a maximal system of linearly independent rows and

$$T = \left[a^{j_1}, \ldots, a^{j_p} \right]$$

is a maximal system of linearly independent columns of A.

Proof Let us show first that S is a maximal system of linearly independent rows of A (Sect. 12.8). We must prove that:

1. S is a linearly independent vector system, and
2. every row a_i of A is linearly dependent on the system S

In order to prove the statement (1) it suffices to verify that the matrix

$$A_n = \begin{bmatrix} a_{i_1 1} & \cdots & a_{i_1 n} \\ \cdots & \cdots & \cdots \\ a_{i_p 1} & \cdots & a_{i_p n} \end{bmatrix}$$

whose rows form S, has rank p (Sect. 13.3). Suppose, by contradiction, that A_n had rank less than p. Then also the matrix $A(i_1, i_2, \ldots, i_p; j_1, j_2, \ldots, j_p)$, which is a submatrix of A_n, should have rank less than p. Therefore, the rows of $A(i_1, i_2, \ldots, i_p; j_1, j_2, \ldots, j_p)$ should be linearly dependent, but this implies (Sect. 14.2) that the determinant $|A(i_1, i_2, \ldots, i_p; j_1, j_2, \ldots, j_p)|$ is equal to zero, what contradicts the hypothesis (14.10).

In order to prove the statement (2) we show that any row a_i of A, with i distinct from $i_1, i_2, \ldots, i_p$, is linearly dependent on the system S. To this aim, let us observe that, fixed such a value for i and for every $j = 1, \ldots, n$, we obtain:

$$\begin{vmatrix} a_{i_1 j_1} & a_{i_1 j_2} & \cdots & a_{i_1 j_p} & a_{i_1 j} \\ a_{i_2 j_1} & a_{i_2 j_2} & \cdots & a_{i_2 j_p} & a_{i_2 j} \\ \cdots & \cdots & \cdots & \cdots & \cdots \\ a_{i_p j_1} & a_{i_p j_2} & \cdots & a_{i_p j_p} & a_{i_p j} \\ a_{i j_1} & a_{i j_2} & \cdots & a_{i j_p} & a_{i j} \end{vmatrix} = 0 \tag{14.12}$$

In fact, equality (14.12) is verified by hypothesis (14.11) if j is distinct from $j_1, \ldots, j_p$; and (14.12) is also verified because two columns of the determinant are equal if j takes one of the values $j_1, j_2, \ldots, j_p$. By (14.12), performing the expansion of the determinant with respect to the last column, we get:

$$k_1 a_{i_1 j} + \cdots + k_p a_{i_p j} + k a_{ij} = 0 \tag{14.13}$$

where $k_1, \ldots, k_p, k$ are the cofactors of $a_{i_1 j}, \ldots, a_{i_p j}, a_{ij}$, respectively. As the cofactors $k_1, \ldots, k_p, k$ do not depend on j, (14.13) implies

$$k_1 a_{i_1} + \cdots + k_p a_{i_p} + k a_i = O$$

where O is the null n-tuple of $\mathbf{R}^n$. Hence, as $k = |A(i_1, i_2, \ldots, i_p; j_1, j_2, \ldots, j_p)| \neq 0$, it follows

$$a_i = -\frac{k_1}{k} a_{i_1} - \ldots - \frac{k_p}{k} a_{i_1}$$

Therefore, a_i depends linearly on S, as we wanted to show.

Similarly, or applying the result to the transpose of A, it can be proved that the system T is a maximal system of linearly independent columns of A. $\qquad\square$

From Theorem 13.3, the statement of the Theorem 14.2 can be expressed in the following form.

Theorem 14.3 *If the matrix A has a non-zero minor of order p non-maximal and, for any choice of i and j, with i distinct from $i_1, i_2, \ldots, i_p$ and j distinct from $j_1, j_2, \ldots, j_p$, the bordered minors of this minor with the i-th row and the j-th column, are null, then the matrix A has rank p.*

14.8.1 Matrix A Has a Non-zero Minor of Maximal Order

Let us suppose that the matrix A (14.9) has a non-zero minor of order $p = m$

$$|A(i_1, i_2, \ldots, i_m; j_1, j_2, \ldots, j_m)| \neq 0$$

By Theorem 13.2 the rows $\left[a_{i_1}, \ldots, a_{i_m} \right]$ form a maximal system of linearly independent rows of A which includes all the rows of A; for this reason $p = m$ is the rank of A. On the other hand, again by virtue of the mentioned theorem, we reach the same conclusion if $p = n$.

Therefore, the statement of theorem 14.3 is completed as follows.

Theorem 14.4 [Theorem of the bordered minors or Kronecker's theorem]. *If matrix A has a non-maximal non-zero minor of order p, $|A(i_1, i_2, \ldots, i_p; j_1, j_2, \ldots, j_m)|$, and, for any choice of i and j, with i distinct from $i_1, i_2, \ldots, i_p$ and j distinct from $j_1, j_2, \ldots, j_p$, the bordered minors of the minor $|A(i_1, i_2, \ldots, i_p; j_1, j_2, \ldots, j_m)|$, with the i-th row and the j-th column, are all null, then the matrix A has rank p. If the matrix A has a non-null minor of maximal order p, then A has rank p.*

14.8.2 Calculating the Rank of a Matrix Via Kronecker's Theorem

Kronecker's theorem comes in handy for computing the rank of a matrix A saving a lot of calculations: as a matter of fact, without Kronecker's theorem, in order to decide that the rank of A is p, once we have found a non-zero minor of A of order p non maximal, we should verify that all the possible minors of order $p + 1$ are zero. Instead, if we apply Kronecker's theorem, having ascertained that there exists a non-zero minor B of order p non maximal, we have to verify that only all the possible bordered minors of B are zero.

Let summarize the steps to compute the rank of a $m \times n$ matrix A:

- find in A a non-null minor B of order $p \geq 1$;
- if $p = \min\{m, n\}$, then the rank of A is equal to p;
- if $p < \min\{m, n\}$, compute all possible bordered minors of B;
- if all bordered minors of B are zero, then the rank of A is equal to p;

- if there exists a non-null bordered minor C of B of order $p + 1$, then the rank of A is greater than or equal to $p + 1$;
- repeat the procedure replacing B with C, i. e., B $\leftarrow$ C, and giving the name p to $p + 1$, i. e., $p \leftarrow p + 1$. (Regarding the symbol $\leftarrow$ see Sect. 11.2.1.)

Example 14.8 Compute the rank of the matrix

$$A = \begin{bmatrix} 1 & 1 & -1 & 1 \\ 3 & 3 & -3 & 5 \\ 0 & 0 & 0 & -1 \end{bmatrix}$$

The minor of the second order $|A(1, 2; 3, 4)| = \begin{vmatrix} -1 & 1 \\ -3 & 5 \end{vmatrix}$ is non-null, and its bordered minors are

$$|A(1, 2; 3, 4)(3; 1)| = \begin{vmatrix} -1 & 1 & 1 \\ -3 & 5 & 3 \\ 0 & -1 & 0 \end{vmatrix} = 0$$

$$|A(1, 2; 3, 4)(3; 2)| = \begin{vmatrix} -1 & 1 & 1 \\ -3 & 5 & 3 \\ 0 & -1 & 0 \end{vmatrix} = 0$$

Since all the bordered minors of $|A(1, 2; 3, 4)|$ are zero the rank of A is $p(A) = 2$.

Example 14.9 Compute the rank of the matrix

$$A = \begin{bmatrix} 1 & 2 & 3 & -1 & 0 \\ 1 & 2 & 0 & 6 & 1 \\ 0 & -1 & 2 & 0 & 0 \end{bmatrix}$$

Since $|A(1, 2; 1, 4)| = 7$ the rank of A is $p(A) \geq 2$. Let us compute the bordered minor:

$$|A(1, 2; 1, 4)(3; 3)| = 14$$

(while $|A(1, 2, 3; 1, 3, 4)| = -14$. Observe that in order to find the rank of A the two minors are both zero or both non-zero (see Remark 14.1). The rank of A is 3 because $|A(1, 2, 3; 1, 3, 4)|$ is a minor of A of maximal order 3.

14.9 Cramer's Rule

Let us consider a system of n linear equations and n unknowns

$$
\begin{aligned}
&a_{11}x_1 + a_{12}x_2 + \cdots + a_{1n}x_n = b_1 \\
&a_{21}x_1 + a_{22}h_2 + \cdots + a_{2n}h_n = b_2 \\
&\cdots\cdots\cdots \\
&a_{n1}x_1 + a_{n2}x_2 + \cdots + a_{nn}x_n = b_n
\end{aligned}
\tag{14.14}
$$

that can be rewritten in the matricial form

$$
\mathrm{A}x = b \tag{14.15}
$$

where A is a square matrix of order n, whose elements are a_{ij} $i = 1, ..., m; j = 1, ..., n$.

If $|\mathrm{A}| \neq 0$ by Theorem 14.4 the rank of A is n. Then the rank of the complete matrix of the system, which has n rows and $n + 1$ columns, is also n. By Theorem 13.3, the system is consistent.

Let us proceed to solve the system.

Rewrite the system in the form (14.15). Since $|\mathrm{A}| \neq 0$, the matrix A has the inverse A^{-1} (Sect. 14.6) and multiplying both sides of (14.15) to the left by A^{-1},

$$
\mathrm{A}^{-1}(\mathrm{A}x) = \mathrm{A}^{-1}b
$$

Therefore, the system (14.15) has the unique solution

$$
x = \mathrm{A}^{-1}b \tag{14.16}
$$

Expanding the right-hand side product (14.16) we have

$$
\begin{aligned}
x_1 &= \frac{A_{11}b_1 + A_{21}b_2 + \cdots + A_{n1}b_n}{|A|} \\
x_2 &= \frac{A_{12}b_1 + A_{22}b_2 + \cdots + A_{n2}b_n}{|A|} \\
&\cdots\cdots\cdots \\
x_n &= \frac{A_{1n}b_1 + A_{2n}b_2 + \cdots + A_{nn}b_n}{|A|}
\end{aligned}
\tag{14.17}
$$

We denote by Cj the matrix obtained from A by replacing the j-th column with the column of known coefficients. The expansion of $|Cj|$ along the j-th column is

$$
|Cj| = b_1 A_{1j} + b_2 A_{2j} + \ldots + b_n A_{nj}
$$

By (14.17) the solution of the system of equations is the n-tuple

$$\left(\frac{|C_1|}{|A|}, \frac{|C_2|}{|A|}, \dots, \frac{|C_n|}{|A|} \right) \tag{14.18}$$

Every linear system with the number of equations equal to the number of unknowns, whose associate matrix has a non-zero determinant is named a *Cramer's system*. The procedure for determining (14.18) the only solution of the system, is called *Cramer's rule*.

Remark 14.3 Cramer's rule also applies to the solution of a compatible system whose associate matrix A has rank p less than the number m of the equations. The rule is applied to a system whose coefficients of the unknowns form a square submatrix of order p of A, whose determinant is non-zero. The assigned system is equivalent to the system of p equations whose coefficients are the elements of the submatrix and only those whose coefficients are in the columns of the submatrix are considered unknowns of the system. The system then takes the form of Cramer, carrying $n - p$ unknowns (which will be considered parameters, or free variables) with the constants. The p unknowns will be expressed in function of the $n - p$ parameters.

14.9.1 Homogeneous Linear Systems

Consider a homogeneous linear system of m equations and n unknowns:

$$\begin{aligned}
a_{11}x_1 + a_{12}x_2 + \cdots + a_{1n}x_n &= 0 \\
a_{21}x_1 + a_{22}x_2 + \cdots + a_{2n}x_n &= 0 \\
&\cdots\cdots\cdots \\
a_{m1}x_1 + a_{m2}x_2 + \cdots + a_{mn}x_n &= 0
\end{aligned} \tag{14.19}$$

The system (14.19) is consistent. Indeed, the null vector of $\mathbf{R}^n$ is a solution of the system. The coefficients a_{ij} are sufficient to define the matrix of the homogeneous system since the complete matrix, obtained by adding the null column of the constants, does not matter.

Let p be the rank of the matrix of the homogeneous system. If $p = n$, the system has only the null solution; if $p < n$, by Theorem 14.1, the system has infinitely many solutions, since the values of $n - p$ unknowns can be fixed arbitrarily (Remark 14.3), and in this case the system also admits solutions other than the null one.

We conclude that a homogeneous linear system admits non-zero solutions if and only if the rank of its matrix is less than the number of unknowns.

The following property holds:

Proposition 14.4 *If p is the rank of the matrix of the homogeneous linear system, the set of solutions of the system is a subspace of dimension $n - p$ of the real vector space $\mathbf{R}^n$.*

Proof If $p = n$, the proposition is trivial because the set of solutions of the system consists of the only null vector of $\mathbf{R}^n$. If $p < n$, it is easy to see that the sum of two solution vectors, however fixed, is a solution of the system and that the product of any scalar by any solution vector is still a solution of the system. By Proposition 12.5 the set of solutions of the homogeneous system is a subspace of the real vector space $\mathbf{R}^n$. To find the dimension of such a subspace, recalling Remark 14.3, the system can be put in the form of p equations, p unknowns and $n - p$ parameters to which attribute real values. Each $(n - p)$-tuple of parameters identifies one and only one solution and the application that associates to each solution of the homogeneous system the vector of $n - p$ components that determines this solution is an isomorphism of the solution set of the system onto the set $\mathbf{R}^{n-p}$ of the $(n - p)$-tuples of real parameters. By Proposition 12.13, the vector space of the solutions of the homogeneous system has dimension $n - p$.   □

14.9.2   Associated Homogeneous Linear System

Let the linear system of linear equations

$$
\begin{aligned}
a_{11}x_1 + a_{12}x_2 + \cdots + a_{1n}x_n &= b_1 \\
a_{21}x_1 + a_{22}x_2 + \cdots + a_{2n}x_n &= b_2 \\
\cdots\cdots\cdots \\
a_{m1}x_1 + a_{m2}x_2 + \cdots + a_{mn}x_n &= b_m
\end{aligned}
\tag{14.20}
$$

be given. The linear system (14.19) is called the *homogeneous linear system associated* to the system (14.20).

Theorem 14.5 *If the system (14.20) is compatible, then the vector h is a solution of the system (14.20) if and only if*

$$
h = h' + h_0
$$

where h' is a fixed solution of (14.20) and h_0 a solution of the homogeneous system associated with (14.20).

Proof Let

$$
h = (h_1, \ldots, h_n)
$$

be a solution of the system (14.20),

$$
h' = \left(h'_1, \ldots h'_n\right)
$$

a fixed solution of the system (14.20). Then the following equalities hold:

$$a^1 h_1 + \cdots + a^n h_n = b$$
$$a^1 h_1' + \cdots + a^n h_n' = b$$

where b is the vector of the known coefficients of (14.20). Subtracting, we get

$$a^1 (h_1 - h_1') + \cdots + a^n (h_n - h_n') = O$$

Then the vector

$$h = h' + h_0$$

is a solution of the homogeneous linear system associated to (14.20), and hence the vector

$$h = h' + h_0$$

is the sum of h' and a solution of the homogeneous linear system associated to (14.20).

Vice versa, if $h_0 = (h_{01}, \ldots, h_{0n})$ is a solution of the homogeneous linear system associated to (14.20), then

$$a^1 h_{01} + \cdots + a^n h_{0n} = O$$

and if $h' = (h_1', \ldots, h_n')$ is a solution of (14.20), then $a^1 h_1' + \cdots + a^n h'n = b$.

By adding, we obtain $h = h' + h_0$ which is a solution of (14.20).   □

14.10 Exercises

1. The system of three equations and three unknowns

$$x - y - z = 1$$
$$2x + z = 0$$
$$x - y = 2$$

is a Cramer's system; in fact the determinant

$$\begin{vmatrix} 1 & -1 & -1 \\ 2 & 0 & 1 \\ 1 & -1 & 0 \end{vmatrix} = -2 \begin{vmatrix} -1 & -1 \\ -1 & 0 \end{vmatrix} - 1 \begin{vmatrix} 1 & -1 \\ -1 & -1 \end{vmatrix} = 2$$

is non-zero. Then the system has only one solution which can be found by Cramer's rule:

$$x = \frac{\begin{vmatrix} 1 & -1 & -1 \\ 0 & 0 & 1 \\ 2 & -1 & 0 \end{vmatrix}}{|A|} = \frac{-\begin{vmatrix} 1 & -1 \\ 2 & -1 \end{vmatrix}}{2} = -\frac{1}{2}$$

$$y = \frac{\begin{vmatrix} 1 & 1 & -1 \\ 2 & 0 & 1 \\ 1 & 2 & 0 \end{vmatrix}}{|A|} = \frac{-2\begin{vmatrix} 1 & -1 \\ 2 & 0 \end{vmatrix} - \begin{vmatrix} 1 & 1 \\ 1 & 2 \end{vmatrix}}{2} = -\frac{5}{2}$$

$$z = \frac{\begin{vmatrix} 1 & -1 & 1 \\ 2 & 0 & 0 \\ 1 & -1 & 2 \end{vmatrix}}{|A|} = \frac{-2\begin{vmatrix} -1 & 1 \\ -1 & 2 \end{vmatrix}}{2} = 1$$

The triple $\left(-\frac{1}{2}, -\frac{5}{2}, 1\right)$ is the solution of the given system.

2. The systems

$$x - y - z = 0$$
$$x - y + 2z = 0$$

of two equations and three unknowns is homogeneous, and then compatible. The matrix of coefficients is

$$A = \begin{bmatrix} 1 & -1 & -1 \\ 1 & -1 & 2 \end{bmatrix}$$

The rank of A is $p(A) = 2$ for there exists a minor of A, $\begin{vmatrix} 1 & -1 \\ 1 & 2 \end{vmatrix} = 3$. Let us call to mind Remark 14.3 and consider as parameter the unknown y; the system is rewritten in the form

$$x - z = y$$
$$x + 2z = y$$

that is a Cramer's system of two equations and unknowns x and z. The parameter y is considered as a constant:

$$x = \frac{\begin{vmatrix} y & -1 \\ y & 2 \end{vmatrix}}{\begin{vmatrix} 1 & -1 \\ 1 & 2 \end{vmatrix}} = \frac{3y}{3} = y$$

$$z = \dfrac{\begin{vmatrix} 1 & y \\ 1 & y \end{vmatrix}}{\begin{vmatrix} 1 & -1 \\ 1 & 2 \end{vmatrix}} = 0$$

Therefore, by Theorem 14.1 the given system has infinitely many solutions that are the triples (x, y, z) such that $x = y$ and $z = 0$, i. e., the triples $(y, y, 0)$, for every real number y. For example, some solutions of the system are the triples $(0, 0, 0)$, $(1, 1, 0)$, $(-5, -5, 0)$.

3. Decide whether the system

$$2x + y - z + t = 1$$
$$3x + y - 2z + t = 2$$
$$x + z = 1$$

is compatible and, if so, find the solutions.

The matrix of the coefficients, and the complete matrix are

$$A = \begin{bmatrix} 2 & 1 & -1 & 1 \\ 3 & 1 & -2 & 1 \\ 1 & 0 & 1 & 0 \end{bmatrix} \qquad A' = \begin{bmatrix} 2 & 1 & -1 & 1 & 1 \\ 3 & 1 & -2 & 1 & 2 \\ 1 & 0 & 1 & 0 & 1 \end{bmatrix}$$

To compute the rank of A, observe that the minor

$$|A(1, 2; 1, 2)| = \begin{vmatrix} 2 & 1 \\ 3 & 1 \end{vmatrix} = -1$$

is non-zero. Then the rank $p(A)$ of A satisfies the inequality $p(A) \geq 2$. Thus we call upon the Kronecker's theorem: the value of the bordered minor of $A(1, 2; 1, 2)$ with third row and third column is

$$|A(1, 2; 1, 2)(3; 3)| = \begin{bmatrix} 2 & 1 & -1 \\ 3 & 1 & -2 \\ 1 & 0 & 1 \end{bmatrix} = \begin{vmatrix} 1 & -1 \\ 1 & -3 \end{vmatrix} + \begin{vmatrix} 2 & 1 \\ 3 & 1 \end{vmatrix} = -2$$

Since the bordered minor $|A(1, 2; 1, 2)(3;3)|$ has maximal order 3, then $p(A) = 3$.

Evidently, also $p(A') = 3$. By Rouché-Capelli's theorem the system is compatible. Furthermore, the value $p(A) = p(A') = 3$ is less than the number of unknowns. Therefore, the system has infinite solutions.

Find the solutions of the system by means of Cramer's rule. Since $|A (1, 2; 1, 2) (3; 3)|$ is different from zero, put the system in the form

$$2x + y - z = 1 - t$$
$$3x + y - 2z = 2 - t$$
$$x + z = 1$$

where t is considered a parameter and move it to the right-hand side. Hence the system has three unknowns x, y, z and its matrix of the coefficients is A (1, 2; 1, 2) (3; 3) whose determinant has value -2.

Apply Cramer's rule to find the solutions. (Choose the expansions along the third rows.)

$$x = \frac{\begin{vmatrix} 1-t & 1 & -1 \\ 2-t & 1 & -2 \\ 1 & 0 & 1 \end{vmatrix}}{\begin{vmatrix} 2 & 1 & -1 \\ 3 & 1 & -2 \\ 1 & 0 & 1 \end{vmatrix}} = \frac{\begin{vmatrix} 1 & -1 \\ 1 & -2 \end{vmatrix} + \begin{vmatrix} 1-t & 1 \\ 2-t & 1 \end{vmatrix}}{-2} = \frac{-1+1-t-(2-t)}{-2} = 1$$

$$y = \frac{\begin{vmatrix} 2 & 1-t & -1 \\ 3 & 2-t & -2 \\ 1 & 1 & 1 \end{vmatrix}}{-2} = \frac{\begin{vmatrix} 1-t & -1 \\ 2-t & -2 \end{vmatrix} - \begin{vmatrix} 2 & -1 \\ 3 & -2 \end{vmatrix} + \begin{vmatrix} 2 & 1-t \\ 3 & 2-t \end{vmatrix}}{-2} = -1 - t$$

$$z = \frac{\begin{vmatrix} 2 & 1 & 1-t \\ 3 & 1 & 2-t \\ 1 & 0 & 1 \end{vmatrix}}{-2} = \frac{\begin{vmatrix} 1 & i-t \\ 1 & 2-t \end{vmatrix} + \begin{vmatrix} 2 & 1 \\ 3 & 1 \end{vmatrix}}{-2} = 0$$

The unknowns x, y, z are expressed as functions of the parameter t. All the solutions of the given system are the 4-tuples $(1, -1 - t, 0, t)$, for every real number t.

4. The system of linear equations

$$x - y + z - t = 0$$
$$x - y + 2z + t = 0 \qquad (14.21)$$
$$3x - 3y + 4z - t = 0$$

is homogeneous and then compatible. Let us state first if it has only one solution or infinite solutions. The matrix of the coefficients

$$A = \begin{vmatrix} 1 & -1 & 1 & -1 \\ 1 & -1 & 2 & 1 \\ 3 & -3 & 4 & -1 \end{vmatrix}$$

has rank $p(A) \geq 2$ because

$$|A(1, 2; 1, 3)| = \begin{vmatrix} 1 & 1 \\ 1 & 2 \end{vmatrix} = 1 \neq 0$$

and the bordered minors of $A(1, 2; 1, 3)$ are

$$\begin{vmatrix} 1 & 1 & -1 \\ 1 & 2 & -1 \\ 3 & 4 & -3 \end{vmatrix} = 0 \quad \begin{vmatrix} 1 & 1 & -1 \\ 1 & 2 & 1 \\ 3 & 4 & -1 \end{vmatrix} = 0$$

There are no more non-null bordered minors of $A(1, 2; 1, 3)$ of the third order.

1. the first two equations are independent and the third is linearly dependent on them;
2. x and z are the formally unknowns and y and t are, as a consequence, to be considered as parameters.

So the system (14.21) has an infinite number of solutions and is equivalent to the system:

$$x - y + z - t = 0$$
$$x - y + 2z + t = 0$$

By statements 1 and 2 let us put the system in the form:

$$x + z = y + t$$
$$x + 2z = y - t$$

that is a Cramer's system that yields:

$$x = \begin{vmatrix} y+t & 1 \\ y-t & 2 \end{vmatrix} = y + 3t, \quad z = \begin{vmatrix} 1 & y+t \\ 1 & y-t \end{vmatrix} = -2t,$$

The set of solutions of the system is the set of 4-tuples $(x, y, z, t) = (y + 3t, y, -2t, t)$, for every y and t in $\mathbf{R}$. For example, if $y = 1$ and $t = 0$ we obtain the solution $(1, 1, 0, 0)$.

5. Find the values of the variable k, such that the system

$$x - ky = 1$$
$$4x + ky = 0$$
$$2x + 3y = -2k$$

is compatible and find the solutions. By Rouché-Capelli's theorem the system is compatible if and only if the matrices

$$A = \begin{bmatrix} 1 & -k \\ 4 & k \\ 2 & 3 \end{bmatrix} \quad A' = \begin{bmatrix} 1 & -k & 1 \\ 4 & k & 0 \\ 2 & 3 & -2k \end{bmatrix}$$

have the same rank. The rank of A is not greater than 2. If the rank of A' is 3, then the system is incompatible. Hence a necessary condition for the compatibility of the system is that $|A'| = 0$, namely

$$\begin{vmatrix} 1 & -k & 1 \\ 4 & k & 0 \\ 2 & 3 & -2k \end{vmatrix} = -10k^2 - 2k + 12 = 0$$

The roots of this equation (Sect. 12.1) are $k = 1$ and $k = -\frac{6}{5}$.
Now let's see what happens if $k = 1$ or $k = -\frac{6}{5}$:

i. $k = 1$. The matrix A has rank 2 because if $k = 1$ the minor

$$\begin{vmatrix} 1 & -k \\ 4 & k \end{vmatrix} = \begin{vmatrix} 1 & -1 \\ 4 & 1 \end{vmatrix} = 5$$

is non-zero. The matrix A', whose determinant is zero when $k = 1$, also has rank 2. The system is compatible and is equivalent to the system of two equations

$$x - y = 1$$
$$4x + y = 0$$

This is a Cramer's system. Its solution is the couple (x, y), where:

$$x = \frac{\begin{vmatrix} 1 & -1 \\ 0 & 1 \end{vmatrix}}{\begin{vmatrix} 1 & -1 \\ 4 & 1 \end{vmatrix}} = \frac{1}{5} \text{ and } y = \frac{\begin{vmatrix} 1 & 1 \\ 4 & 0 \end{vmatrix}}{\begin{vmatrix} 1 & -1 \\ 4 & 1 \end{vmatrix}} = \frac{4}{5}$$

ii. $k = -\frac{6}{5}$. The matrix A has rank 2 since

$$\begin{vmatrix} 1 & -k \\ 4 & k \end{vmatrix} = \begin{vmatrix} 1 & \frac{6}{5} \\ 4 & -\frac{6}{5} \end{vmatrix} = -6$$

and the matrix A' has rank 2. The system is compatible and equivalent to the Cramer's system

$$x + \frac{6}{5}y = 1$$
$$4x - \frac{6}{5}y = 0$$

that has the unique solution $\left(\frac{1}{5}, \frac{2}{3}\right)$.

6. Consider the homogeneous system with two non-identical equations and unknowns x, y, z:

$$ax + by + cz = 0$$
$$a'x + b'y + c'z = 0$$

Since the equations are not identical the two triples of coefficients (a, b, c) and (a', b', c') are nonnull. The system is compatible because it is homogeneous. Let the rank of the matrix be 1. Then the system is equivalent to one of its equations, for example $ax + by + cz = 0$. Therefore, the system has infinite solutions. For example, all the solutions of the system

$$x - y + z = 0$$
$$-2x + 2y + 2z = 0$$

are the triples of real numbers $(y - z, y, z)$, with y and z real numbers.

Suppose that the system

$$ax + by + cz = 0$$
$$a'x + b'y + c'z = 0$$

has rank 2, i. e., the matrix

$$A = \begin{bmatrix} a & b & c \\ a' & b' & c' \end{bmatrix}$$

has a non-zero minor of order 2, for example,

$$\begin{vmatrix} a & b \\ a' & b' \end{vmatrix} \neq 0$$

Let z be as a free variable

$$ax + by = -cz$$
$$a'x + b'y = c'z$$

It is possible to apply Cramer's rule:

$$x = \frac{\begin{vmatrix} -cz & b \\ -c'z & b' \end{vmatrix}}{\begin{vmatrix} a & b \\ a' & b' \end{vmatrix}} = \frac{\begin{vmatrix} -c & b \\ -c' & b' \end{vmatrix}}{\begin{vmatrix} a & b \\ a' & b' \end{vmatrix}} z$$

$$y = \frac{\begin{vmatrix} a & -cz \\ a' & -c'z \end{vmatrix}}{\begin{vmatrix} a & b \\ a' & b' \end{vmatrix}} = \frac{\begin{vmatrix} c & a \\ c' & a' \end{vmatrix}}{\begin{vmatrix} a & b \\ a' & b' \end{vmatrix}} z$$

Hence, setting

$$h = \begin{vmatrix} b & c \\ b' & c' \end{vmatrix}, \quad k = \begin{vmatrix} c & a \\ c' & a' \end{vmatrix}, \quad l = \begin{vmatrix} a & b \\ a' & b' \end{vmatrix}$$

the solutions of the system are the triples

$$\left(\frac{h}{l}z, \frac{k}{l}z, z \right) = \left(\frac{h}{l}, \frac{k}{l}, 1 \right) z$$

for every $z \in \mathbf{R}$. Any two of these non-zero triples are proportional to each other. If we set $z = l$, a solution of the system is (h, k, l). It follows that the set of non-zero solutions of the system are, the triples proportional to (h, k, l), i. e., the minors of the second order, extracted from the matrix A with alternate signs:

$$\begin{vmatrix} b & c \\ b' & c' \end{vmatrix}, \begin{vmatrix} c & a \\ c' & a' \end{vmatrix}, \begin{vmatrix} a & b \\ a' & b' \end{vmatrix}$$

The result is generalized to the homogeneous systems of n linear equations and $n + 1$ unknowns, having rank n: all the non-zero solutions of such a system are proportional to the n-tuple of order n minors of the matrix of the system, obtained by deleting the first, second, ..., the n-th column and taken with alternate signs.

7. Let the planes α and β be given:

$$\alpha)\ x - y + z - 1 = 0, \quad \beta)\ 2y - z + 4 = 0 \tag{14.22}$$

The planes are non-parallel (Sect. 10.4) and then they have a line r in common, represented by the system of Eqs. (14.22). The line s parallel to r and passing through the origin of the coordinates $O(0, 0, 0)$ has equations

$$x - y + z = 0$$
$$2y - z = 0$$

In fact, each of the planes representing s passes through the origin of the coordinates and is parallel to a plane of the representation (14.22) of r.

Let us find a triple of direction numbers of r. If $P(x, y, z)$ is a point of s distinct from $O(0, 0, 0)$, the vector $OP = (x, y, z)$ is parallel to the line r. Since the point of coordinates $(-1, 1, 2)$ belongs to s, then $(-1, 1, 2)$ is a triple of direction numbers of r.

It is easy to generalize. The direction numbers of the line r of equations

$$ax + by + cz + d = 0$$
$$a'x + b'y + c'z + d' = 0$$

are the direction numbers of the line s of equations

$$ax + by + cz = 0$$
$$a'x + b'y + c'z = 0$$

and vice versa. Then a triple of direction numbers of s coincides with a non-zero solution of the system representing s whose non-zero solutions, i. e., the coordinates of the points of s distinct from the origin O $(0, 0, 0)$, are proportional to second order minors (see Exercise 14.6)

$$\begin{vmatrix} b & c \\ b' & c' \end{vmatrix}, \begin{vmatrix} c & a \\ c' & a' \end{vmatrix}, \begin{vmatrix} a & b \\ a' & b' \end{vmatrix}$$

of the matrix $\begin{bmatrix} a & b & c \\ a' & b' & c' \end{bmatrix}$.

8. Solve the following problems

a. State if the following vectors of $\mathbf{R}^4$ are linearly dependent or independent:

$$(-1, -2, 1, 1), (0, 2, 1, 0), (1, 0, -2, -2);$$

b. find at least one non-null vector belonging to the subspace generated by the system

$$T = [(-1, -2, 1, 1), (0, 2, 1, 0), (1, 0, -2, -2)];$$

c. find the set of non-null vectors of $\mathbf{R}^4$ orthogonal to the vectors of T.

Solution

a. The vectors are linearly independent if the unique linear combination equal to the null vector has the coefficients all null. Examine the equation:

$$h_1[(-1, -2, 1, 1) + h_2(0, 2, 1, 0) + h_3(1, 0, -2, -2) = (0, 0, 0, 0).$$

Perform the products and equate the components in the same places. The following linear system of four equations and three unknowns is obtained:

$$-h_1 + h_3 = 0$$
$$-2h_1 + 2h_2 = 0$$
$$h_1 + h_2 - 2h_3 = 0$$
$$h_1 - 2h_3 = 0$$

The matrix of the coefficients

$$\begin{bmatrix} -1 & 0 & 1 \\ -2 & 2 & 0 \\ 1 & 1 & -2 \\ 1 & 0 & -2 \end{bmatrix}$$

has rank 3, since there is a minor non-null of the third (maximum) order:

$$\begin{vmatrix} -1 & 0 & 1 \\ 1 & 1 & -2 \\ 1 & 0 & -2 \end{vmatrix} = 1$$

The system is compatible, has the unique solution $(0, 0, 0)$ and the vectors of T are linearly independent.

b. It suffices to consider any linear combination of the three vectors of T with coefficients h_1, h_2, h_3 not all null. For example, choose $h_1 = 0, h_2 = -1, h_3 = 1$ and perform the linear combination:

$$0(-1, -2, 1, 1) - 1(0, 2, 1, 0) + (1, 0, -2, -2)$$
$$= (0, -2, -1, 0) + (1, 0, -2, -2) = (1, -2, -3, -2),$$

which is a non-null vector belonging to the subspace generated by the system T.

c. A vector $(x_1, x_2, x_3, x_4) \in \mathbf{R}^4$ is orthogonal to each of three given vectors if the following condition are fulfilled (remind that two vectors are orthogonal if the scalar product is null. (See Sect. 12.10):

$$(x_1, x_2, x_3, x_4) \times (-1, -2, 1, 1) = 0$$
$$(x_1, x_2, x_3, x_4) \times (0, 2, 1, 0) = 0$$
$$(x_1, x_2, x_3, x_4) \times (1, 0, -2, -2) = 0$$

Let us perform the products:

$$-x_1 - 2x_2 + x_3 + x_4 = 0$$
$$2x_2 + x_3 = 0$$
$$x_1 - 2x_3 - 2x_4 = 0$$

The matrix of the coefficients has a minor non-null of maximum order 3:

$$\begin{vmatrix} -2 & 1 & 1 \\ 2 & 1 & 0 \\ 0 & -2 & -2 \end{vmatrix} = 4$$

Then system of equations is compatible (see Sect. 13.5). In order to find the solutions consider $x_1 = m =$ parameter:

$$-2x_2 + x_3 + x_4 = m$$
$$2x_2 + x_3 = 0$$
$$-2x_3 - 2x_4 = -m$$

and apply Cramer's rule (see Sect. 14.9):

$$x_2 = \frac{\begin{vmatrix} m & 1 & 1 \\ 0 & 1 & 0 \\ -m & -2 & -2 \end{vmatrix}}{4} = -\frac{m}{4}$$

$$x_3 = \frac{\begin{vmatrix} -2 & m & 1 \\ 2 & 0 & 0 \\ 0 & -m & -2 \end{vmatrix}}{4} = \frac{m}{2}$$

$$x_4 = \frac{\begin{vmatrix} -2 & 1 & m \\ 2 & 1 & 0 \\ 0 & -2 & -m \end{vmatrix}}{4} = 0$$

The general solution consists of the set of 4-tuples $\left(m, -\frac{m}{4}, \frac{m}{2}, 0\right)$, for every $m \in \mathbf{R}$. As a result, for every non zero m, any vector $\left(m, -\frac{m}{4}, \frac{m}{2}, 0\right)$ is orthogonal to each of the three vectors of T.

9. Solve the following problems

   a. Find the rank of the matrix

   $$A = \begin{bmatrix} 1 & 2 & -1 & 1 \\ t & 0 & 1 & 1 \\ 1 & t+1 & 0 & t+1 \end{bmatrix}$$

   in function of the parameter $t \in \mathbf{R}$.

   b. Consider the rows of A for $t = 0$ and find a linear combination of them that equals the vector $(2, 5, -1, 4)$.

Solution

a. In order to calculate the rank of a matrix follow the procedure (see Sect. 14.8.2.):

   - There is in A a non-null minor of order 2:

$$|B| = \begin{vmatrix} 2 & -0 \\ 0 & 1 \end{vmatrix} = 20$$

Then the rank of A is $p \geq 2$, for every value of t.

   - As $p < \min\{3, 4\}$, compute all possible bordered minors of B. There are two bordered minors of B:

$$|C| = \begin{vmatrix} 1 & 2 & -1 \\ t & 0 & 1 \\ 1 & t+1 & 0 \end{vmatrix} = -t \begin{vmatrix} 2 & -1 \\ t+1 & 0 \end{vmatrix} - 1 \begin{vmatrix} 1 & 2 \\ 1 & t+1 \end{vmatrix}$$
$$= -t^2 - 2t + 1,$$

where $|C| = 0$ if and only if $t = -1 \pm \sqrt{2}$, and

$$|D| = \begin{vmatrix} 2 & -1 & 1 \\ 0 & 1 & 1 \\ t+1 & 0 & t+1 \end{vmatrix} = 0$$

for every $t \in \mathbf{R}$.

   Therefore, if $t = -1 + \sqrt{2}$ and $t = -1 - \sqrt{2}$ all the bordered minors of $|B|$ vanish and by the Kronecker's theorem, the rank of A is equal to 2; if $t \neq -1 \pm \sqrt{2}$, then $|C| \neq 0$ and the rank of A is equal to 3.

b. If $t = 0$, matrix A becomes:

$$A = \begin{bmatrix} 1 & 2 & -1 & 1 \\ 0 & 0 & 1 & 1 \\ 1 & 1 & 0 & 1 \end{bmatrix}$$

In order to express the vector $(2, 5, -1, 4)$ as a linear combination of the rows of A, it is necessary to determine coefficients a, b and c such that

$$(2, 5, -1, 4) = a(1, 2, -1, 1) + b(0, 0, 1, 1) + c(1, 1, 0, 1)$$

$$(2, 5, -1, 4) = (a, 2a, -a, a) + (0, 0, b, b) + (c, c, 0, c)$$

$$(2, 5, -1, 4) = (a + c, 2a + c, -a + b, a + b + c)$$

Equating the components,

$$a + c = 2$$
$$2a + c = 5$$
$$-a + b = -1$$
$$a + b + c = 4$$

and substituting,

$$c = 2 - a$$
$$2a + c = 5$$
$$-a + b = -1$$
$$a + b + c = 4$$

Observe that the last equation can be cancelled as it is the sum of the second and third. So we obtain:

$$c = -1$$
$$a = 3$$
$$b = 2$$

Therefore, $(a, b, c) = (3, 2, -1)$ is the unique solution. As a consequence, the required linear combination is

$$(2, 5, -1, 4) = 3(1, 2, -1, 1) + 2(0, 0, 1, 1) - (1, 1, 0, 1).$$

Bibliography

Aitken, A.C.: Determinants and matrices. Oliver and Boyd LTD, Edinburgh (1965)

Lang, S.: Linear algebra. Addison-Wesley Publishing Company, Reading Massachusetts (1966)

Lipschutz, S.: Theory and problems of linear algebra. McGraw-Hill Book Company, New York (1968)

Schwartz, J.T.: Introduction to matrices and vectors. Dover Publications Inc., New York (1971)

Strang, G.: Linear algebra and applications. Academic Press Inc., New York (1976)

Ventre, A.: Algebra lineare e algoritmi. Zanichelli, Bologna (2021)

Chapter 15
Lines and Planes

15.1 Introduction

We will study some applications of the linear algebra to analytic geometry; in particular, we will apply the results on systems of linear equations to the study of the reciprocal positions concerning lines, planes and vectors in the space. We will extend to the space some concepts developed in a coordinate system of the plane xy. The geometric problems are solved from both the synthetic point of view, based on the Euclidean axioms (Chap. 4), and the analytic point of view, by means of equations.

Problems of parallelism and perpendicularity concerning lines and planes, coplanarity, intersections of planes and lines, bundles of planes are solved.

15.2 Parallel Lines

Let us consider the lines r and s of equations in the coordinate system of the space xyz (Sect. 10.5)

$$r) \quad \begin{aligned} x &= x_1 + mt \\ y &= y_1 + nt \\ z &= z_1 + pt \end{aligned}$$

$$s) \quad \begin{aligned} x &= x_2 + m't \\ y &= y_2 + n't \\ z &= z_2 + p't \end{aligned}$$

Direction numbers of r and s are the triples proportional to (m, n, p) and (m', n', p'), respectively. Like in the plane geometry the following statement holds.

Property 1 [Parallelism of lines]. The lines r and s are parallel if and only if the triples of respective direction numbers are proportional.

© The Author(s), under exclusive license to Springer Nature Switzerland AG 2023 243
A. G. S. Ventre, *Calculus and Linear Algebra*,
https://doi.org/10.1007/978-3-031-20549-1_15

Example 15.1 The line s passing through the point P $(3, 2, -5)$ and parallel to the line r of parametric equations

$$r) \quad \begin{aligned} x &= -1 - 4t \\ y &= 7 + 9t \\ z &= t \end{aligned}$$

has direction numbers proportional, in particular equal, to $(-4, 9, 1)$. Then s has parametric equations

$$s) \quad \begin{aligned} x - 3 &= -4t \\ y - 2 &= 9t \\ z + 5 &= t \end{aligned}$$

15.3   Coplanar Lines and Skew Lines

Let us consider the lines r and s:

$$r) \quad \begin{aligned} x &= x_1 + mt \\ y &= y_1 + nt \\ z &= z_1 + pt \end{aligned}$$

$$s) \quad \begin{aligned} x &= x_2 + m't \\ y &= y_2 + n't \\ z &= z_2 + p't \end{aligned}$$

One of the two following cases occurs (see Chap. 4):

1.   the lines are *coplanar*,
2.   the lines are *skew*.

If the lines r and s are coplanar, then

- either the lines are *parallel*, in particular coincident, and the respective triples of direction numbers are proportional,
- or they are *incident* at a single point; then the system of the six equations of r and s and the unknowns t, x, y, z is compatible.

If the lines are skew, however two distinct points in r and two distinct points in s are taken, there is no plane that contains the four points; the system of six equations of r and s and the unknowns t, x, y, z is incompatible and the two triples of direction numbers of r and s are not proportional.

Observe that also the system of equations of two distinct and parallel lines is incompatible, but the triples of direction numbers are proportional.

15.4   Line Parallel to Plane. Perpendicular Lines. Perpendicular Planes

Let the plane α

$$\alpha) \quad ax + by + cz + d = 0$$

and the line

$$r) \quad \begin{array}{l} x = x_1 + mt \\ y = y_1 + nt \\ z = z_1 + pt \end{array}$$

be given.

One of the two cases occurs:

1. The plane and the line are *incident in a unique point*. Then the system of the four equations of α and r and the unknowns x, y, z and t has a unique solution.
2. The plane and the line are *parallel*. Then

   2.1   either the line lies on the plane (the line and the plane are *improperly parallel*); the system of equations of α and r has infinitely many solutions,
   2.2   or the line and the plane have no point in common (the line and the plane are *properly parallel*); the system of equations of α and r is incompatible.

Let us plug the right-hand sides of the equations of r in the equation of α. We obtain an equation in the unknown t:

$$(am + bn + cp)t + ax_1 + by_1 + cz_1 + d = 0 \qquad (15.1)$$

We can draw some conclusions from the inspection of Eq. (15.1). If the coefficient of t is zero

$$am + bn + cp = 0 \qquad (15.2)$$

then the line and the plane are improperly parallel or properly parallel, according to whether $ax_1 + by_1 + cz_1 + d = 0$ (when (15.1) is the identical equation), or $ax_1 + by_1 + cz_1 + d \neq 0$ (when (15.1) is the incompatible equation). Therefore, we obtain:

Property 2 [Parallelism of a line and a plane]. The line r and the plane α are parallel if and only if

$$am + bn + cp = 0$$

Since every line s with direction numbers (a, b, c) is perpendicular to the plane $\alpha)$ $ax + by + cz + d = 0$ (Sect. 10.4, any line r parallel to α, will be perpendicular to s. Therefore,

Property 3 [Perpendicularity of lines]. A line with direction numbers (m, n, p) and a line with direction numbers (m', n', p') are perpendicular if and only if $mm' + nn' + pp' = 0$.

As the coefficients of the equation of the plane are proportional to the direction numbers of any perpendicular line, we obtain:

Property 4 [Perpendicular planes]. The planes of equations $ax + by + cz + d = 0$ and $a'x + b'y + c'z + d' = 0$ are perpendicular if and only if $aa' + bb' + cc' = 0$.

Given in the space xyz a point and a plane there are infinitely many lines that pass through the point and are parallel to the plane.

Exercise 15.1 Find the equations of the lines parallel to the plane

$$\alpha)\quad x - y + 2z = 0$$

and passing through the point A$(-1, 0, 4)$.

Solution We observe that there exists a set of the required lines that form a bundle of lines (Chap. 4). By (15.2) the direction numbers of any line parallel to the plane α are the triples (m, n, p) such that $m - n + 2p = 0$, namely, $m = n - 2p$. Then the lines having equations

$$
\begin{aligned}
x &= -1 + (n - 2p)t \\
y &= nt \\
z &= 4 + pt
\end{aligned}
\tag{15.3}
$$

$t \in \mathbf{R}$, form the bundle of lines (Sect. 4.6) passing through A and parallel to α. The system (15.3) depends on two parameters n and p.

15.5   Intersection of Planes

We dealt with parallel planes in (Sects. 4.2.4 and 10.4).

Two non-parallel planes α and α' identify one and only one line, which is the intersection of α and α'. Let

$$
\begin{aligned}
\alpha)\quad & ax + by + cz + d = 0 \\
\alpha')\quad & a'x + b'y + c'z + d' = 0
\end{aligned}
$$

be the equations of the planes α and α'. The triples (a, b, c) and (a', b', c') are non-proportional. The system of equations of α and α', along with the condition (a, b, c) and (a', b', c') non proportional, *represents* a line.

Such a way to define a line as the intersection of two planes we name *representation* $\alpha\alpha'$.

Let us describe by means of an example how from the parametric equations of the line we pass to the system $\alpha\alpha'$.

The equations

$$
\begin{aligned}
x &= 1 + 6t \\
y &= -7 + 2t \\
z &= -5t
\end{aligned}
\tag{15.4}
$$

define the line r with direction numbers $(6, 2, -5)$ and passing through the point $(1, -7, 0)$. If we eliminate the parameter t, we obtain a system $\alpha\alpha'$. In fact, from the third equation we have $t = -\frac{1}{5} z$ that substitute in the remaining equations:

$$
\begin{aligned}
x &= 1 - \tfrac{6}{5}z \\
y &= -7 - \tfrac{2}{5}z
\end{aligned}
$$

that doing the calculations take the form:

$$
\begin{aligned}
5x + 6z - 5 &= 0 \\
5y + 2z + 35 &= 0
\end{aligned}
$$

These equations form a representation $\alpha\alpha'$ of the line r.

Set $z = 0$ in the system above; we obtain $x = 1$, $y = -7$, whence the point $(1, -7, 0)$; another point is obtained by assigning z a new value, for example $z = 5$; then $x = -5$, $y = -9$ and the point $(-5, -9, 5)$ is got. From these two points we obtain the direction numbers of r, as differences of the coordinates: $(-5, -9, 5) - (1, -7, 0) = (-6, -2, 5)$. Operating over the system $\alpha\alpha'$ we obtain the parametric equations

$$
\begin{aligned}
x &= 1 - 6t \\
y &= -7 - 2t \\
z &= 5t
\end{aligned}
\tag{15.5}
$$

With (15.5) we have not gone back exactly to the parametric Eqs. (15.4), but to an equivalent system. In fact, the Eqs. (15.5) define a line passing through $(1, -7, 0)$ with direction numbers $(-6, -2, 5)$ proportional to $(6, 2, -5)$. So, also the Eqs. (15.5) represent r.

Exercise 15.2 Find a triple of direction numbers of the line r represented by the equations

$$x - y - z + 1 = 0$$
$$x - y + z - 1 = 0$$

and give parametric equations of r.

Solution It is easy to check that A(0, 0, 1) and B(1, 1, 1) are distinct points of r. Therefore, the numbers

$$m = 0 - 1 = -1$$
$$n = 0 - 1 = -1$$
$$p = 1 - 1 = 0$$

form a triple of direction numbers of r. Parametric equations of r are:

$$x = t$$
$$y = t$$
$$z = 1$$

$t \in \mathbf{R}$.

15.6 Bundle of Planes

The set of planes parallel to a given plane α is called an *improper bundle of planes* or a *bundle of parallel planes*. By Theorem 10.1, for each point P of the space, there exists a unique plane parallel to α and containing P.

Let a line r of the space be given. A *bundle of planes of axis r* is, by definition, the set of the planes containing r.

Since two non-parallel planes identify the common line r, the bundle of planes of axis r is determined by two planes passing through r. In fact, let α and α' be non-parallel planes

$$\alpha) \quad ax + by + cz + d = 0$$
$$\alpha') \quad a'x + b'y + c'z + d' = 0$$

and r the common line. Let us call *system $\alpha\alpha'$* the above system.

We prove that any linear combination of the equations α) and α')

$$m(ax + by + cz + d) + m'(a'x + b'y + c'z + d') = 0 \qquad (15.6)$$

with coefficients m and m' not both null, represents a plane that contains r. The statement is obvious if one of the two coefficients, m or m', is zero. Then suppose that m and m' are both non-zero and rewrite (15.6) in the form

$$(ma + m'a)x + (mb + m'b')y + (mc + m'c')z + md + m'd' = 0 \qquad (15.7)$$

Suppose, by contradiction,

$$ma + m'a' = mb + m'b' = mc + m'c' = 0 \qquad (15.8)$$

Then

$$a = -\frac{m'}{m}a'$$
$$b = -\frac{m'}{m}b'$$
$$c = -\frac{m'}{m}c'$$

Therefore, the triples (a, b, c) and (a', b', c') should be proportional and the planes α and α' parallel, contrary to the hypothesis. Thus, the equalities (15.8) are not acceptable and (15.7) is the equation of a plane β. We now verify that (15.6), or the equivalent (15.7), is the equation of a plane containing the line r represented by the system $\alpha\alpha'$. In fact, if the point $P_0(x_0, y_0, z_0)$ belongs to r, then the following equalities hold:

$$ax_0 + by_0 + cz_0 + d = 0$$
$$a'x_0 + b'y_0 + c'z_0 + d' = 0$$

and P_0 belongs to the plane (15.6). Hence, each plane (15.6) passes through r.

Vice versa, every plane β passing through r falls within the form (15.6), i.e., it is represented by (15.6), for suitable values of m and m'. Among the planes containing r the plane β is determined by a point $Q(x_1, y_1, z_1)$ not belonging to r. The point Q lies in the plane (15.6) if and only if there exist values of m and m' such that

$$m(ax_1 + by_1 + cz_1 + d) + m'(a'x_1 + b'y_1 + c'z_1 + d') = 0$$

As Q does not belong to r, both equalities

$$ax_1 + by_1 + cz_1 + d = 0$$
$$a'x_1 + b'y_1 + c'z_1 + d' = 0$$

are not satisfied. Then suppose, for example, $a'x_1 + b'y_1 + c'z_1 + d' \neq 0$. The values of m and m' satisfying (15.6) are tied by:

$$m' = -\frac{ax_1 + by_1 + cz_1 + d}{a'x_1 + b'y_1 + c'z_1 + d'}m$$

Summarizing, a plane belongs to the bundle determined by the non-parallel planes α and α' if and only if it has an Eq. (15.6) with both non-zero values of m and m'.

Remark 15.1 Obviously, if the planes α and α' are parallel, any linear combination of the equations

$$\alpha) \quad ax + by + cz + d = 0$$
$$\alpha') \quad a'x + b'y + c'z + d' = 0$$

is the equation of a plane parallel to α and α' and then any linear combination of the equations represents an improper bundle of planes.

Exercise 15.3 Determine the plane through the origin $O(0, 0, 0)$ and perpendicular to the line r defined by the system

$$2x - y + z - 1 = 0$$
$$3x + z + 2 = 0$$

Solution The direction numbers of the line r are obtained (see Exercise 14.6) from the matrix

$$\begin{bmatrix} 2 & -1 & 1 \\ 3 & 0 & 1 \end{bmatrix}$$

and are proportional to

$$m = \begin{vmatrix} -1 & 1 \\ 0 & 1 \end{vmatrix} = -1; \ n = \begin{vmatrix} 1 & 2 \\ 1 & 3 \end{vmatrix} = 1; \ p = \begin{vmatrix} 2 & -1 \\ 3 & 0 \end{vmatrix} = 3$$

A plane containing the origin has the equation $ax + by + cz = 0$ whose coefficients a, b, c, by Theorem 10.1, are direction numbers of r. Therefore, the required plane has equation $x - y - 3z = 0$.

Exercise 15.4 Find the plane passing through the origin and parallel to the line r of equations

$$x - 5z - 1 = 0$$
$$y - z + 2 = 0$$

and to the line s of equations

$$x = -1 + 4t$$
$$y = 1 + 2t$$
$$z = t$$

Solution From the matrix

$$\begin{bmatrix} 1 & 0 & -5 \\ 0 & 1 & -1 \end{bmatrix}$$

are computed the direction numbers of r (see Exercise 14.6)

$$m = \begin{vmatrix} 0 & -5 \\ 1 & -1 \end{vmatrix} = 5; n = \begin{vmatrix} -5 & 1 \\ -1 & 0 \end{vmatrix} = 1; p = \begin{vmatrix} 1 & 0 \\ 0 & 1 \end{vmatrix} = 1;$$

from Property 2, a plane parallel to r satisfies the condition: $5a + b + c = 0$. The direction numbers of s are 4, 2, 1 and a plane parallel to s satisfies the condition: $4a + 2b + c = 0$. The system of the conditions

$$5a + b + c = 0$$
$$4a + 2b + c = 0$$

has a solution (a, b, c)

$$a = \begin{vmatrix} 1 & 1 \\ 2 & 1 \end{vmatrix} = -1; b = \begin{vmatrix} 1 & 5 \\ 1 & 4 \end{vmatrix} = -1; c = \begin{vmatrix} 5 & 1 \\ 4 & 2 \end{vmatrix} = 6$$

So the required plane, that contains the origin, has equation: $-x - y + 6z = 0$.

Exercise 15.5 Find the plane α parallel to the line r of equations

$$x = t$$
$$y = -2 + 3t$$
$$z = t$$

perpendicular to the plane $x - 5y + z - 4 = 0$ and passing through the point $(1, 2, -1)$.

Solution Notice that $(1, 3, 1)$ is a triple of direction numbers of r and the equation of the plane α we want to determine is $ax + by + cz = 0$. By Properties 2 and 4 the following conditions must be fulfilled by the plane α:

$$a + 3b + c = 0$$
$$a - 5b + c = 0$$

So the solution (a, b, c) is proportional to the triple

$$\begin{vmatrix} 3 & 1 \\ -5 & 1 \end{vmatrix} = 8; \begin{vmatrix} 1 & 1 \\ 1 & 1 \end{vmatrix} = 0; \begin{vmatrix} 1 & 3 \\ 1 & -5 \end{vmatrix} = -8$$

We choose the triple $(a, b, c) = (1, 0, -1)$ and get the equation $x - z + d = 0$. The coefficient d is determined by imposing that the required plane passes through the point $(1, 2, -1)$:

$$1 - (-1) + d = 0.$$

Thus, the plane α has equation: $x - z - 2 = 0$.

Exercise 15.6 Represent the line r passing though the point P(1, 0, 1) and parallel to the line s

$$2x - y + z + 1 = 0$$
$$4x - 7y + 2z + 3 = 0$$

Solution The direction numbers m, n, p of s are proportional to

$$\begin{vmatrix} -1 & 1 \\ -7 & 2 \end{vmatrix} = 5; \quad \begin{vmatrix} 1 & 2 \\ 2 & 4 \end{vmatrix} = 0; \quad \begin{vmatrix} 2 & -1 \\ 4 & -7 \end{vmatrix} = -10$$

We choose $(m, n, p) = (1, 0, -2)$ and get the equations of r

$$x = 1 + t$$
$$y = 0$$
$$z = 1 - 2t$$

Exercise 15.7 Find the equation of the plane containing the line r

$$x = 1 - t$$
$$y = 3t - 2$$
$$z = t$$

and the point P(2, -1, 0).

Solution We first observe that the problem is determined, i.e., it admits a unique solution. In fact, the point P does not belong to r, because the coordinates $(x, y, z) = (2, -1, 0)$ does not satisfy all the equations of r. We must now find, in the set of the planes containing r, the one that passes through P. Let us eliminate the parameter t, setting $z = t$ in the first two equations of r:

$$x + z - 1 = 0$$
$$y - 3z + 2 = 0$$

The line r is now defined as the intersection of two planes. The equation of the bundle of planes of axis r is

$$m(x + z - 1) + m'(y - 3z + 2) = 0 \tag{15.9}$$

where the coefficients m and m' are not both zero.

The plane in the bundle that contains P(2, -1, 0) satisfies (15.9), i.e., $m(2 + 0 - 1) + m'(-1 - 0 + 2) = m + m' = 0$. Non-null values m and m' such that $m + m' = 0$ are $m = 1$ and $m' = -1$; from the values replaced in (15.9) we obtain the equation

$x + z - 1 - (y - 3z + 2) = 0$, which yields

$$x - y + 4z - 3 = 0$$

that represents the plane rP.

Exercise 15.8 Verify that the line

$$r) \quad \begin{aligned} x &= 1 \\ y &= t - 2 \\ z &= -2t \end{aligned}$$

and the line

$$s) \quad \begin{aligned} x &= 9 + t \\ y &= 4 \\ z &= 1 - t \end{aligned}$$

are not parallel. Then determine the equation of the plane α that contains r and is parallel to s.

Solution A triple of direction numbers of r is $(0, 1, -2)$ and a triple of direction numbers of s is $(1, 0, -1)$. The triples are not proportional, which means that the lines are not parallel.

Let us eliminate the parameter t from the equations of r: two planes passing through r are obtained:

$$\begin{aligned} x - 1 &= 0 \\ 2y + z + 4 &= 0 \end{aligned}$$

The bundle of planes of axis r has equation

$$m(x - 1) + m'(2y + z + 4) = 0 \tag{15.10}$$

rewritten as follows

$$mx + 2m'y + m'z - m + 4m' = 0 \tag{15.11}$$

By Property 2 the plane parallel to s in the bundle satisfies the equation $1m + 0(2 m') - 1 m' = m - m' = 0$. Then the values $m = m' = 1$ are replaced in (15.10) or (15.11) to get the equation of the plane

$$x + 2y + z + 3 = 0$$

that contains r and is parallel to s.

Exercise 15.9 Given the point O(0, 0, 0) and the line r

$$x = 1 - t$$
$$y = t$$
$$z = 2t$$

a.   verify that the point O does not belong to the line;
b.   determine the equation of the plane that passes through the point and the line.

Solution a. The point O belongs to the line r if and only if there exists a real number t such that the following equalities are satisfied:

$$0 = 1 - t$$
$$0 = t$$
$$0 = 2t$$

It is easily seen that the system of equations is incompatible: $t = 0$ satisfies the second equation, but not the first. Thus, O does not belong to r.

Solution b. The problem can be solved following the procedure of Exercise 15.5. Here we prefer to choose another way. Consider two distinct points of r: for example, A(1, 0, 0) and B(0, 1, 2), obtained in correspondence of the values $t = 0$ and $t = 1$ in the equations of r, respectively. The points O, A, B are non-collinear, then there exists a unique plane α passing through O, A, B. Let

$$ax + by + cz + d = 0$$

be the generic equation of the plane α. If O belongs to α, then $d = 0$ and α) $ax + by + cz = 0$. If A belongs to the plane α, then $a1 + b0 + c0 = 0$, i.e., $a = 0$, and, therefore, the equation of α has the form $by + cz = 0$. The plane α passes through B(0, 1, 2), then $b1 + c2 = 0$, i.e., $b = -2c$ and the equation of α becomes $-2cy + cz = 0$. Since $c \neq 0$ (otherwise the equation $ax + by + cz + d = 0$ would reduce to identity $0 = 0$) we can divide the equation of α by c. Therefore, the equation of the plane passing through O, A, B is $2y - z = 0$.

Let us remark that a point $P_1(x_1, y_1, z_1)$ belongs to the plane α) $ax + by + cz + d = 0$ if and only if $ax_1 + by_1 + cz_1 + d = 0$. This obvious consideration suggests a way to solve the exercise. Indeed, the points P(x, y, z), O(0, 0, 0), A(1, 0, 0) and B(0, 1, 2) belong to the plane α, if and only if the following equalities hold:

$$ax + by + cz + d = 0$$
$$a0 + b0 + c0 + d = 0$$
$$a1 + b0 + c0 + d = 0$$
$$a0 + b1 + c2 + d = 0$$

The equations form a homogeneous system in the unknowns a, b, c, d which shows the coplanarity of $P(x, y, z)$ with the points O, A and B. The system admits solutions distinct from the null solution, $(a, b, c, d) \neq (0, 0, 0, 0)$, then (Sect. 114.10 the determinant of the coefficients is equal to zero:

$$\begin{vmatrix} x & y & z & 1 \\ 0 & 0 & 0 & 1 \\ 1 & 0 & 0 & 0 \\ 0 & 1 & 2 & 0 \end{vmatrix} = 0$$

Expand the above determinant with respect to the first row:

$$\begin{vmatrix} 0 & 0 & 1 \\ 0 & 0 & 0 \\ 1 & 2 & 0 \end{vmatrix} x - \begin{vmatrix} 0 & 0 & 1 \\ 1 & 0 & 0 \\ 0 & 2 & 0 \end{vmatrix} y + \begin{vmatrix} 0 & 0 & 1 \\ 1 & 0 & 0 \\ 0 & 1 & 0 \end{vmatrix} z - \begin{vmatrix} 0 & 0 & 0 \\ 1 & 0 & 0 \\ 0 & 1 & 2 \end{vmatrix} = 0$$

to obtain α) $-2y + z = 0$.

Bibliography

Aitken, A.C.: Determinants and Matrices. Oliver and Boyd LTD, Edinburgh (1965)
Albert, A.A.: Solid Analytic Geometry, Dover Publications Inc., New York (2016)
D'Apuzzo, L., Ventre, A.: Algebra lineare e Geometria Analitica, Cedam, Padova (1995)
Lang, S.: Linear Algebra. Addison-Wesley Publishing Company, Reading Massachusetts (1966)
Lipschutz, S.: Theory and Problems of Linear Algebra. McGraw-Hill Book Company, New York (1968)

Chapter 16
Algorithms

16.1 Introduction

While we have not explicitly named them, even we encountered and applied numerous *algorithms*. We called them *procedures* or *methods* and they concerned various topics, such as: key codes, adding natural numbers in a range (Chap. 2), performing a reduction of a system of linear equations (Chap. 11), searching for solutions of systems of equations (Chap. 14).

The four arithmetic operations are among the simplest algorithms we know.

It is possible to build algorithms for processes considered complicated. The practical difficulties associated with the execution of these processes derive from the fact that algorithms often require an enormous number of operations, even if every single operation is simple. The more complex operations are performed by repetition of the simpler ones. This is one of the reasons why algorithms are associated with automatic calculating machines.

Consider now an elementary procedure which, however, contains some characteristics that contribute to formulate the a concept of algorithm.

We want to find the maximum value in a *n*-tuple of numerical data. We have an executor at our disposal to communicate our instructions to solve the problem.

Here is the list of *instructions*:

1. take the first component of the *n*-tuple with your left hand; go to instruction 2,
2. take the next component with the right hand; go to instruction 3,
3. if the component in the right hand is greater than the component in the left hand, throw away the number in the left hand and transfer the component in your right hand to the left hand and go to instruction 2 above. If not, go to step 4 below,
4. throw away the number in your right hand and go to instruction 2.

The executor has to compare two numbers at a time, transfer a number from one hand to the other, choose at a crossroads: from instruction 3 the executor can go to instruction 2 or 4. Each instruction or statement translates into an elementary operation or choice.

© The Author(s), under exclusive license to Springer Nature Switzerland AG 2023   257
A. G. S. Ventre, *Calculus and Linear Algebra*,
https://doi.org/10.1007/978-3-031-20549-1_16

16.2   Greatest Common Divisor: The Euclidean Algorithm

We will not give a formal definition of algorithm. We will rather explain some examples in order to dwell upon properties and features that an algorithm should meet.

Let us find the *greatest common divisor* (GCD) of the numbers 210 and 45; the computation of the GCD(210, 45) is obtained in two steps:

> *First step.* We factorize the given numbers into prime factors: $210 = 2 \times 3 \times 5 \times 7$ and $45 = 3^2 \times 5$;
>
> *Second step.* GCD(210, 45) is equal to the product of the common factors to 210 and 45 each with the least exponent, $\text{GCD}(210, 45) = 3 \times 5 = 15$.

We describe now a different method to find the GCD(210, 45):

- *First step.* Calculate the Euclidean division (Sect. 2.1) 210:45 where the dividend is 210 and the divisor is 45. Obtain $210 = 4 \times 45 + 30$; the *quotient* is 4 and the *remainder* is 30.
- *Second step.* Perform the division of the divisor by the remainder, 45:30. Obtain $45 = 1 \times 30 + 15$. The *quotient* is 1 and the *remainder* is 15.
- *Third step.* Perform the division of the divisor by the remainder, 30:15. Obtain $30 = 2 \times 15 + 0$.
- *Fourth step.* At the first remainder null, the procedure ends. Obtain the result $\text{GCD}(210, 45) = 15$.

The procedure extends to the search for the GCD(a, b) of any two natural numbers a and b and it is known as *Euclid's algorithm* for calculating the greatest common divisor.

There are as many different problems as there are the pairs of natural numbers. The algorithm for finding GCD(a, b) solves each of these problems by building a decreasing sequence of natural numbers, the first of which is the greater between a and b, the second is the smaller; the third is the remainder of the division of the first by the second; the fourth is the remainder of the division of the second by the third and so on. The procedure ends when a division with zero remainder is reached: the divisor of this last division is the greatest common divisor of a and b.

Obviously, if $a = b$ this number is the greatest common divisor of a and b.

The Euclidean algorithm does not explicitly involve decompositions into prime factors, which generally deals with onerous operations, but consists of a list of comparison and division statements. This consideration leads us to observe that the statements of an algorithm, expressed in natural language or through formulas, are made simple and split into elements so that they can be executed regardless of the executor. Let us clarify the concept by introducing such kind of simplifications or refinements in the Euclidean algorithm, based on the fact that a division can be reduced to the execution of successive subtractions. The algorithm is expressed by the following list of statements, addressed to a hypothetical performer.

1.   Read the initial data, i. e., the numbers a and b.

Go to the next statement.
2. Compare the two numbers just read, i. e., determine if the first is greater than, equal to or less than the second.
Go to the next statement.
3. If the two numbers are equal each of them gives the required result, that is GCD(a, b).
The procedure ends. Otherwise, go to the next statement.
4. If the first number is less than the second, swap the two numbers together.
Go to the next statement.
5. Subtract the second number from the first and replace the two numbers read with the subtrahend and the difference, respectively.
Go to statement 2.

After having carried out the 5 statements you go back to the second and then to the third, fourth, fifth; and again to the second, the third, etc., until the condition of the third statement is fulfilled, that is, until the two numbers are equal. When this occurs, the problem is solved and the calculation ends, as required by statement 3.

The algorithm contains an automatic ending procedure.

We calculate the GCD(210, 45) by applying the algorithm just described:

Start
statement 1: $a = 210, b = 45$
statement 2: $a \geq b$
statement 3: $a \neq b$
statement 4: $a \geq b$
statement 5: $210 - 45 = 165$; $a = 45$ subtrahend, $b = 165$ difference
statement 2: a $\leq$ b
statement 4: a $= 165$, b $= 45$
statement 5: a $= 45$, b $= 120$
statement 2: a $\leq$ b
statement 4: a $= 120$, b $= 45$
statement 5: a $= 45$, b $= 75$
statement 2: a $\leq$ b
statement 4: a $= 75$, b $= 45$
statement 5: a $= 45$, b $= 30$
statement 2: a $\geq$ b
statement 4: a $= 30$, b $= 15$
statement 5: a $= 15$, b $= 15$
statement 2: a $= b = 15$
statement 3: GCD(210, 45) $= 15$
End

The statements include commands: start, read, compare, swap, go to, subtract, replace, end. The meaning of the statement is immediate and simulates a conversation with the executor. Statement 4 contains an interchange command; statement 5 contains a replacement command. In the passage from statement 4, $a = 120, b = 45$

to statement 5, $a = 45$, $b = 30$, the number 45 changes its name from b to a. The equalities, for example b $= 45$, are *assignments* and must be read from right to left: "set 45 in b" or "give the name b to 45".

The Euclidean algorithm is detailed. The fundamental operations consist in subtracting two numbers, comparing and interchanging them: the problem is fragmented into elementary operations. The situation would have been different if we had had to communicate to a performer to factorize each given natural number into prime numbers; the statements would have been much more complex and the communication difficult.

The above considerations provide the first elements of the concept of algorithm.

To summarize, an algorithm is a single finite sequence of statements which

i.   acts on a set of initial data. In the case of the example described above, the initial data are the natural numbers a and b;
ii.   is made of unambiguous statements; for example, arithmetic operations, comparisons, substitutions, jumps to some statement other than the next one in the sequence;
iii. is deterministic: once accomplished a non-final statement, the next statement is determined; and if the problem has been solved, then the algorithm provides an automatic ending procedure;
iv.   is a procedure for solving a class of similar problems and not a particular problem.

Let us accept the properties listed above as a plausible approach to the concept of algorithm. There remains some margin of ambiguity: for example, it should be specified what is meant by "class of similar problems". However, we understand that the GCD example generalizes to a class of similar problems.

If a performer, a human or a machine, is able to accomplish each of the listed statements and communicate us the results of the statements, then we will know how to solve the problem, while the performer may not know which problem has been solved by the sequence of statements.

16.3   Regular Subdivision of a Segment

The parametric equations of the line passing through the points $A(x_1, y_1)$, $B(x_2, y_2)$ are (see Eqs. 7.3 and 7.4)

$$x = x_1 + t(x_2 - x_1), \quad y = y_1 + t(y_2 - y_1)$$

$t \in \mathbf{R}$. The points $A(x_1, y_1)$ and $B(x_2, y_2)$ are determined by the values 0 and 1 of the parameter t, respectively. For every t such that $0 < t < 1$ there is one and only one point P of the segment $\overline{AB}$, between A and B and distinct from A and B, whose coordinates are $(x_1 + t(x_2 - x_1), y_1 + t(y_2 - y_1))$. Vice versa, for each point P belonging to the segment $\overline{AB}$ and distinct from the endpoints, there is one and only one value of t,

$0 < t < 1$, such that P has coordinates $(x_1 + t\,(x_2 - x_1),\ y_1 + t\,(y_2 - y_1))$. Then the parametric representation of the segment $\overline{AB}$ is as follows:

$$x = x_1 + t(x_2 - x_1), \quad y = y_1 + t(y_2 - y_1)$$

$t \in [0, 1]$. Let us now construct a procedure for subdividing the segment $\overline{AB}$ into 10 segments by means of the points $P_0 = A$, P_1, P_2, ..., P_9, $P_{10} = B$, such that P_i follows P_{i-1} in the orientation from A to B, for $i = 0, 1, ..., 10$, so that the segments $\overline{P_0 P_1}$, $\overline{P_1 P_2}$, ..., $\overline{P_9 P_{10}}$ are congruent. Let us define the procedure, or algorithm, as a *regular subdivision of the segment* $\overline{AB}$. The values of t to find the points P_1, P_2, ..., P_9, obtained by Thales' theorem, are $\frac{1}{10}$, $\frac{2}{10}$, ..., $\frac{9}{10}$, respectively. Therefore, the coordinates of the point P_i, $i = 0, 1, ..., 10$, are:

$$x_1 + \frac{i}{10}(x_2 - x_1), \quad y_1 + \frac{i}{10}(y_2 - y_1)$$

For example, let us implement the regular subdivision of the segment with endpoints A(4, 2), B(7, 4), having the parametric representation

$$x = 4 + 3t, \quad y = 2 + 2t$$

$t \in [0, 1]$. Specifically, we have to organize the *steps* of the procedure, performing a list of the operations, or statements in order to determine step by step the points $P_0 = A$, P_1, P_2, ..., P_9, $P_{10} = B$. The coordinates of the point P_i are $\left(4 + 3\frac{i}{10},\ 2 + 2\frac{i}{10}\right)$, $i = 0, 1, ..., 10$.

A list of statements follows.

Start

1. Write the parametric equations of segment $\overline{AB}$.
   Go to statement 2 below.
2. Activate the index or *counter i*. Set $i = 0$. The point $P_0 = A(4, 2)$ is obtained.
   Go to statement 3 below.
3. Set $i = 1$. The point $P_1\left(4 + \frac{3}{10},\ 2 + \frac{2}{10}\right)$ is obtained.
   Go to statement 4.
4. Increase the value of counter i by 1, i. e., put $i + 1$ in place of i. Continue to denote the new index value with i. In symbols, apply the statement $i \leftarrow i + 1$, which reads "put $i + 1$ in i", or "call i the value $i + 1$". (To the left of $\leftarrow$ there is the name of the quantity on the right).
   The point $\left(4 + 3\frac{i}{10},\ 2 + 2\frac{i}{10}\right)$ is obtained.
   Go to statement 5.
5. If $i > 10$, execute statement 6. Otherwise, go to statement 4.
6. The problem is solved having found the points $P_0 = A$, P_1, P_2, ..., P_9, $P_{10} = B$.
   Execute statement 7.
7. End

In the statement 3 the point P_1 is found; then go to execute statement 4, where i = 2: then $P_2 \left(\frac{46}{10}, \frac{24}{10}\right)$ is obtained. Then go to the statement 5 wondering if this value of i is greater than 10. Not yet. Then go back to the statement 4: now i becomes 3 and carry out $P_3 \left(\frac{49}{10}, \frac{26}{10}\right)$. Afterwards go to the statement 5 and then to statement 4, and so on, until $i = 11$. Now, the point P_1, P_2, ..., P_9 are got and the procedure ends, as commented by statement 6 and required by statement 7.

The algorithms can be often represented and controlled by means of a *block diagram* that highlights the sequence of the statements. A block diagram is often useful whenever an algorithm needs to be translated into a programming language. Let now exhibit the block diagram of the algorithm for the regular subdivision of a segment into n segments (Fig. 16.1).

Arrows connecting boxes indicate the statement path. The frames and the contained statements are called *blocks*: the operations and statements are written

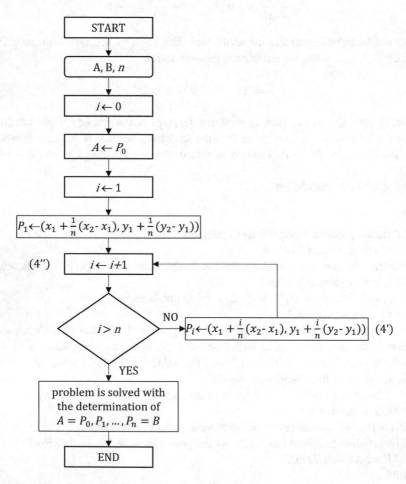

Fig. 16.1 Block diagram for the regular subdivision of a segment

inside rectangular frames, a diamond indicates any sort of decision, the round frames contain reading instructions or the definitions of the initial data or calculation elements (input) or exit instructions (for example, "print result"). The use of the imperative for statement of the type $a \leftarrow b$ is effective: "call a the element b", or "set b in a" (see block (4'')). For example, the block (4') contains the instruction "call P_i the couple of coordinates $(x_1 + i(x_2 - x_1), y_1 + i(y_2 - y_1))$". The statement of the block (4'') allows the *iteration*, i. e., the reoccurrence of the calculation.

Remark 16.1 The symbol $\leftarrow$ was introduced to describe the actions of the elementary operations in the procedure of *reduction* of the system of linear equations (Chap. 11). Besides, the symbol $=$ in the statements of Euclidean algorithm plays the role of $\leftarrow$.

16.4 Gauss Elimination

Gauss elimination is an algorithm for examining and solving systems of linear equations. The algorithm transforms a system of linear equations into a particular reduced system (Sect. 11.3).

An example may clarify how the algorithm works.

Consider the system of linear equations:

$$
\begin{aligned}
E_1 : &\quad x + y + z = 1 \\
E_2 : &\quad 2x + 3y - z = 2 \\
E_3 : &\quad x - y - z = 3
\end{aligned} \tag{16.1}
$$

Multiply equation E_1 by 2, the coefficient of x in equation E_2, and subtract equation E_2 from equation $2E_1$, then replace equation $2E_1 - E_2$ with E_2 (i. e., $E_2 \leftarrow 2E_1 - E_2$, that means "call E_2 the equation $2E_1 - E_2$); finally replace $E_1 - E_3$ with E_3 ($E_3 \leftarrow E_1 - E_3$). Obtain the system, equivalent to (16.1):

$$
\begin{aligned}
E_1 : &\quad x + y + z = 1 \\
E_2 : &\quad -y + 3z = 0 \\
E_3 : &\quad 2y + 2z = -2
\end{aligned} \tag{16.2}
$$

Then in (16.2) do $E_3 \leftarrow 2E_2 + E_3$ and obtain the following system equivalent to (16.1):

$$
\begin{aligned}
E_1 : &\quad x + y + z = 1 \\
E_2 : &\quad -y + 3z = 0 \\
E_3 : &\quad 8z = -2
\end{aligned} \tag{16.3}
$$

To solve the system (16.3) we first find the value of z in E_3: $z = -1/4$; then we proceed by "back substitution", i. e., we substitute $z = -1/4$ in E_2, and determine the value $y = -3/4$ and finally substitute the values $y = -3/4$ and $z = -1/4$ in E_1 to find $x = 1 + 3/4 + 1/4 = 2$. So the solution of system (16.1) is the triple $(2, -3/4, -1/4)$. (The system (16.3) is called a *triangular system* from its matrix of coefficients).

Let us retrace the above procedure to focus the main objective at any step. To make the unknown x disappear in the second and third equations of the system it is necessary that the coefficient of x in the first equation is different from 0. The non-zero coefficient of x in E_1 is called the *pivot* of the first elimination step. Similarly, in order to eliminate y from equation E3 in (16.2) it is necessary that the coefficient of y in E_2 be non-zero (therefore, the coefficient of y in E_2 is the pivot of the second elimination step; so the pivot of the first step is 1, the pivot of the second step is -1 $= a_{22}$ in the system (16.3).

Let us describe the algorithm in general. Let a system of m linear equations and n unknowns $x_1, x_2, ..., x_n$ be given.

The *first step* consists in constructing a system, equivalent to the given system, such that x_1 appears with non-zero coefficient in the first equation. To this aim perform the following two operations:

Operation 1. If necessary, rearrange the equations of the system so that in the first equation the first unknown x_1 has a non-zero coefficient a_{11}. So a_{11} is the pivot of the first step in the elimination procedure:

$$E_1 : \quad a_{11}x_1 + a_{12}x_2 + \cdots + a_{1n}x_n = b_1$$
$$E_2 : \quad a_{21}x_1 + a_{22}x_2 + \cdots + a_{2n}x_n = b_2 \qquad (16.4)$$
$$\cdots$$
$$E_m : \quad a_{m1}x_1 + a_{m2}x_2 + \cdots + a_{mn}x_n = b_m$$

Operation 2. Proceed to eliminate the unknown x_1 in all the equations following the first. For $i = 2, 3, ..., m$, carry out the substitution:

$$E_i \leftarrow a_{11}E_i - a_{i1}E_1 \qquad (16.5)$$

or the equivalent substitution:

$$E_i \leftarrow E_i - d_i E_1 \qquad (16.6)$$

with $d_i = a_{i1}/a_{11}$. Then the system (16.4) is transformed into the equivalent system

$$E_1: \quad a_{11}x_1 + a_{12}x_2 + \cdots + a_{1j}x_j + \cdots + a_{1n}x_n = b_1$$
$$E_2: \quad 0 + a_{22}^{(1)}x_2 + \cdots + a_{2j}^{(1)}x_j + \cdots + a_{2n}^{(1)}x_n = b_2^{(1)}$$
$$\cdots$$
$$E_i: \quad 0 + a_{i2}^{(1)}x_2 + \cdots + a_{ij}^{(1)}x_j + \cdots + a_{in}^{(1)}x_n = b_i^{(1)} \qquad (16.7)$$
$$\cdots$$
$$E_m: \quad 0 + a_{m2}^{(1)}x_2 + \cdots + a_{mj}^{(1)}x_j + \cdots + a_{mn}^{(1)}x_n = b_m^{(1)}$$

where, by (16.5) and (16.6),

$$a_{ij}^{(1)} = a_{ij} - d_i a_{1j}$$
$$b_i^{(1)} = b_i - d_i b_1$$

for $i = 2, \ldots, m$ and $j = 1, \ldots, n$.

The *second step* has the objective of building a system, equivalent to the given system, such that the coefficients of the unknown x_2 in the equations that follow the second equation are all zeros. To this aim we consider the subsystem of system (16.7) made of the equations $E_2, \ldots, E_i, \ldots, E_m$, which has fewer unknowns. This subsystem is treated similarly to the original system (16.4) by performing the operations 1. and 2., in which x_2 plays the role of x_1 and the first equation of the subsystem takes the place of E_1.

Remark 16.2 It may happen that all the coefficients $a_{22}^{(1)}, \ldots, a_{i2}^{(1)}, \ldots, a_{m2}^{(1)}$ of the unknown x_2 in $E_2, \ldots, E_i, \ldots, E_m$, are null. Then, the second step must not be referred to the unknown x_2, but to the first of the unknowns following x_2 which appear with a non-zero coefficient in one of the equations $E_2, \ldots, E_i, \ldots, E_m$ in the system (16.7) (the pivot must be chosen among the non-zero coefficients of an unknown $x_j, j = 3, \ldots, n$). For example, the circumstance occurs in the system.

$$x + 3y - z = 1$$
$$2x + 6y + z = 0 \qquad (16.8)$$
$$3x + 9y - 2z = 2$$

where eliminating x in the second and third equation, by means of (16.5): $E_2 \leftarrow E_2 - 2E_1$, $E_3 \leftarrow E_3 - 3E_1$, the system, equivalent to (16.8), is obtained:

$$x + 3y - z = 1$$
$$3z = -2$$
$$z = -1$$

Then the second step consists in eliminating z from the third equation, applying the substitution $E_3 \leftarrow 3E_3 - E_2$; thus the system

$$x + 3y - z = 1$$
$$3z = -2$$
$$0 = -1$$

is incompatible along with the equivalent system (16.8). The algorithm ends.

At the k-th step the unknown x_k should be eliminated from the equations following the k-th equation. Proceed similarly:

The equations are rearranged from the k-th to the last, so as to call E_k the equation with the non-zero coefficient $a_{kk}^{(k-1)}$ of the unknown x_k . The coefficient $a_{kk}^{(k-1)}$ becomes the pivot of the k-th step. If a non-zero pivot among the coefficients of x_k does not exists, search among the coefficients of x_{k+1}, and so on (see Remark 16.2).

If a non-zero coefficient can become the pivot $a_{kk}^{(k-1)}$, then proceed to the operation $E_i \leftarrow E_i - d_i^{(k-1)}E_k$, for $i = k + 1, \ldots, m$, where $d_i^{(k-1)} = a_{ik}^{(k-1)}/a_{kk}^{(k-1)}$. Then the coefficients of the subsystem $E_{k+1}, \ldots, E_m$, when the unknown x_k has been eliminated, are

$$a_{ij}^{(k)} = a_{ij}^{(k-1)} - d_{kj}^{(k-1)}, \quad b_i^{(k)} = b_i^{(k-1)} - d_i^{(k-1)}b_k^{(k-1)}$$
$$i = k + 1, \ldots, m; \quad j = k, \ldots, n.$$

Remark 16.3 If during the procedure the incompatible equation $0x_1 + 0x_2 + \ldots + 0x_n = b$, $b \neq 0$, is obtained, then the given system is incompatible (see Remark 16.2).

Remark 16.4 If the identical equation $0x_1 + 0x_2 + \ldots + 0x_n = 0$ is obtained, then it is canceled.

16.5   Conclusion

The reader has certainly studied various algorithms since primary school. Euclid built the oldest algorithm in Western culture to calculate the greatest common divisor of two natural numbers. Algorithms are used in programming and constitute the dynamics of the software and the net itself. The theory of algorithms goes to the deepest foundations of mathematics and technology.

Bibliography

Adler, I.: Thinking Machines. The John Day Co., NewYork (1961)

Adler, I.: New Look at Geometry. Dover Publications Inc., New York (2012)

Aho, A.V., Hopcroft, J.E., Ullman, J.D.: The Design and Analysis of Computer Algorithms. Addison-Wesley Publishing Company, Reading, Massachusetts (1974)

McCracken, D.D., Dorn, W.S.: Numerical Methods and Fortran Programming. John Wiley and Sons Inc., New York (1966)

Trakhtenbrot, B.A.: Algorithms and Automatic Computing Machines. D. C. Heath and Company, Boston (1963)

Ventre, A.: Algebra lineare e algoritmi. Zanichelli, Bologna (2021)

Chapter 17
Elementary Functions

17.1 Introduction

We introduced some concepts concerning real-valued functions of a single real variable. We also dealt with the linear function (Chap. 7) and some circular functions (Chap. 8). Let us now consider some properties of the real-valued functions, such as monotonicity and invertibility, and define other classes of real-valued functions of a single real variable: the power, the exponential function and the logarithm which, along with the linear and circular functions, are known as *elementary functions*. Of course, we will define the domain, the range of the functions in each class, and the basic analytic and geometric properties.

17.2 Monotonic Functions

Let f be a real-valued function of a real variable whose domain is a subsct A, not reduced to a single point. The function f is said to be:

- *increasing* in A if $f(x_1) \leq f(x_2)$ whenever x_1 and x_2 belong to A and $x_1 < x_2$;
- *decreasing* in A if $f(x_1) \geq f(x_2)$ whenever x_1 and x_2 belong to A and $x_1 < x_2$.

A real-valued function of a real variable is called a *monotonic function* in A if it is increasing in A or decreasing in A.

In particular, the function f is called *strictly increasing* in A if $f(x_1) < f(x_2)$ whenever $x_1 < x_2$; the function f is called *strictly decreasing* in A if $f(x_1) > f(x_2)$ whenever $x_1 < x_2$.

For example, the absolute value function $f(x) = |x|$ is strictly decreasing in $(-\infty, 0]$ and strictly increasing in $[0, +\infty)$.

The function f is called *strictly monotonic* in A if it is strictly increasing or strictly decreasing in A (Fig. 17.1).

Example 17.1 Let us verify that the linear function $f(x) = 2x - 3$ is strictly increasing.

© The Author(s), under exclusive license to Springer Nature Switzerland AG 2023   269
A. G. S. Ventre, *Calculus and Linear Algebra*,
https://doi.org/10.1007/978-3-031-20549-1_17

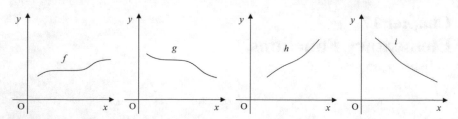

Fig. 17.1 Graphs of monotonic functions: f increasing, g decreasing, h strictly increasing, i strictly decreasing

Indeed, if $x_1 < x_2$, then $f(x_1) = 2x_1 - 3 < 2x_2 - 3 = f(x_2)$. In general, the linear function $f(x) = kx + n$ is strictly increasing if $k > 0$, strictly decreasing if $k < 0$; the constant function $f(x) = n$ is increasing and decreasing in **R**. Therefore, the linear function is monotonic in **R**.

17.3 Invertible Functions and Inverse Functions

We will state some connections between invertible and strict monotonic functions (Sects. 5 and 7.7). The following theorem applies:

Theorem 17.1 *Let $f: A \to B$ be a real-valued function whose domain is the subset $A \subseteq$ **R** and whose range is B. If f is strictly monotonic in A, then f is invertible in A.*

Proof We show that for every $y \in B$ there exists one and only one $x \in A$ such that $f(x) = y$. Indeed, there exists an $x \in A$ such that $f(x) = y$ since B is the range of f; moreover, as f is strictly monotonic there are no two distinct x_1 and x_2 such that $f(x_1) = y$ and $f(x_2) = y$.   □

The following statements can be easily proved.

Theorem 17.2 *A strictly increasing (decreasing) function is invertible and its inverse is strictly increasing (decreasing).*

Theorem 17.3 *A composite function of invertible functions is invertible.*

17.4 The Power

Let us get in touch with some known notions. When it comes to natural numbers, repeated multiplication by the same number is the operation of *raising to a power*: the symbol 2^6 means "2 raised to the 6th power", and denotes the operation $2 \times 2 \times 2 \times 2 \times 2 \times 2 = 64$. The word "raised" is usually omitted. Therefore, we write: $2^6 = 2 \times 2 \times 2 \times 2 \times 2 \times 2 = 64$, $1^{43} = 1$, $3^3 = 3 \times 3 \times 3 = 27$.

Operations 2^6, 1^{43}, 3^3 are called *powers*. The expression 3^2 also reads: *the square of* 3, the expression 6^3: *the cube* of 6. The power 2^6 has *base* 2 and *exponent* 6, the power $(-3)^4$ has base -3 and exponent 4. The power with negative integer base and natural exponent is defined. For example, $(-3)^3 = (-3) \times (-3) \times (-3) = -27$.

We will give a meaning to the powers whose exponent is a natural, relative integer, rational and a real number.

17.4.1 Power with Natural Exponent

Whatever the natural number n is, the function of the variable x

$$f(x) = x^n$$

called the *power function with natural* (or positive integer) *exponent n*, or the *n*-th power of x, has domain **R** and its value x^n, for each $x \in \mathbf{R}$, is calculated by means of the multiplications:

$$x^n = x \times x \times \ldots \times x$$

where the factor x occurs $n > 1$ times at the right-hand side; if $n = 1$, by definition $x^1 = x$ and the power reduces to the identical function (Sect. 7.3.2).

Let us compute, for example, the following powers with natural exponent:

$$0^n = 0; \quad 7^1 = 7; \quad (-3)^2 = (-3) \times (-3) = 9; \quad (0.5)^2 = 0.5 \times 0.5 = 0.25;$$

$$3^2 = 3 \times 3 = 9; \quad -3^2 = -(3 \times 3) = -9; \quad (-0.5)^2 = (-0.5) \times (-0.5) = 0.25;$$

$$-3^3 = -\left(3^3\right) = -27; \quad (-3)^3 = (-3) \times (-3) \times (-3) = -27$$

The following properties of the range of x^n, $n \in \mathbf{N}$, hold:

- if n is even, then the number x^n is non-negative, for every $x \in \mathbf{R}$,
- if n is odd and x is positive, then x^n is positive;
- if n is odd and x is negative, then x^n is negative.

Therefore, the range of x^n depends on being n even or odd. Since $0^n = 0$, for every natural n, the range of x^n is $[0, +\infty)$, if n is even; the range of x^n is **R**, if n is odd.

The symbol $\mathbf{R}^+$ denotes the set of non-negative real numbers, namely $\mathbf{R}^+ = [0, +\infty) = \{x \in \mathbf{R} : x \geq 0\}$.

Let us state the following propositions.

Proposition 17.1 The function x^n is strictly increasing in $\mathbf{R}^+$.

Let's verify that the function x^2 is strictly increasing in $\mathbf{R}^+$; indeed, if $0 \leq x_1 < x_2$, multiplying the inequality $x_1 < x_2$ first by x_1 and afterwards by x_2 we get:

$$x_1^2 \leq x_1 x_2, x_1 x_2 < x_2^2$$

Therefore,

$$x_1^2 \leq x_2^2$$

On this basis the proposition may be proved by induction on the exponent n.

Proposition 17.2 If n is even, x^n is strictly decreasing in $(-\infty, 0]$ (Fig. 17.2).

Let us calculate the coordinates of some points of the graph of $f(x) = x^2$:

$$f(0) = 0$$
$$f\left(\tfrac{1}{2}\right) = \left(\tfrac{1}{2}\right)^2 = \tfrac{1}{4}$$
$$f(1) = 1^2 = 1$$
$$f(2) = 2^2 = 4$$
$$f\left(-\tfrac{1}{2}\right) = \left(-\tfrac{1}{2}\right)^2 = \tfrac{1}{4}$$
$$f(-1) = (-1)^2 = 1$$
$$f(-2) = (-2)^2 = 4$$

The graph of $f(x) = x^2$ is a curve named *parabola*.

Proposition 17.3 If n is odd, x^n is strictly increasing in **R** (Fig. 17.3).

Let us calculate the coordinates of some points of the graph of $f(x) = x^3$ (Fig. 17.3):

Fig. 17.2 Graph of $f(x) = x^2$

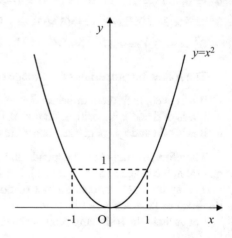

Fig. 17.3 Graphs of the powers x, x^2 and x^3

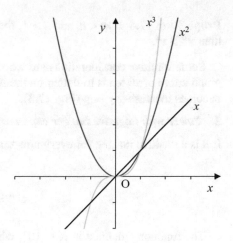

$$f(0) = 0$$
$$f\left(\tfrac{1}{2}\right) = \left(\tfrac{1}{2}\right)^3 = \tfrac{1}{8}$$
$$f(1) = 1^3 = 1$$
$$f(2) = 2^3 = 8$$
$$f\left(-\tfrac{1}{2}\right) = \left(-\tfrac{1}{2}\right)^3 = -\tfrac{1}{8}$$
$$f(-1) = (-1)^3 = -1$$
$$f(-2) = (-2)^3 = -8$$

For every $x \in \mathbf{R}$, $x^2 = (-x)^2$, $-x^3 = (-x)^3$.

17.4.2 Power with Non-Zero Integer Exponent

1. *Power with natural exponent*

Further properties of the power with natural exponent are shown by the following identities.

Proposition 17.4 For every real numbers x_1 and x_2, the following equality holds:

$$(x_1 x_2)^n = x_1^n x_2^n$$

Proposition 17.5 For every natural numbers m and n and the real number x we have:

$$x^m x^n = x^{m+n}$$
$$\left(x^m\right)^n = x^{mn}$$

Proposition 17.6 If $m < n$, and $x > 1$, then $x^m < x^n$, while, if $m < n$ and $0 < x < 1$, then $x^m > x^n$.

Some intuitive considerations are worth noticing. If $0 < x < 1$, as n increases the graph curve of x^n tends to flatten on the interval $[0, 1]$, whereas, if $x > 1$, the curve becomes increasingly step (Fig. 17.3).

2. *Power with negative integer exponent*

If n is a positive integer, for every non-zero real number x, the power x^{-n} is defined by:

$$x^{-n} = \frac{1}{x^n}$$

The function f defined in $\mathbf{R} - \{0\}$, which associates the real number x^{-n} to $x \in \mathbf{R} - \{0\}$ is called the *power with negative integer exponent*. The function $f(x) = x^{-n}$ with natural odd n (Fig. 17.4), has domain $\mathbf{R}-\{0\}$ and range $\mathbf{R}-\{0\}$. The function $f(x) = x^{-n}$, with even natural n, has domain $\mathbf{R}-\{0\}$ and range $(0, +\infty)$ (Fig. 17.5). The coordinates of some points of the graph of $f(x) = x^{-1}$ follow:

$$f\left(\tfrac{1}{2}\right) = 2$$
$$f(1) = 1$$
$$f(2) = \tfrac{1}{2}$$
$$f\left(-\tfrac{1}{2}\right) = -2$$
$$f(-1) = -1$$
$$f(-2) = -\tfrac{1}{2}$$

The graph of the function $f(x) = x^{-1}$ is a curve called *equilateral hyperbola* (Fig. 17.4).

The coordinates of some points of the graph of $f(x) = \frac{1}{x^2}$ follow:

Fig. 17.4 The graph of $f(x) = \frac{1}{x}$

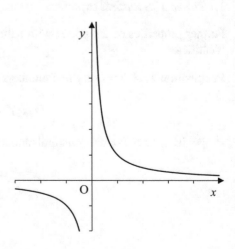

Fig. 17.5 The graph of
$f(x) = \frac{1}{x^2}$

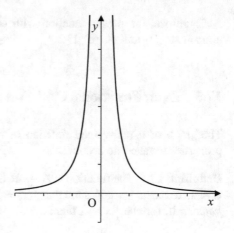

$$f\left(\tfrac{1}{2}\right) = 4$$
$$f(1) = 1$$
$$f(2) = \tfrac{1}{4}$$
$$f\left(-\tfrac{1}{2}\right) = 4$$
$$f(-1) = 1$$
$$f(-2) = -\tfrac{1}{4}$$

The graph of the function $f(x) = x^{-2}$ is drawn in Fig. 17.5.

3. *Further properties of the power*

For every non-null real numbers x, x_1, x_2 and integers p, q the following identities hold:

$$(x_1 x_2)^p = x_1^p x_2^p$$
$$x^p x^q = x^{p+q}$$
$$(x^p)^q = x^{pq}$$
$$x^{-p} = \tfrac{1}{x^p}, \quad x \neq 0$$

The power with exponent $-n$, for every $n \in \mathbf{N}$, is strictly decreasing in $(0, +\infty)$. If n is even (odd) the power is strictly increasing (strictly decreasing) in $(-\infty, 0)$ (Figs. 17.4 and 17.5).

17.4.3   Null Exponent

Whatever the non-zero real number x is, by definition it is assumed $x^0 = 1$. For reasons that will be clarified, it is not convenient to attribute meaning to the symbol 0^0.

Therefore, the power function with exponent 0 is the constant function with domain $\mathbf{R}-\{0\}$ and range $\{1\}$.

17.5 Even Functions, Odd Functions

The graph of a real-valued function of a real variable sometimes has symmetry properties to take into account.

Definition 17.1 The function $f: A \to \mathbf{R}$ is said to be an *even function* if, for every x $\in$ A there exists $-x \in$ A such that $f(x) = f(-x)$; the function f is said to be an *odd function* if, for every $x \in$ A there exists $-x \in$ A such that $-f(x) = f(-x)$.

As a consequence of the definition, if the point $(x, y) = (x, f(x))$ belongs to the graph of an even function f, then also the point $(-x, y) = (-x, f(x))$ belongs to the graph of f; therefore, the graph of an even function is symmetrical with respect to the y axis (Sect. 7.1) (Fig. 17.6).

Similarly, if the point $(x, y) = (x, f(x))$ belongs to the graph of an odd function f, then also the point $(-x, -y) = (-x, -f(x)) = (-x, f(-x))$ belongs to the graph of f; therefore, the graph of an odd function is symmetrical with respect to the origin of the coordinates (Sect. 7.1.1).

Examples The function cosine (Sect. 8.1.3) is even, the functions sine and tangent are odd.

If n is an even positive integer, then the function $f(x) = x^n$ is even (Fig. 17.6); if n is odd, the function $f(x) = x^n$ is odd (Fig. 17.7).

If n is an even positive integer, the function $f(x) = x^{-n}$ is even (Fig. 17.5); if n is odd, the function $f(x) = x^{-n}$ is odd (Fig. 17.4).

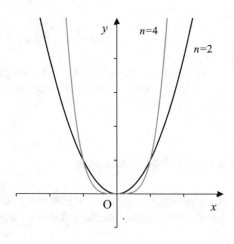

Fig. 17.6 Even functions. The graphs of f(x) = x^n, n = 2, 4

Fig. 17.7 Odd functions.
The graphs of f(x) = x^n, n =
1, 3, 5

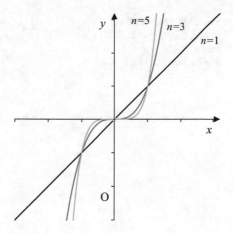

Remark 17.1 For m and n even positive integers and $m < n$, if $x \in (-1, 1)$ then x^m $> x^n$; for x not belonging to the interval $[-1, 1]$, $x^m < x^n$ (Fig. 17.6).

Remark 17.2 For m and n odd positive integers and $m < n$, if $x \in (0, 1)$ then $x^m > x^n$; if $x > 1$, then $x^m < x^n$ (Fig. 17.7).

17.6 The Root

Definition 17.2 For every natural number n, the *root function of index n*, or the *n*-th *root* of x, denoted $\sqrt[n]{x}$, is defined as the function which has domain **R** and range **R** if n is odd, has domain $[0, +\infty)$ and range $[0, +\infty)$ if n is even and whose value is calculated in this way:

– if n is even, for each x belonging to $[0, +\infty)$, the symbol $\sqrt[n]{x}$ denotes the unique nonnegative real number whose n-th power is equal to x;
– if n is odd, for each real number x the symbol $\sqrt[n]{x}$ denotes the unique real number whose n-th power is equal to x.

Given the function $\sqrt[n]{x}$, the symbol $\sqrt[n]{}$ is called the *radical symbol* and the number x underneath the radical symbol is called *radicand*.

Remark 17.3 If n is even, the equation $x^n = a$, with positive real a, admits in addition to the positive solution $\sqrt[n]{x}$, called the n-th *arithmetic root* of the positive real number a, also the negative solution $-\sqrt[n]{x}$.

The root of index 2 is also called the square root and is indicated by omitting to write the index, i.e., $\sqrt{x} = \sqrt[2]{x}$; the root of index 3, $\sqrt[3]{x}$, is called the *cube root*.
Whatever the natural numbers m and n are, the following equalities hold:

Fig. 17.8 Graphs of
$x, \sqrt{x}, \sqrt[3]{x}$. If $0 < x < 1$,
then $x < \sqrt{x} < \sqrt[3]{x}$; if
$x > 1$, then $\sqrt[3]{x} < \sqrt{x} < x$

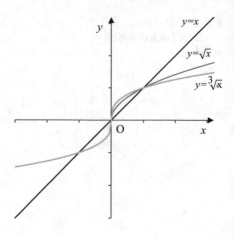

$$\left(\sqrt[n]{x}\right)^m = \sqrt[n]{x^m}$$
$$\sqrt[m]{\sqrt[n]{x}} = \sqrt[mn]{x}$$

The two equalities above are valid for any real x such that both sides have meaning. The equality

$$\sqrt[n]{x} = \sqrt[nm]{x^m}$$

applies only if $x \geq 0$. Moreover, for every natural n, the following equality holds:

$$\sqrt[n]{xy} = \sqrt[n]{x}\,\sqrt[n]{y}$$

whatever $x, y \in \mathbf{R}$, such that both sides have meaning. Furthermore, if m and n are natural numbers and $m < n$, if $0 < x < 1$, then $\sqrt[m]{x} < \sqrt[n]{x}$; if $x > 1$, then $\sqrt[n]{x} < \sqrt[m]{x}$ (Fig. 17.8).

If $x > 0$ and $r = \frac{m}{n}$ is a rational number with integer m and natural n, it is assumed by definition,

$$x^r = x^{\frac{m}{n}} = \sqrt[n]{x^m} \tag{17.1}$$

In particular

$$x^{\frac{1}{n}} = \sqrt[n]{x}$$

Equation (17.1) defines the power function with rational exponent.

Remark 17.4 The rational number r may be represented in infinite ways by means of a fraction. It is proved that definition (17.1) is independent of the particular fractional representation of r.

If f and $g(x) = x^n$, then $f(g(x)) = \sqrt[n]{g(x)} = \sqrt[n]{x^n} = x^{\frac{n}{n}} = x$. Also:

Fig. 17.9 Graphs of x, x^2
and $\sqrt{x}$

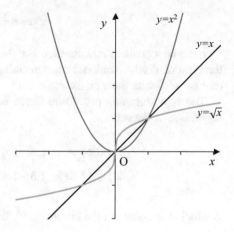

$$g(f(x)) = g(\sqrt[n]{x}) = \left(x^{\frac{1}{n}}\right)^n = x.$$

Thus, for every x and for every natural n, such that the function $\sqrt[n]{x}$ is defined, the functions $\sqrt[n]{x}$ and x^n are inverse of each other (Sect. 7.6). The graphs of x, x^2, $\sqrt{x}$ are drawn in (Fig. 17.9).

17.6.1 Further Properties of the Power with Rational Exponent

For every real positive a, whatever the rational numbers r and s are, the following equalities hold:

$$a^{r+s} = a^r a^s$$
$$\left(a^r\right)^s = a^{rs}$$
$$a^{-r} = \frac{1}{a^r}$$

If a is a real number greater than 1 and r and s are two rational numbers, if $r > s$, then $a^r > a^s$. If the positive real number a is less than 1 and r and s are rational numbers with $r > s$, then $a^r < a^s$.

17.7 Power with Real Exponent

In (Sect. 17.6) the power function with rational exponent $r = \frac{m}{n}$ has been defined:

$$a^r = a^{\frac{m}{n}} = \sqrt[n]{a^m}$$

Let a be a positive real number. For the completeness of the real field (Sect. 6.7) the symbol a^r with irrational r has meaning. To get an idea of the construction of the number a^r, let us show an example.

The approximation procedure (Sect. 6.7) of the value of $\sqrt{2}$, as the element of separation of the sets

$$A\left(\sqrt{2}\right) = 1;\ 1.4;\ 1.41;\ 1.414;\ 1.4142;\ \dots$$
$$B\left(\sqrt{2}\right) = 2;\ 1.5;\ 1.42;\ 1.415;\ 1.4143;\ \dots$$

is adopted to construct the number $a^{\sqrt{2}}$. Indeed, the two sets:

$$A = \left\{ a^1, a^{\frac{14}{10}}, a^{\frac{141}{100}}, a^{\frac{1414}{1000}}, a^{\frac{14142}{10000}},\ \dots \right\}$$
$$B = \left\{ a^2, a^{\frac{15}{10}}, a^{\frac{142}{100}}, a^{\frac{1415}{1000}}, a^{\frac{14143}{10000}},\ \dots \right\}$$

are separate and contiguous (Sect. 6.9) and $a^{\sqrt{2}}$ is the element of separation of the sets A and B.

Definition 17.3. Let γ be a non-integer real number. Then the function of x.

$$x^\gamma$$

which has domain $[0, +\infty)$ if γ is positive and has domain $(0, +\infty)$ if γ is negative, is called *power function with non-integer real exponent*. The value x^γ is calculated as the element of separation of two contiguous classes like described above.

For every positive real x and for $\gamma \neq 0$, it is assumed:

$$x^{-\gamma} = \frac{1}{x^\gamma}$$

17.8   The Exponential Function

The expression a^r gives rise to two different classes of functions, depending on whether the independent variable is the base or the exponent. In the first case, already (Sects. 17.4.1 to 17.5) studied, the power function is defined.

Let us now consider the real-valued function f of the real variable x

$$f(x) = a^x$$

with real positive a different from 1, called the *exponential function* with base a. As indicated in the (Sect. 17.7), the domain of the exponential function is **R** and the range is the set of the positive real numbers.

17.8.1 Properties of the Exponential Function

If a and b are positive real numbers different from 1, the following identities hold:

$$a^0 = 1$$
$$a^{x+y} = a^x a^y$$
$$a^{-x} = \frac{1}{a^x}$$
$$a^{x-y} = \frac{a^x}{a^y}$$
$$(a^x)^y = a^{xy}$$
$$(ab)^x = a^x b^x$$

whatever the real numbers x and y are.

We illustrate with examples the graphs of a^x in the cases: $a > 1$ and $0 < a < 1$.

i. $a > 1$. For example, let $a = 2$. The graph of $f(x) = 2^x$ is the set of the points $(x, 2^x)$, $x \in \mathbf{R}$.

Let us get some points of the graph:

$$f(0) = 2^0 = 1$$
$$f\left(\tfrac{1}{2}\right) = 2^{\frac{1}{2}} = \sqrt{2}$$
$$f(1) = 2$$
$$f(2) = 2^2$$
$$f(-1) = \tfrac{1}{2}$$
$$f(-2) = 2^{-2} = \tfrac{1}{4}$$
$$f(-3) = \tfrac{1}{8}$$

The graph of 2^x is drawn in Fig. 17.10.

If $a > 1$, the exponential function a^x is strictly increasing in **R**. Namely,

$$\text{if } a > 1 \text{ and } p < q, \text{ then } a^p < a^q.$$

In particular,

$$a^{-1} < a^0 = 1 < a < a^2$$

For example, as a^x is strictly increasing in **R**, the solutions of the inequality

$$2^x < 2^3$$

Fig. 17.10 Graph of the exponential function $f(x) = 2^x$

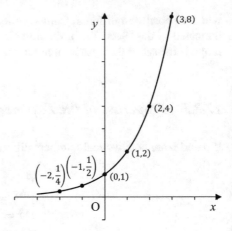

are the real numbers $x < 3$.

ii.　$0 < a < 1$. For example, let $a = \frac{1}{2}$. The graph of $f(x) = \left(\frac{1}{2}\right)^x$ is the set of points $\left(x, \left(\frac{1}{2}\right)^x\right)$, $x \in \mathbf{R}$. Let us get some points of the graph:

$$f(0) = \left(\tfrac{1}{2}\right)^0 = 1$$
$$f\left(\tfrac{1}{2}\right) = \left(\tfrac{1}{2}\right)^{\frac{1}{2}} = \tfrac{1}{\sqrt{2}}$$
$$f(1) = \tfrac{1}{2}$$
$$f(2) = \left(\tfrac{1}{2}\right)^2 = \tfrac{1}{4}$$
$$f(-1) = \left(\tfrac{1}{2}\right)^{-1} = 2$$
$$f(-2) = \left(\tfrac{1}{2}\right)^{-2} = 4$$
$$f(-3) = 8$$

The graph of $\left(\frac{1}{2}\right)^x$ is drawn in Fig. 17.11.

If $0 < a < 1$, the exponential function a^x is strictly decreasing in $\mathbf{R}$. This means:

$$\text{if } 0 < a < 1 \text{ and } p < q, \text{ then } a^p > a^q$$

In particular,

$$a^{-1} > a^0 = 1 > a > a^2$$

If we want to find the solutions of the inequality

$$\left(\frac{1}{2}\right)^x < \left(\frac{1}{2}\right)^3$$

Fig. 17.11 Graph of the exponential function $\left(\frac{1}{2}\right)^x$

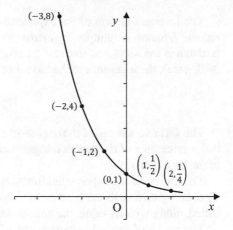

let us just consider that the function $\left(\frac{1}{2}\right)^x$ is strictly decreasing in **R** to conclude that the solutions of the inequality are the real numbers $x > 3$.

For every $a > 0$ and $a \neq 1$, the graph of the exponential function $f(x) = a^x$ contains the point $(0, a^0) = (0, 1)$.

17.8.2 The Number of Napier

John Napier (1550–1617) was a Scottish mathematician. *Napier's number* is an irrational number denoted by the letter e, whose first digits are

$$e = 2.7182818284\ldots$$

The exponential function with base e, namely the function e^x, is simply called the *exponential function* and is also denoted by the symbol *exp(x)*.

Remark 17.5 *Famous numbers.* In the interval $(1, 2)$ the irrational number $\sqrt{2}$ had long been found. Since ancient times, π had found its place in the interval $(3, 4)$. A famous number was therefore expected in the interval $(2, 3)$. This eventually came: the number e, of which we will learn about later.

17.9 The Logarithm

We have seen that the exponential function a^x is strictly increasing in **R** if $a > 1$, and is strictly decreasing in **R** if $0 < a < 1$. In each of the two cases then the exponential function is invertible.

The inverse function of the exponential function with base a is called the *logarithmic function* or, simply, *logarithm* to the base a. The domain of the logarithm function is the set of positive real numbers $(0, +\infty)$, and the range is **R**. For each $x \in (0, +\infty)$, the logarithm to the base a of x, is denoted

$$\log_a x$$

The variable x is called the *argument* of the logarithm. The number $\log_a x$, which is the value on x of the function logarithm to the base a, is still called the *logarithm* to the base a of x.

The inverse of the exponential function e^x, being e the number of Napier, is called *logarithm function* to the base e, or, simply, *logarithm* and is denoted $\log x$, or $\ln x$, called, along with its value, the *natural logarithm* of x.

The graph of $\log_a x$ is symmetrical of the graph of $y = a^x$ with respect to the bisector of the first and third quadrant (Sect. 7.7).

If $a > 1$, the logarithm function to the base a is strictly increasing in $(0, +\infty)$. This means that, if $a > 1$ and $p < q$, then $\log_a p < \log_a q$ (Fig. 17.12). In particular,

$$0 = \log_a 1 < \log_a 2$$

Let us solve the inequality: $\log_a x < \log_a 3$ with $a > 1$, in the unknown x. Since $\log_a x$ is strictly increasing, the solutions x are such that: $0 < x < 3$.

If $0 < a < 1$, the logarithm to the base a is strictly decreasing in **R**. This means that if $p < q$, then $\log_a p > \log_a q$ (Fig. 17.12). In particular, $0 = \log_a 1 > \log_a 2$.

Let us solve the inequality: $\log_a x < \log_a 3$ with $0 < a < 1$, in the unknown x. Since $\log_a x$ is strictly decreasing, the set of the solutions of $\log_a x < \log_a 3$ is $\{x \in$ **R**$: x > 3\}$.

We have stated that the exponential function and the logarithm are inverse of each other:

Fig. 17.12 The function $g(x) = \log_a x, a > 1$, is strictly increasing; the function $f(x) = \log_a x, 0 < a < 1$, is strictly decreasing

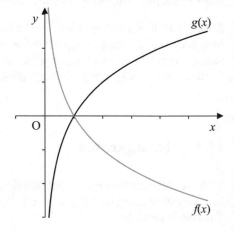

$$y = f(x) = a^x \text{ if and only if } x = f^{-1}(y) = \log_a y \qquad (17.2)$$

In particular

$$y = f(x) = e^x \text{ if and only if } x = f^{-1}(y) = \ln y \qquad (17.3)$$

For example, if $a = 2$ the following equivalence holds:

$$3 = \log_2 8 \text{ if and only if } 8 = 2^3$$

(3 is the logarithm to the base 2 of 8 means that 3 is the exponent to give to 2 to obtain 8). Furthermore

$$-3 = \log_2 \frac{1}{8} \text{ if and only if } \frac{1}{8} = 2^{-3}$$

Propositions (17.2) and (17.3) imply:

$$a^{\log_a x} = x, \quad e^{\ln x} = x$$

for every $x > 0$.
The following property holds:

1. $\log_a 1 = 0$, for every $a > 0$ and $a \neq 1$,
    For every positive x, x_1 and x_2, operations within logarithm arguments apply:
2. $\log_a(x_1 x_2) = \log_a x_1 + \log_a x_2$.
    $$\log_a \frac{x_1}{x_2} = \log_a x_1 - \log_a x_2$$
3.
    $$\log_a \frac{1}{x} = -\log_a x$$
4. for every positive x, for every h in $\mathbf{R}$ and a, b positive and different from 1, the equalities hold:

$$\log_a x^h = h \log_a x \qquad (17.4)$$

$$\log_a a^x = x \quad \ln e^x = x$$

$$\log_b x = \frac{\log_a x}{\log_a b} \qquad (17.5)$$

5. $a^{k \log_a x} = x^k \quad e^{k \ln x} = x^k$
    $\ln e^x = x$
    Equation (17.5) is called *logarithm change of base formula*.

Applying properties 1 to 4 and (17.4) evaluate:

$$\log_2 8^{-1} = -\log_2 8 = -\log_2 2^3 = -3\log_2 2 = -3$$
$$\log_2(8 \times 32 \times 64) = \log_2 8 + \log_2 32 + \log_2 64 = 3 + 5 + 6 = 14$$
$$\log_4 \sqrt[5]{64} = \tfrac{1}{5}\log_4 64 = \tfrac{3}{5}$$
$$\ln e^7 = 7$$

17.10   Conclusion

In this chapter we have introduced some noteworthy properties of the real-valued functions of a real variable. In particular, we have studied the power, exponential and logarithm functions. In Chap. 8 the sine, cosine and tangent functions were studied. The mentioned functions belong to the class of *elementary functions*. There are other functions named elementary. They are the inverse functions of the circular functions, that we will see below.

17.11   Exercises

1. Find the domain of the following functions and determine the values $y = f(x)$ such that $y = 0$.

   (a) $f(x) = 8 - x^3$ *Ans*. (a) $\mathbf{R}$; $f(2) = 0$
   (b) $f(x) = \frac{x-1}{x+2}$ *Ans*. (b)$\mathbf{R}$; $f(1) = 0$
   (c) $f(x) = \frac{1}{x^2+1}$ *Ans*. (c)$\mathbf{R}$; $f(x) > 0, \forall x \in \mathbf{R}$
   (d) $f(x) = \sqrt{x^2-4}$ *Ans*.   (d) $(-\infty, -2] \cup [2, +\infty)$; $f(-2) = 0$, $f(2) = 0$
   (e) $f(x) = \sqrt{-7+x}$ *Ans*. (e)$[7, +\infty)$; $f(7) = 0$

2. Find the domain of the function $f(x) = \ln \frac{x+2}{x-3}$

   Solution The argument of logarithm is positive: $\frac{x+2}{x-3} > 0$ This implies:

   $$x > -2 \text{ and } x > 3 \quad \text{or} \quad x < -2 \text{ and } x < 3$$

   Then either $x > 3$ or $x < -2$. Thus, Dom$(f) = (-\infty, -2) \cup (3, +\infty)$
3. Find the domain of the function:

   $$f(x) = \frac{1}{1 - 2^x}$$

   Solution The denominator must be non-null: $2^x \neq 1$. Then: Dom$(f) = \mathbf{R}-\{0\}$.
4. Find the inverse of the functions:

   a.   The function $f(x) = \frac{2x+1}{3x-1}$ is defined in the set $\mathbf{R} - \{\frac{1}{3}\}$.

Let us consider the equation with the unknown x:

$$y = \frac{2x+1}{3x-1}$$

Solving for x we obtain:

$$x = \frac{y+1}{3y-2}. \qquad (17.6)$$

Then, for every $y \neq \frac{2}{3}$ Eq. (17.6) admits a unique solution, while if $y = \frac{2}{3}$ it is incompatible. This means that the range of f is $\mathbf{R} - \{\frac{2}{3}\}$ and f is invertible and its inverse is

$$f^{-1} : y \in \mathbf{R} - \left\{\frac{2}{3}\right\} \rightarrow \frac{y+1}{3y-2}$$

b. $y = f(x) = x - 3$ Ans. $x = y - 7$
c. $y = 5x - 7$ Ans. $x = \frac{1}{5}y + \frac{7}{5}$
d. $y = 2^{x+1}$ Ans. $x + 1 = \log_2 y \Leftrightarrow x = -1 + \log_2 y$
e. $y = \ln(x - 1)$ Ans. $x - 1 = e^y \Leftrightarrow x = 1 + e^y$

5. Find the domain of the function

$$f(x) = \frac{6\sqrt{x+1} + 7\sqrt{4-x}}{x^2 - 7x + 6}$$

By Definition 17.2, since the indices of the roots of $\sqrt{x+1}$ and $\sqrt{4-x}$ are even and the denominator is null if and only if $x = 1$ or $x = 6$ (Sect. 12.1), the domain of $f(x)$ is the set of the real numbers x that are the solutions of the system of inequalities:

$$x \geq -1$$
$$x \leq 4$$
$$x \neq 1 \text{ and } \neq 6$$

Therefore, the domain of $f(x)$ the set $\{x \in \mathbf{R} : -1 \leq x < 1, 1 < x \leq 4\}$.

6. Find the domain of the function

$$f(x) = \frac{\sqrt[3]{x} + 3}{1 - \sqrt{x+2}}$$

By definition 17.2 we have $x > -2$ and as $1 - \sqrt{x+2} \neq 0$ if and only if $x \neq -1$, the domain of $f(x)$ is the set $\{x \in \mathbf{R} : -2 < x < -1, -1 < x < +\infty\}$.

7. Solve the exponential equation:

$$6^{2-x} \times 3^{x+1} = 864$$

By the properties of the exponential function (Sect. 17.8), the equation may be written as:

$$\frac{6^2}{6^x} \times 3^x \times 3 = 864$$

Then

$$\frac{3^x}{6^x} \times 6^2 \times 3 = 864$$

or

$$\left(\frac{3}{6}\right)^x \times 6^2 \times 3 = 864$$

i.e.

$$\left(\frac{1}{2}\right)^x \times 108 = 864$$

and

$$\left(\frac{1}{2}\right)^x = 8$$

Hence, $x = -3$.

8. Solve the exponential equation:

$$3^{2x} + 3^x - 6 = 0 \tag{17.7}$$

Let us put

$$y = 3^x \tag{17.8}$$

to obtain the equation $y^2 + y - 6 = 0$ whose solution are: $y_1 = -3$ and $y_2 = 2$. By (17.8) we get: $-3 = 3^x$ that does not yield any solution of (17.7), and $2 = 3^x$ that admits the solution $x = \log_3 2$.

9. Solve the logarithmic equation:

$$\ln(x^2 + 1) = \ln(6x - 6) \tag{17.9}$$

Observe that the necessary condition for the Eq. (17.9) to be compatible is that the equation

$$x^2 + 1 = 6x - 6, \text{ i.e., the equation } x^2 - 6x + 5 = 0 \qquad (17.10)$$

be compatible. The solutions of Eq. (17.10) are $x_1 = 1$ and $x_2 = 5$, but only x_2 is acceptable since the logarithm is not defined at 0 (indeed, $(6x-6)_{x=1} = 0$). Therefore, the Eq. (17.9) is compatible and has the unique solution $x_1 = 1$.

10. Solve the logarithmic equation:

$$\ln(x^2 + 1) = \ln(6x - 7)$$

The equation $x^2 + 1 = 6x - 7$, i.e., $x^2 - 6x + 8 = 0$, has solutions $x_1 = 2$ and $x_2 = 4$, that are both acceptable.

Bibliography

Anton, H.: Calculus. John Wiley & Sons Inc., New York (1980)

Lax, P., Burnstein, S., Lax, A.: Calculus with Applications and Computing. Springer-Verlag, New York (1976)

Royden, H.L., Fitzpatrick, P.M.: Real Analysis. Pearson, Toronto (2010)

Spivak, M.: Calculus. Cambridge University Press (2006)

Ventre, A.: Matematica. Fondamenti e calcolo. Wolters Kluwer Italia, Milan (2021)

Chapter 18
Limits

18.1 Introduction

The concepts of accumulation point and limit are basic. The idea of movement is introduced. In fact, we observed (Sect. 6.6) that if a is an accumulation point of the numerical set A, every neighborhood of a contains infinite points of A. This infinite set of points is being formed by successive choices, each of which is made by ascertaining that:

- in a neighborhood I' of a there exists a point a' of A and a' is different from a,
- in a neighborhood I'' of a properly contained in I' there exists a point a'' of A and a'' is different from a and a', and so on.

The way of narrowing, approaching, accumulating the points a', a'', …, near a given point a, suggests the idea of movement.

18.2 Definition

In Sect. 6.6 the extended set of real numbers $\mathbf{R}^* = \mathbf{R} \cup \{-\infty, +\infty\}$ was introduced. The elements $+\infty$ and $-\infty$ are called *points at infinity* of $\mathbf{R}^*$. In Sect. 6.5 the notions of neighborhood of a point a of $\mathbf{R}$ and neighborhood of $+\infty$ and $-\infty$ have been introduced. Moreover, the elements $+\infty$ and $-\infty$, which belong to $\mathbf{R}^*$, are accumulation points of $\mathbf{R}$ that do not belong to $\mathbf{R}$.

Let f be a real-valued function of a real variable whose domain is the set $\mathrm{Dom}(f) = A \subseteq \mathbf{R}$. Let c be an accumulation point of A belonging to the set $\mathbf{R}^*$. The behavior of f in a neighborhood of c allows to introduce the concept of *limit of the function f as x approaches c*.

Definition 18.1 [Definition of limit]. Let f be a real-valued function of a real variable having domain $\mathrm{Dom}(f) = A \subseteq \mathbf{R}$ and let $c \in \mathbf{R}^*$ be an accumulation point of A. The element $\ell \in \mathbf{R}^*$ is called *limit of the function f as x approaches* (or *tends to*) c, if for

© The Author(s), under exclusive license to Springer Nature Switzerland AG 2023   291
A. G. S. Ventre, *Calculus and Linear Algebra*,
https://doi.org/10.1007/978-3-031-20549-1_18

every neighborhood J of ℓ there exists a neighborhood $I(c)$ of c, such that $f(x) \in$ J, whenever $x \neq c$ and $x \in \mathrm{Dom}(f) \cap I(c)$. Then we write

$$\lim_{x \to c} f(x) = \ell \tag{18.1}$$

An expression equivalent to (18.1) is "$f(x) \to \ell$ as $x \to c$" ($f(x)$ approaches ℓ as x approaches c) that stresses the relation between the two movements $x \to c$ and $f(x) \to \ell$.

Remark 18.1 The definition of limit does not require that c belongs to $\mathrm{Dom}(f)$ and even though $c \in \mathrm{Dom}(f)$ it is not required $f(c) \in$ J.

If $\ell \in \mathbf{R}$ and $c \in \mathbf{R}^*$, then the function f is said *to have finite limit* ℓ as x approaches c, or f *converges* to ℓ as x approaches c.

If $c \in \mathbf{R}^*$ and $\lim_{x \to c} f(x) = +\infty$ ($\lim_{x \to c} f(x) = -\infty$), then f is said to positively (negatively) diverge; also, f is said to be positively (negatively) divergent as $x \to c$.

If f satisfies the definition of limit as x tends to $c \in \mathbf{R}^*$, then the function is said to be *regular* at c. Other possible ways to convey the concept of the regularity of f at c are: f is regular if it is convergent or divergent, positively or negatively, as $x \to c$; or f is regular if and only if it is endowed with limit at c. A regular function f at c is said to *admit* limit at c.

If ℓ and c belong to $\mathbf{R}$, the neighborhood J may be replaced with a suitable neighborhood of radius $\varepsilon > 0$, i.e., $(\ell - \varepsilon, \ell + \varepsilon)$, and centered on ℓ; hence, Definition 18.1 may be restated as follows:

Definition 18.2 The number $\ell \in \mathbf{R}$ is called *limit* of $f(x)$ as x approaches $c \in \mathbf{R}$, written

$$\lim_{x \to c} f(x) = \ell$$

if for every positive real number ε there exists a neighborhood $I(c)$ of c, such that

$$\ell - \varepsilon < f(x) < \ell + \varepsilon \tag{18.2}$$

for every $x \in I(c) \cap \mathrm{Dom}(f)$, $x \neq c$.

Let us remark that by (6.15) the inequalities (18.2) assume the equivalent form:

$$|f(x) - \ell| < \varepsilon$$

So the definition of limit when $\ell \in \mathbf{R}$ and $c \in \mathbf{R}$ may be reformulated in this way: the number $\ell \in \mathbf{R}$ is called *limit* of $f(x)$ as $x \to c \in \mathbf{R}$, written $\lim_{x \to c} f(x) =$, if for any $\varepsilon > 0$ a positive number δ exists such that $|f(x) - \ell| < \varepsilon$, whenever $0 < |x - c| < d$.

Remark 18.2 We observe that if the function f admits finite limit ℓ as x approaches the point $c \in \mathbf{R}$, only the values taken by f at the points in a neighborhood of c, but different from c, contribute to the determination of the limit ℓ of f as x approaches c. Therefore, if c belongs to the domain of f, the value $f(c)$ and the limit ℓ are numbers completely independent of each other.

18.2.1 Specific Applications of Definition 18.1

Recall (Sect. 6.6.1) that a left neighborhood of $c \in \mathbf{R}$ is a half-open interval $(a, c]$, $a < c$, and a right neighborhood of c is a half-open interval $[c, b)$, $c < b$. We denote by $I^-(c)$ and $I^+(c)$ a left neighborhood and a right neighborhood of the point c, respectively. Evidently, the union $I^-(c) \cup I^+(c)$ is a neighborhood of c.

Let us comment on specific applications of Definition 18.1.

(a) ℓ and c are finite (Fig. 18.1).

The function f converges to ℓ as x approaches $c \in \mathbf{R}$. Consider the graph of a function f (Fig. 18.1). For every neighborhood $J(\ell)$ of ℓ there exists a neighborhood $I(c)$ of c, such that if $x \neq c$ and $x \in \mathrm{Dom}(f) \cap I(c)$, then $f(x) \in J(\ell)$.

Example 18.1 Let us consider the function

$$f(x) = \frac{x + 6}{x - 2}$$

whose domain is $\mathrm{Dom}(f) = \mathbf{R} - \{2\}$. The point $c = 4$ is an accumulation point of $\mathrm{Dom}(f)$. Replacing x with 4, we suspect that $f(x)$ will approach $\frac{4+6}{4-2} = 5$. To verify our conjecture we need to prove that

$$\lim_{x \to 4} \frac{x + 6}{x - 2} = 5$$

By Definition 18.2, the inequalities (18.2) become:

$$\frac{x + 6}{x - 2} > 5 - \varepsilon \tag{A}$$

Fig. 18.1 $\lim\limits_{x \to c} f(x) = \ell, c$
and ℓ real
numbers, $I(c) = I^-(c) \cup I^+(c)$

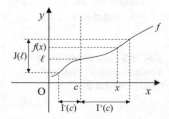

$$\frac{x+6}{x-2} < 5 + \varepsilon \tag{B}$$

Let us find the solutions of inequality (A). Let us assume that x is in a fairly narrow neighborhood of 4 so that the number $x-2$ is positive. Then, multiplying by $x-2$ the two sides of (A), we get (Sect. 6.4, property 4), $(5-\varepsilon)(x-2) < x+6$, and

$$(5-\varepsilon)x - 10 + 2\varepsilon < x + 6$$

Then

$$(4-\varepsilon)x < 16 - 2\varepsilon$$

Consider ε small enough, e.g., $4 > \varepsilon > 0$, to obtain:

$$x < \frac{16 - 2\varepsilon}{4 - \varepsilon}$$

Observe that

$$\frac{16 - 2\varepsilon}{4 - \varepsilon} > \frac{16 - 4\varepsilon}{4 - \varepsilon} = 4$$

Therefore, the solutions of inequality (A) are the numbers x smaller than a number greater than 4, and define a right neighborhood $I^+(4)$ of 4 (Fig. 18.2).

Let us find the solutions of inequality (B). Similarly, we obtain

$$x + 6 < (5+\varepsilon)x - 10 - 2\varepsilon$$

and

$$\frac{16 + 2\varepsilon}{4 + \varepsilon} < x$$

Observe that

$$\frac{16 - 2\varepsilon}{4 - \varepsilon} > \frac{16 - 4\varepsilon}{4 - \varepsilon} = 4$$

Therefore, the solutions of (B) are the numbers x greater than a number smaller than 4, and define a left neighborhood $I^-(4)$ of 4 (Fig. 18.2). To summarize, it has

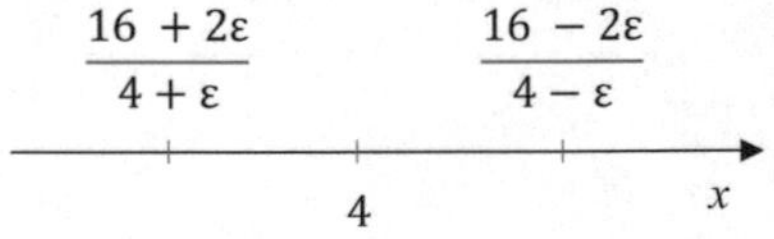

Fig. 18.2 Neighborhood of 4 containing the points that satisfy (A) and (B)

$$\frac{16 + 2\varepsilon}{4 + \varepsilon} \qquad \frac{16 - 2\varepsilon}{4 - \varepsilon}$$

been shown that, for every $\varepsilon > 0$, there exists a neighborhood $I(4) = I^-(4) \cup I^+(4)$ such that, if $x \neq 4$ and x belongs to the intersection $\text{Dom}(f) \cap I(4)$, then $f(x)$ belongs to the interval $(5-\varepsilon, 5 + \varepsilon)$. Hence, (18.2) is verified.

Let us continue to illustrate specific applications of Definition 18.1, those in which at least one of the two elements, ℓ or c, is $+\infty$ or $-\infty$. Remind that a neighborhood of $+\infty$ is the set of the real numbers greater than a fixed real number h, a neighborhood of $-\infty$ is the set of the real numbers smaller than a fixed real number k.

Let the function f be defined in a neighborhood of $+\infty$. Consider the case:

(b) ℓ finite, $c = + \infty$:

$$\lim_{x \to +\infty} f(x) = \ell \tag{18.3}$$

In case (18.3) the function f converges to ℓ as x approaches $+\infty$. The Definition 18.1 is specified as follows:

- for every neighborhood $J(\ell)$, there exists a neighborhood $I(+\infty)$ such that $f(x) \in J(\ell)$, whenever $x \in I(c) \cap \text{Dom}(f)$, $x \neq c$ or, rewritten in terms of ε and h,
- for every $\varepsilon > 0$, there exists $h \in \mathbf{R}$, such that, for every $x > h$, the inequalities $\ell - \varepsilon < f(x) < \ell + \varepsilon$ hold (Fig. 18.3).

Let the function f be defined in a neighborhood of $-\infty$. Consider the case:

(c) ℓ finite, $c = -\infty$:

$$\lim_{x \to -\infty} f(x) = \ell$$

The function f is convergent to ℓ as x approaches $-\infty$. Definition 18.1 takes the form:

- for every $\varepsilon > 0$, there exists $k \in \mathbf{R}$, such that for every $x < k$, the inequalities $\ell - \varepsilon < f(x) < \ell + \varepsilon$ hold (Fig. 18.4).

Let us now consider the case

(d) $\ell = +\infty$, c finite:

$$\lim_{x \to c} f(x) = +\infty$$

Fig. 18.3 ℓ is limit of f as x approaches $+\infty$, $J(\ell) = (\ell - \varepsilon, \ell + \varepsilon)$, $I(+\infty) = (h, +\infty)$

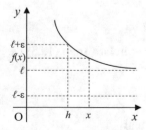

Fig. 18.4 ℓ is limit of f as x approaches $-\infty$, $J(\ell) = (\ell - \varepsilon, \ell + \varepsilon)$, $I(-\infty) = (-\infty, k)$

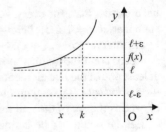

Fig. 18.5 $+\infty$ is limit of f as x approaches c, $J(+\infty) = (k, +\infty)$, $I(c) = I^- U I^+$

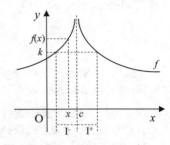

Definition 18.1 takes the form:

– for every neighborhood of $+\infty$, $J(+\infty)$, there exists a neighborhood $I(c)$ such that $f(x) \in J(+\infty)$, whenever $x \in I(c) \cap \text{Dom}(f)$, $x \neq c$.

The function f is positively divergent as $x \to c$ (Fig. 18.5).

Example 18.2 Let us verify the limit

$$\lim_{x \to 1} \frac{1}{(1-x)^2} = +\infty. \tag{18.4}$$

To this aim we first consider that the number 1 is an accumulation point for the domain $\mathbf{R} - \{1\}$ of the function $f(x) = \frac{1}{(1-x)^2}$. We must verify that for every neighborhood of $+\infty$, $J(+\infty)$, there exists a neighborhood I of 1, such that if $x \in I$ and $x \neq 1$, then $\frac{1}{(1-x)^2} \in J(+\infty)$. In other words, we have to check that for every positive real number k, there exists a neighborhood I of 1, such that if $x \in I$ and $x \neq 1$, then

$$\frac{1}{(1-x)^2} > k. \tag{18.5}$$

In fact, from (18.5) we get

$$(1-x)^2 < \frac{1}{k}$$

and

$$|1 - x| < \frac{1}{\sqrt{k}}$$

Thus, by a property of absolute value (Sect. 6.5)

$$1 - \frac{1}{\sqrt{k}} < k < 1 + \frac{1}{\sqrt{k}}$$

These inequalities define a neighborhood of $x = 1$ such that every $x \neq 1$ belonging to the neighborhood satisfies (18.5). Therefore, the limit (18.4) is verified.

Remark 18.3 The verification of (18.4) rests on a property of the fractions: fixed the value of the numerator, if the absolute value of the denominator decreases, then the absolute value of the fraction increases. For example,

$$\frac{1}{0.1} = \frac{1}{\frac{1}{10}} = 10 \qquad \frac{1}{0.01} = \frac{1}{\frac{-1}{100}} = -100 \qquad \frac{1}{0.001} = \frac{1}{\frac{1}{1000}} = 1000$$

The case

(e) $\ell = -\infty$, c finite, i.e., $\lim_{x \to c} f(x) = -\infty$.

is similar to the case (d). The function f is negatively divergent as $x \to c$ (Fig. 18.6).

Consider the case

(f) $\ell = +\infty$, $c = +\infty$:

$$\lim_{x \to +\infty} f(x) = +\infty,$$

Definition 18.1 takes the form:

- for every neighborhood of $+\infty$, $J(+\infty)$, there exists a neighborhood $I(+\infty)$, such that $f(x) \in J(+\infty)$, whenever $x \in I(+\infty) \cap \text{Dom}(f)$, or

$$\forall k > 0, \ \exists h : \ x > h \Rightarrow f(x) > k$$

Fig. 18.6 $-\infty$ is limit of f as $x \to c$, $J(-\infty) = (-\infty, k)$, $I(c) = I^-UI^+$

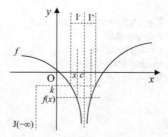

Fig. 18.7 $+\infty$ is limit of f as
x approaches
$+\infty$, $J(+\infty) =$
$(k, +\infty)$, $I(+\infty) =$
$(h, +\infty)$

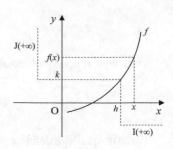

The function f is positively divergent as x approaches $+\infty$ (Fig. 18.7).

Consider the case

(g) $\ell = +\infty$, $c = -\infty$:

$$\lim_{x \to -\infty} f(x) = +\infty$$

The Definition 18.1 takes the form:

– for every neighborhood $J(+\infty)$ there exists a neighborhood $I(-\infty)$ such that if x belongs to $I(-\infty)$, then $f(x) \in J(+\infty)$, or in other terms:

$$\forall k > 0, \quad \exists h : \quad x < h \Rightarrow \quad f(x) > k$$

The function f is positively divergent as x approaches $-\infty$ (Fig. 18.8).

Consider the case

(h) $\ell = -\infty$, $c = +\infty$:

$$\lim_{x \to +\infty} f(x) = -\infty$$

Let us apply Definition 18.1:

– for every neighborhood $J(-\infty)$ there exists a neighborhood $I(+\infty)$ such that if x belongs to $I(+\infty)$, then $f(x) \in J(-\infty)$, or:

Fig. 18.8 $+\infty$ is limit of f as
x approaches
$-\infty$, $J(+\infty) =$
$(k, +\infty)$, $I(-\infty) =$
$(-\infty, h)$

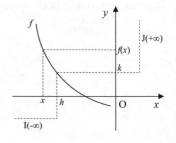

Fig. 18.9 −∞ is limit of f
as x approaches +
∞,J(−∞) =
(−∞, k), I(+∞) =
(h, +∞)

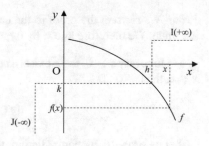

Fig. 18.10 −∞ is limit of f
as x approaches −∞, J(−∞)
= (−∞, k), I(−∞) = (−∞,
h)

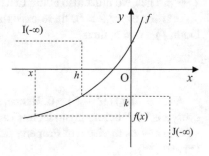

$$\forall k < 0, \quad \exists h : \quad x > h \Rightarrow \quad f(x) < k$$

The function f is negatively divergent as x approaches +∞ (Fig. 18.9).

Consider the case

(i) $\ell = -\infty, \ c = -\infty$:

$$\lim_{x \to -\infty} f(x) = -\infty$$

The Definition 18.1 takes the form:

− for every neighborhood J(−∞) there exists a neighborhood I(−∞) such that if x
belongs to I(−∞), then $f(x) \in$ J(−∞),

$$\forall k < 0, \quad \exists h : \quad x < h \Rightarrow \quad f(x) < k$$

The function f is negatively divergent as x approaches -∞ (Fig. 18.10).

18.2.2 Uniqueness of the Limit

Theorem 18.1 [Uniqueness of the limit]. *If* $\lim_{x \to c} f(x) = \ell$ *exists, then it is unique.*

Proof We restrain the proof to the case that c and ℓ are real numbers, i.e., c and ℓ are finite. We therefore know, by hypothesis, that:

(*) for every $\varepsilon > 0$, there exists a neighborhood I of c such that if $x \in I \cap \text{Dom}(f)$, $x \neq c$, then

$$|f(x) - \ell| < \varepsilon.$$

Let us suppose, by contradiction, that the function f converges also to $\ell' \neq \ell$, as $x \to c$. Then we must also admit that:

(**) for every $\varepsilon > 0$, there exists a neighborhood I' of c such that if $x \in I' \cap \text{Dom}(f)$, $x \neq c$, then

$$\left| f(x) - \ell' \right| < \varepsilon.$$

As $\ell' \neq \ell$, it is $|\ell - \ell'| > 0$. Hence, for every $\varepsilon > 0$, such that $2\varepsilon < |\ell - \ell'|$, and for every $x \in I \cap I'$ belonging to $\text{Dom}(f)$ and $x \neq c$, we obtain, by adding and subtracting $f(x)$ to $\ell - \ell'$, in virtue of Property 6.3 of the absolute value, and by the hypotheses (*) and (**),

$$2\varepsilon < \left| \ell - \ell' \right| = \left| \ell - f(x) + f(x) - \ell' \right| \leq \left| \ell - f(x) \right| + \left| f(x) - \ell' \right| < 2\varepsilon$$

But these inequalities lead to the contradiction $2\varepsilon < 2\varepsilon$. Then we cannot admit the hypothesis (**), that is the existence of limit ℓ'. What claims the uniqueness of the limit ℓ.   □

The theorem holds even if c or ℓ or both are infinite. Therefore, a regular function at c cannot tend to two distinct limits therein.

18.3   Limits of Elementary Functions

In the present section we tell in advance some notions that will be developed in the next chapter. We have already remarked (Sect. 18.2) that if the function f converges to ℓ as x approaches the point c belonging to the domain of f, then the value $f(c)$ and the limit ℓ are numbers completely independent of each other. But if f is an elementary function or the composite function of elementary functions, the computation of the limit $\lim_{x \to c} f(x) = \ell$ when c is a point in the domain of f, is a conceptually simple operation: indeed, we will show that $\lim_{x \to c} f(x) = f(c)$.

Examples

i.   The function x^2 approaches $3^2 = 9$ as x approaches $3 \Leftrightarrow \lim_{x \to 3} x^2 = 9$.

ii. $\lim_{x \to 5} \sqrt{x-1} = \sqrt{5-1} = 2$.

iii. $\lim\limits_{x \to 0} e^x = e^o = 1$

iv. $\lim\limits_{x \to -7} \frac{1}{x} = -\frac{1}{7}$

v. $\lim\limits_{x \to 1} \ln x = \ln 1 = 0$, where ln is the natural logarithm function.

vi. $\lim\limits_{x \to c} k = k$, whatever the real constants c and k are.

vii. The following equalities hold:

$$\lim_{x \to +\infty} 1 = 1$$

$$\lim_{x \to +\infty} x = +\infty$$

$$\lim_{x \to -\infty} x = -\infty$$

$$\lim_{x \to +\infty} \frac{1}{x} = 0$$

$$\lim_{x \to -\infty} \frac{1}{x} = 0$$

$$\lim_{x \to +\infty} \frac{1}{x^2} = 0$$

$$\lim_{x \to -\infty} \frac{1}{x^2} = 0$$

Furthermore, the limits of the power function with natural exponent n, as x approaches $+\infty$ or $-\infty$, can be checked:

if n is even, then

$$\lim_{x \to +\infty} x^n = +\infty \qquad \lim_{x \to -\infty} x^n = +\infty$$

if n is odd, then

$$\lim_{x \to +\infty} x^n = +\infty \qquad \lim_{x \to -\infty} x^n = -\infty$$

if n is natural:

$$\lim_{x \to +\infty} \frac{1}{x^n} = 0 \qquad \lim_{x \to -\infty} \frac{1}{x^n} = 0$$

ix. The limits of exponential function a^x as x approaches $+\infty$ or $-\infty$, are given by:

if $a > 1$, then

$$\lim_{x \to +\infty} a^x = +\infty \qquad \lim_{x \to -\infty} a^x = 0$$

if $0 < a < 1$, then

$$\lim_{x \to +\infty} a^x = 0 \qquad \lim_{x \to -\infty} a^x = +\infty$$

The functions sin$x$, cos$x$ and tan$x$ do not admit limits as x approaches $+\infty$ or $-\infty$.

18.4 Properties of Limits

We have given the definition of limit, finite or infinite, of a function f as x approaches c, finite or infinite (see Sect. 18.2). We have verified the existence of the limit in some cases. The limit does not always exist, as noted at the end of (Sect. 18.3) regarding circular functions. In the following we will illustrate some notable properties of the limits.

18.4.1 Operations

If f and g are real-valued functions defined in a subset D of $\mathbf{R}$, we consider the functions:

$$
\begin{array}{ll}
\text{sum} & f(x) + g(x) \\
\text{difference} & f(x) - g(x) \\
\text{product} & f(x)g(x) \\
\text{ratio} & \frac{f(x)}{g(x)}, g(x) \neq 0
\end{array}
$$

The following theorem, we state without the proof, allows to computing with limits.

Theorem 18.2 *If f and g converge as $x \to c \in \mathbf{R}$, then the sum $f(x) + g(x)$, the difference $f(x) - g(x)$ and the product $f(x)\,g(x)$ converge as $x \to c$ and the following equalities hold*:

$$\lim_{x \to c}(f(x) + g(x)) = \lim_{x \to c} f(x) + \lim_{x \to c} g(x) \qquad (18.6)$$

$$\lim_{x \to c}(f(x) - g(x)) = \lim_{x \to c} f(x) - \lim_{x \to c} g(x) \qquad (18.7)$$

$$\lim_{x \to c}(f(x)g(x)) = \lim_{x \to c} f(x) \lim_{x \to c} g(x) \qquad (18.8)$$

In particular, if $f(x) = h$, $h \in \mathbf{R}$, then

$$\lim_{x \to c} hg(x) = h \lim_{x \to c} g(x)$$

Furthermore,

i.   if h and k are real numbers and $h \leq f(x) \leq k$, for every $x \in \text{Dom}(f)$, and $\lim_{x \to c} g(x)$
$= 0$, then $\lim_{x \to c} (f(x)g(x)) = 0$.

Example 18.3 $\lim_{x \to 0} x \sin \frac{1}{x} = 0$, since $-1 \leq \sin \frac{1}{x} \leq 1$ and $\lim_{x \to 0} x = 0$.

ii.   If f and g converge as $x \to c \in \mathbf{R}$, and if $\lim_{x \to c} g(x) \neq 0$, then also the ratio $\frac{f(x)}{g(x)}$
converges at c and

$$\lim_{x \to c} \frac{f(x)}{g(x)} = \frac{\lim_{x \to c} f(x)}{\lim_{x \to C} g(x)}$$

Equalities (18.6), (18.7) and (18.8) can be expressed in the abbreviated form so: the limit of the sum, difference and product are equal to the sum, difference and product of the limits, respectively. As well as the proposition (ii) can be expressed so: the limit of the ratio is equal to the ratio of the limits, provided that $\lim_{x \to c} g(x) \neq 0$.

As far as the sum is concerned, it is easy to realize that if f is convergent and g positively (negatively) divergent, then $f + g$ is positively (negatively) divergent; similarly, if f and g are positively (negatively) divergent, then also $f + g$ is positively (negatively) divergent. (Let us remember Hotel Hilbert (Sect. 5.3), where $+\infty + 1 = +\infty$, and $+\infty + \infty = +\infty$.)

But, if $\lim_{x \to c} f(x) = +\infty$ and $\lim_{x \to c} g(x) = -\infty$, we are not able to say, without further investigation, what is the meaning of the result of the addition $\lim_{x \to c} f(x) + \lim_{x \to c} g(x) = +\infty - \infty$: it could be 0, or 1, or any number or infinity, positive or negative, or it might not exist. In order to express this circumstance, we say that $+\infty - \infty$ is an *indeterminate form*.

Let now examine when the theorem 18.2 and proposition (ii) can be extended to the case that the functions f and g despite being regular at c, are not both convergent at c. If $\hat{l}$ is a real number and $c \in \mathbf{R}^*$, then:

–   if $\lim_{x \to c} f(x) = \ell$ and $\lim_{x \to c} g(x) = \pm\infty$, then $\lim_{x \to c} (f(x) + g(x)) = \pm\infty$ $[\ell \pm \infty = \pm\infty]$

–   if $\lim_{x \to c} f(x) = +\infty$ and $\lim_{x \to c} g(x) = +\infty$, then $\lim_{x \to c} (f(x) + g(x)) = +\infty$ $[+\infty + \infty = +\infty]$

–   if $\lim_{x \to c} f(x) = -\infty$ and $\lim_{x \to c} g(x) = -\infty$, then $\lim_{x \to c} (f(x) + g(x)) = -\infty$ $[-\infty - \infty = -\infty]$

–   if $\lim_{x \to c} f(x) = \ell \neq 0$ and $\lim_{x \to c} g(x) = \pm\infty$, then $\lim_{x \to c} |(f(x)g(x))| = +\infty$ $[|\ell(\pm\infty)| = +\infty, \ell \neq 0]$

–   if $\lim_{x \to c} f(x) = \pm\infty$ and $\lim_{x \to c} g(x) = \pm\infty$, then $\lim_{x \to c} |(f(x)g(x))| = +\infty$ $[|(\pm\infty)(\pm\infty)| = +\infty]$

–   if $\lim_{x \to c} f(x) = \ell$ and $\lim_{x \to c} g(x) = \pm\infty$, then $\lim_{x \to c} \frac{f(x)}{g(x)} = 0$ $[\frac{\ell}{\pm\infty} = 0]$

– if $\lim_{x \to c} f(x) = \ell$ and $\lim_{x \to c} g(x) = \pm\infty$, then $\lim_{x \to c} \left| \frac{g(x)}{f(x)} \right| = +\infty$ $[\left| \frac{\pm\infty}{\ell} \right| = +\infty]$

– if $\lim_{x \to c} f(x) = \ell \neq 0$ and $\lim_{x \to c} g(x) = 0$, then $\lim_{x \to c} \left| \frac{f(x)}{g(x)} \right| = +\infty$ $[\left| \frac{\ell}{0} \right| = +\infty, \ell \neq 0]$

Some cases, called indeterminate forms and denoted

$$+\infty - \infty, \quad 0 \cdot \infty, \quad \frac{\infty}{\infty}, \frac{0}{0} \qquad (18.9)$$

are excluded from the previous list. Let us stress: the fact that a limit results in an indeterminate form does not necessarily mean that the limit does not exist, but merely a supplement of information is needed to ascertain the existence of the limit and, in case, its value. Sometimes, through suitable transformation or simplification, the uncertainty is eliminated.

Example 18.4 Consider the functions $f(x) = (x + 2)^2$ and $g(x) = x^2$. The limit

$$\lim_{x \to +\infty} (f(x) - g(x))$$

is an indeterminate form $+\infty - \infty$. But, performing a simple calculation we get $f(x) - g(x) = 4x + 4$. Hence, $\lim_{x \to +\infty} (f(x) - g(x)) = \lim_{x \to +\infty} (4x + 4) = +\infty$.

Example 18.5 Consider the functions $f(x) = -2x + 3$ and $g(x) = x^2 + 4x$. The limit

$$\lim_{x \to +\infty} \frac{f(x)}{g(x)} = \lim_{x \to +\infty} \frac{-2x + 3}{x^2 + 4x}$$

is an indeterminate form of the type $\frac{\infty}{\infty}$. To eliminate the indeterminateness let us divide the numerator and denominator by x^2 (the power of the variable x having maximum degree in the fraction). By Theorem 18.2,

$$\lim_{x \to +\infty} \frac{\frac{-2x+3}{x^2}}{\frac{x^2+4}{x^2}} = \lim_{x \to +\infty} \frac{\frac{-2}{x} + \frac{3}{x^2}}{1 + \frac{4}{x}} = \frac{0+0}{1+0} = 0$$

Example 18.6 Consider the functions $f(x) = 5x + 3$ and $g(x) = 7x - 2$. The limit

$$\lim_{x \to +\infty} \frac{5x + 3}{7x - 2}$$

is an indeterminate form $\frac{\infty}{\infty}$. Let us divide the numerator and denominator by x (the power of the variable x having maximum degree in the fraction) to eliminate the indeterminateness. By Theorem 18.2,

$$\lim_{x \to +\infty} \frac{\frac{5x+3}{x}}{\frac{7x-2}{x}} = \lim_{x \to +\infty} \frac{\frac{5x}{x} + \frac{3}{x}}{\frac{7x}{x} - \frac{2}{x}} = \frac{5+0}{7+0} = \frac{5}{7}$$

Example 18.7 Consider the functions $f(x) = 2x^2 + 1$ and $g(x) = x + 4$. The limit

$$\lim_{x \to +\infty} \frac{2x^2 + 1}{x + 4}$$

is an indeterminate form $\frac{\infty}{\infty}$. Let us divide the numerator and denominator by x^2 (the power of the variable x having maximum degree in the fraction). By Theorem 18.2, and (Sect. 18.3 vii)

$$\lim_{x \to +\infty} \frac{2x^2 + 1}{x + 4} = \frac{2 + \lim_{x \to +\infty} \frac{1}{x^2}}{\lim_{x \to +\infty} \frac{1}{x} + 4 \lim_{x \to +\infty} \frac{1}{x^2}} = +\infty.$$

Example 18.8 $\lim_{x \to 1} \frac{\sqrt{x}-1}{x-1}$ is an indeterminate form of the type $\frac{0}{0}$.

Observe that $x - 1 = (\sqrt{x}-1)(\sqrt{x}+1)$. Hence,

$$\lim_{x \to 1} \frac{\sqrt{x}-1}{x-1} = \lim_{x \to 1} \frac{\sqrt{x}-1}{(\sqrt{x}-1)(\sqrt{x}+1)} = \lim_{x \to 1} \frac{1}{\sqrt{x}+1} = \frac{1}{2}$$

18.4.2 Permanence of the Sign

Theorem 18.3 [Permanence of the sign]. *Let $f(x)$ be a real-valued function defined in the set $A \subseteq \mathbf{R}$ that converges at the finite non-zero limit ℓ as x approaches c. Then there exists a neighborhood I of c such that $f(x)$ takes values of the same sign as the limit, in every $x \in A \cap I$, $x \neq c$.*

Proof Suppose

$$\lim_{x \to c} f(x) = \ell > 0$$

By Definition 18.2 for every $\varepsilon > 0$ there exists a neighborhood $I(c)$ such that $\ell - \varepsilon < f(x) < \ell + \varepsilon$ for every $x \in A \cap (I(c) - \{c\}) = H$. Therefore, it is sufficient to choose $\varepsilon < \ell$. in order that $f(x) > 0$ in H (Fig. 18.11). The proof is similar if $\ell < 0$. □

Theorem 18.3 still holds in case of infinite limit.

Fig. 18.11 Permance of the sign

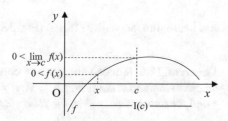

Remark 18.4 The inverse of the Theorem 18.3 does not hold: indeed,

$f(x) > 0$ for every $x \in A \cap (I(c) - \{c\})$, and

$f(x) \to \ell$ as $x \to c$.

do not imply $\ell > 0$. For instance, the function x^2 approaches 0 as $x \to 0$, even if $x > 0$ for every $x \neq 0$.

By Theorem 18.3 reasoning by contradiction, we deduce:

Theorem 18.4 *If ℓ is the limit of $f(x)$ as x approaches c and the values of f in a neighborhood of c are not negative (positive), then*

$$\lim_{x \to c} f(x) = \ell \geq 0 \quad \left(\lim_{x \to c} f(x) = \ell \leq 0 \right)$$

18.4.3   Comparison

We prove some theorems that from the comparison between functions in a neighborhood yield a comparison between the respective limits.

Theorem 18.5 [First theorem of comparison]. *If the functions f and g are convergent as x approaches c and if $f(x) \leq g(x)$ in a neighborhood of c, excluding at most the point c, then*

$$\lim_{x \to c} f(x) \leq \lim_{x \to c} g(x) \tag{18.10}$$

Proof By Theorem 18.2 the function $g(x) - f(x)$ admits limit as x approaches c and its values are not negative in the points x distinct from c and belonging to a neighborhood of c. From Theorem 18.4 we obtain

$$\lim_{x \to c} (g(x) - f(x)) \geq 0$$

and

$$\lim_{x \to c} g(x) - \lim_{x \to c} f(x) \geq 0$$

and finally the inequality (18.10) (Fig. 18.12).   □

Theorem 18.6 [Second theorem of comparison]. *Let f, g and h be real-valued functions defined in the subset $A \subseteq \mathbf{R}$. If $f(x)$ and $h(x)$ converge to ℓ as x approaches c and for any $x \in A$, $x \neq c$, the following inequalities hold*

Fig. 18.12 First theorem of comparison

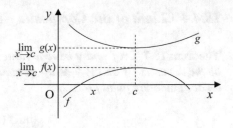

$$f(x) \le g(x) \le h(x)$$

then the function g(x) also converges to ℓ as x → c.

Proof From Definition 18.2, for every $\varepsilon > 0$ there exists a neighborhood of c, $I(c)$, such that for every $x \in I(c)$, $x \ne c$, $x \in A$, we obtain

$$\ell - \varepsilon < f(x) < \ell + \varepsilon$$
$$\ell - \varepsilon < h(x) < \ell + \varepsilon$$

Then, by hypothesis, for every $x \in I(c)$, $x \ne c$, $x \in A$, the following inequalities hold:

$$\ell - \varepsilon < f(x) \le g(x) \le h(x) < \ell + \varepsilon$$

Therefore,

$$\lim_{x \to c} g(x) = \ell$$

as we wanted to show. □

Exercise 18.1 Let us check the limit

$$\lim_{x \to +\infty} \frac{\sin x}{x} = 0. \tag{18.11}$$

In fact, observe that, for every $x > 0$,

$$-\frac{1}{x} \le \frac{\sin x}{x} \le \frac{1}{x}$$

and

$$\lim_{x \to +\infty} \left(-\frac{1}{x} \right) = \lim_{x \to +\infty} \frac{1}{x}$$

Therefore, by the second theorem of the comparison, the limit (18.11) is verified.

18.4.4   Limit of the Composite Function

Theorem 18.7 *Let f and g be real-valued functions of a real variable. Let* $A \subseteq \mathbf{R}$ *be the domain of f and* $B \subseteq \mathbf{R}$ *the domain of g, such that* $f(A) \subseteq B$. *Let* $c \in \mathbf{R}$ *be an accumulation point for A. If*

$$\lim_{x \to c} f(x) = \ell \tag{18.12}$$

$\ell \in \mathbf{R}$, *and there exists a neighborhood H of c, such that for every* $x \in H \cap A$, $x \neq c$, *it results*

$$f(x) \neq \ell \tag{18.13}$$

If

$$\lim_{y \to \ell} g(y) = h \tag{18.14}$$

$h \in \mathbf{R}$, *then*

$$\lim_{x \to c} g(f(x)) = h \tag{18.15}$$

Proof Let us first observe that the composite function $g(f(x))$ has meaning (Sect. 5.4). Let M be a neighborhood of h. Then, by (18.14), there exists a neighborhood J of ℓ such that for every y belonging to $(J \cap B) - \{\ell\}$, it is $g(y) \in M$; then assigned the neighborhood J, from (18.12) there exists a neighborhood I' of c, such that for every $x \in (A \cap I') - \{c\}$ it is $f(x) \in J$. Since (18.13) holds for every $x \in (H \cap A) - \{c\}$, set $I = I' \cap H$, for every $x \in (A \cap D) - \{c\}$ we have $f(x) \in (J \cap f(A)) - \{\ell\}$, so $g(f(x)) \in M$ and (18.15) holds. (Observe that the intersection of two neighborhoods of c, $I = I' \cap H$, is a neighborhood of c; and that if $x \in (H \cap A) - \{c\}$, then $0 < |f(x) - \ell|$). $\qquad \square$

Remark 18.5 Theorem 18.7 still holds if c, ℓ, h are elements of $\mathbf{R}^*$. Furthermore, the theorem stays on in absence of hypothesis (18.13), provided that (18.14) be replaced with the equality.

$$\lim_{y \to \ell} g(y) = g(\ell)$$

Remark 18.6 The hypothesis (18.13) is essential to the validity of Theorem 18.7. The following example shows that in the mere absence of hypothesis (18.13) the theorem no longer holds. In fact, consider the functions

$$f(x) = \frac{sinx}{x}$$

And

$$g(y) = \begin{cases} 1 \ if \ \ y = 0 \\ 0 \ if \ \ y \neq 0 \end{cases}$$

We have:

$$\lim_{x+\infty} \frac{sin x}{x} = 0 \quad \text{and} \quad \lim_{y \to 0} g(y) = 0$$

For every non-null integer $k, f(k\pi = 0$ and $g(f(k\pi)) = 1$. Then the function $g(f(x))$ does not admit limit as $x \to +\infty$.

18.4.5 Right and Left Limits

Given a non-empty subset A of **R** and a point $c \in$ **R** we denote $A(c-)$ the subset of A whose elements are the real numbers smaller than c, and $A(c+)$ the subset of A whose elements are the real numbers greater than c. The sets $A(c-)$ and $A(c+)$ are called *the part of* A *to the* *left* of c and *the part of* A *to the right* of c, respectively. One of the two sets $A(c-)$ and $A(c+)$ can be empty. If c is an accumulation point for A, then c is an accumulation point for at least one of the two sets. The notion of *left* (*right*) *neighborhood* is introduced in (Sect. 6.6.1).

Definition 18.3 A point $c \in$ **R** is called a *right* (*left*) *accumulation point* for a set $A \subseteq$ **R** if each right (left) neighborhood of c contains a point of A distinct from c.

Definition 18.4 Let f be a real-valued function with domain $A \subseteq$ **R**. Let $c \in$ **R** be a right (left) accumulation point for A. The element ℓ of **R*** is said to be the *right* (*left*) *limit* of $f(x)$ as x approaches c, in symbols.

$$\lim_{x \to c^+} f(x) = \ell \quad \left(\lim_{x \to c^-} f(x) = \ell \right) \tag{18.16}$$

if, for every neighborhood J of ℓ there exists a right neighborhood I^+ (left neighborhood I^-) of c, such that for every $x \in A \cap I^+$ ($x \in A \cap I^-$) and $x \neq c, f(x)$ belongs to J. Alternative symbols to (18.16) are $f(c +)$ and $f(c-)$, respectively.

The uniqueness theorems of right and left limits hold in analogy with Theorem 18.1.

The symbols (18.16) mean: "limit of $f(x)$ as x approaches c from the right (from the left) is equal to ℓ" (Figs. 18.13and 18.14).

Example 18.9 The function $f(x)$ defined in **R**:

Fig. 18.13 Right limit of f
(x) at c, $\lim\limits_{x\to c^+} f(x)$

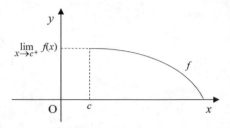

Fig. 18.14 Left limit of f
(x) at c, $\lim\limits_{x\to c^-} f(x)$

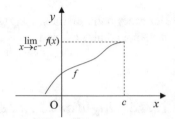

$$f(x) = 1 \quad if \quad x \geq 0$$
$$f(x) = -1 \quad if \quad x < 0$$

is called *sign* (or *signum*) of x, denoted $f(x) = \text{sgn}(x)$. It is convergent at the right and at the left of the point 0 and we have: $f(0^+) = 1$, $f(0^-) = -1$.

Example 18.10 The function $f(x) = \frac{1}{x}$ has no limit in $x = 0$; in fact, it diverges negatively to the left of 0 and positively to the right of 0: $f(0^-) = -\infty$, $f(0^+) = +\infty$.

If there is an accumulation point only on the right (left) for the set A, the concept of right (left) limit is identified with that of limit.

Theorem 18.8 *Let the real-valued function f be defined in* $A \subseteq \mathbf{R}$. *If c is both a right and left accumulation point of A, the limit of f at c exists if and only if the left and right limits exist and are equal. Then the following equalities hold*:

$$\lim_{x\to c} f(x) = \lim_{x\to c-} f(x) = \lim_{x\to c+} f(x)$$

In (Fig. 18.15) the graph of a function f, with $f(c+) = +\infty$, is drawn.

Example 18.11 The function $f(x) = \sin\frac{1}{x}$ defined in $\mathbf{R}-\{0\}$ does not admit limit as x approaches zero. In fact, $\sin\frac{1}{x} = 0$ if and only if $\frac{1}{x} = k\pi$, for every integer k. Then $\sin\frac{1}{x} = 0$ if and only if $x = \frac{1}{k\pi}$, for every integer k. Furthermore, $\sin\frac{1}{x} = 1$ if and only if $\frac{1}{x} = \frac{\pi}{2} + 2k\pi$, i.e., $x = \frac{2}{(4k+1)\pi}$, for every integer k. Therefore in each neighborhood of 0 there are points on which $\sin\frac{1}{x}$ takes value 0 and other points where $\sin\frac{1}{x}$ takes value 1. In conclusion, $\sin\frac{1}{x}$ does not admit limit as $x \to 0$.

Fig. 18.15 $\lim\limits_{x \to c-} f(x) = \ell$; $\lim\limits_{x \to c+} f(x) = +\infty$

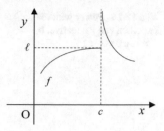

18.4.6 More on the Limits of Elementary Functions

The values of the limits of the logarithm function to the base a, as $x \to 0$ and as $x \to +\infty$, are:

a. if $a > 1$, then

$$\lim_{x \to 0+} \log_a x = \lim_{x \to 0} \log_a x = +\infty, \quad \lim_{x \to +\infty} \log_a x = +\infty$$

b. if $0 < a < 1$, then

$$\lim_{x \to 0+} \log_a x = \lim_{x \to 0} \log_a x = +\infty, \quad \lim_{x \to +\infty} \log_a x = -\infty$$

The values of the limits of the function $\tan x$, as x approaches $\frac{\pi-}{2}$ and $\frac{\pi+}{2}$ are:

$$\lim_{x \to \frac{\pi-}{2}} \tan x = +\infty, \quad \lim_{x \to \frac{\pi+}{2}} \tan x = -\infty,$$

Since $\tan x$ is π periodic (Chap. 8),

$$\lim_{x \to \left(\frac{\pi}{2}+k\pi\right)-} \tan x = +\infty, \quad \lim_{x \to \left(\frac{\pi}{2}+k\pi\right)+} \tan x = -\infty,$$

for any integer k.

The limits of the power function with real exponent r, as x approaches 0 or $+\infty$, follow (Fig. 18.16):

if $r < 0$, then $\lim\limits_{x \to 0+} x^r = +\infty$ and $\lim\limits_{x \to +\infty} x^r = 0$,

if $0 < r < 1$, then $\lim\limits_{x \to 0+} x^r = 0$ and $\lim\limits_{x \to +\infty} x^r = +\infty$,

if $r > 1$, then $\lim\limits_{x \to 0+} x^r = 0$ and $\lim\limits_{x \to +\infty} x^r = +\infty$,

Example 18.12 The function $f(x) = \sqrt{x-1}$, is defined in the interval $[1, +\infty)$ and has limits as $x \to 1$ and $x \to +\infty$:

Fig. 18.16 Power with real
exponent r: **a** r negative, **b**
$r > 1$, **c** $0 < r < 1$

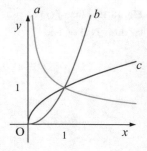

$$\lim_{x \to 1+} \sqrt{x-1} = \lim_{x \to 1} \sqrt{x-1} = 0, \quad \lim_{x \to +\infty} \sqrt{x-1} = +\infty$$

Example 18.13 The function $f(x) = \frac{1}{\sqrt{1-x}}$, whose domain is the open interval $(-\infty, 1)$, has the following limits as $x \to 1$ and $x \to -\infty$:

$$\lim_{x \to 1-} \frac{1}{\sqrt{1-x}} = \lim_{x \to 1} \frac{1}{\sqrt{1-x}} = +\infty \quad \lim_{x \to -\infty} \frac{1}{\sqrt{1-x}} = 0$$

Example 18.14 The function $f(x) = \frac{\sqrt{1-x}}{x}$, whose domain is the set $(-\infty, 1] - \{0\}$ has the following limit as $x \to 1$:

$$\lim_{x \to 1-} \frac{\sqrt{1-x}}{x} = \lim_{x \to 1} \frac{\sqrt{1-x}}{x} = 0$$

The function $\frac{\sqrt{1-x}}{x}$ has right (left) limit as $x \to 0^- \ (x \to 0^+)$

$$\lim_{x \to 0-} \frac{\sqrt{1-x}}{x} = -\infty \quad \lim_{x \to 0+} \frac{\sqrt{1-x}}{x} = +\infty$$

and the limit

$$\lim_{x \to -\infty} \frac{\sqrt{1-x}}{x} = 0$$

Furthermore $f(1) = 0$. The graph of f is drawn in Fig. 18.17.

18.4.7 Solved Problems

Calculate the following limits using the results in (Sect. 18.4.1):

a. $\lim\limits_{x \to +\infty} \frac{x + 2\sqrt{x}}{3x - 1}$

Fig. 18.17 Graph of the function $\frac{\sqrt{1-x}}{x}$

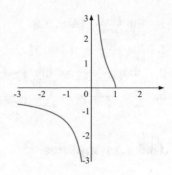

The limit is an indeterminate form $\frac{\infty}{\infty}$.

$$\lim_{x \to +\infty} \frac{x + 2\sqrt{x}}{3x - 1} = \lim_{x \to +\infty} \frac{x + 2x^{\frac{1}{2}}}{3x - 1} = \lim_{x \to +\infty} \frac{x\left(1 + \frac{2}{\sqrt{x}}\right)}{x\left(3 - \frac{1}{x}\right)} = \lim_{x \to +\infty} \frac{x}{3x} = \frac{1}{3}$$

b. $\lim\limits_{x \to 0} \frac{4x^3 - 2x^2 + 5x}{2x^4 + x^3 - 3x}$

The limit is an indeterminate form $\frac{0}{0}$. Factorize numerator and denominator:

$$\lim_{x \to 0} \frac{x\left(4x^2 - 2x + 5\right)}{x\left(2x^3 + x^2 - 3\right)} = \lim_{x \to 0} \frac{4x^2 - 2x + 5}{2x^3 + x^2 - 3} = -\frac{5}{3}$$

c. $\lim\limits_{x \to 7} \frac{7-x}{\sqrt{7} - \sqrt{x}}$

The limit is an indeterminate form $\frac{0}{0}$. Multiply numerator and denominator by $(\sqrt{7} + \sqrt{x})$ to obtain:

$$\lim_{x \to 7} \frac{(7-x)\left(\sqrt{7} + \sqrt{x}\right)}{7 - x} = \lim_{x \to 7}\left(\sqrt{7} + \sqrt{x}\right) = 2\sqrt{7}$$

18.4.8 *Supplementary Problems*

Calculate the limits:

a. $\lim\limits_{x \to 1} \frac{\sqrt{x} - 1}{x - 1}$ [Ans. $\frac{1}{2}$]

b. $\lim\limits_{x \to 2} \frac{\sqrt{2} - \sqrt{x}}{2 - x}$ [Ans. $\frac{\sqrt{2}}{4}$]

c. $\lim\limits_{x \to +\infty} \left(\sqrt{x^2 + 8} - x\right)$ [Ans. 0]

d. $\lim\limits_{x \to +\infty} \left(\sqrt{x^2 + 1} - \sqrt{x^2 - 1}\right)$ [Ans. 0]

e. $\lim\limits_{x \to 2} \frac{16-x^4}{2-x}$ [Ans. 32]

f. $\lim\limits_{x \to 0} \frac{\sqrt{x+4}-2}{x}$ [Ans. $\frac{1}{4}$]

g. $\lim\limits_{x \to 1+} \frac{e^x}{x^2-3x+2} = \lim\limits_{x \to 1+} \frac{e^x}{(x-2)(x-1)}$ [Ans. $-\infty$]

h. $\lim\limits_{x \to 1-} \frac{e^x}{x^2-3x+2} = \lim\limits_{x \to 1-} \frac{e^x}{(x-2)(x-1)}$ [Ans. $+\infty$]

18.5   Asymptotes

Given a real-valued function of a real variable f, it is useful for sketching the graph of f to know if there exists any line in the plane, whose distance from the curve $y = f(x)$, approaches zero when the point $P(x, f(x))$ of the graph indefinitely moves away from the origin of the coordinates.

Such lines are called *asymptotes*, which we study in the following sections.

18.5.1   Vertical Asymptotes

Given the function f, a vertical line $x = c$ such that $f(x)$ approaches $+\infty$ or $-\infty$ as x approaches c either from the right or from the left is said to be a *vertical asymptote* for the graph of f. Therefore, if the line $x = c$ is a vertical asymptote for the graph of f, at least one of the following cases occurs (Sect. 18.2.1):

$$\lim\limits_{x \to c+} f(x) = +\infty, \quad \lim\limits_{x \to c+} f(x) = -\infty, \quad \lim\limits_{x \to c-} f(x) = +\infty, \quad \lim\limits_{x \to c-} f(x) = -\infty$$

For example, the line $x = 2$ is a vertical asymptote (Fig. 18.18) of the graph of the function

$$f(x) = \frac{1}{x-2}$$

because

$$\lim\limits_{x \to 2-} \frac{1}{x-2} = -\infty \quad \text{and} \quad \lim\limits_{x \to 2+} \frac{1}{x-2} = +\infty$$

18.5.2   Horizontal Asymptotes

Let f be a function defined in an unbounded interval, let's say $(-\infty, a)$, $(b, +\infty)$, or **R**. If either

Fig. 18.18 $x = 2$ is a vertical asymptote of the graph of $f(x) = \frac{1}{x-2}$

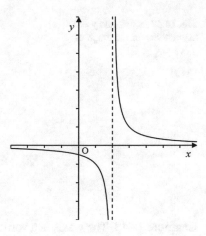

$$\lim_{x \to -\infty} f(x) = \ell$$

or

$$\lim_{x \to +\infty} f(x) = \ell,$$

$\ell \in \mathbf{R}$, then the horizontal line $y = \ell$ is called a *horizontal asymptote* for the graph of f.

Example 18.15 The line $y = 1$ is a horizontal asymptote for the graph of

$$f(x) = \frac{x+1}{x-2}$$

In fact,

$$\lim_{x \to -\infty} \frac{x+1}{x-2} = 1 \quad \text{and} \quad \lim_{x \to +\infty} \frac{x+1}{x-2} = 1$$

Besides, the line $x = 2$ is a vertical asymptote for the graph of f (Fig. 18.19).

Example 18.16 The x axis is a horizontal asymptote for the graph of $f(x) = \frac{1}{x-2}$. In fact,

$$\lim_{x \to \pm\infty} \frac{1}{x-2} = 0$$

Example 18.17 The x axis is a horizontal asymptote for the graph of $f(x) = a^x$, for every a positive and different from 1. In fact, if $a > 1$, then $\lim_{x \to -\infty} a^x = 0$; if $0 < a < 1$, then $\lim_{x \to +\infty} a^x = 0$.

Fig. 18.19 $x = 2$ and $y = 1$ asymptotes for the graph of $f(x) = \frac{x+1}{x-2}$

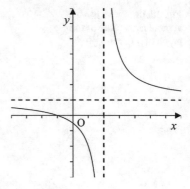

Example 18.18 The y axis is a vertical asymptote for the graph of $f(x) = \log_a x$, for every a positive and different from 1. In fact, if $a > 1$, then $\lim\limits_{x \to 0+} \log_a x = +\infty$; if $0 < a < 1$, then $\lim\limits_{x \to 0+} \log_a x = +\infty$.

Example 18.19 The lines $x = -\frac{\pi}{2}$ and $x = \frac{\pi}{2}$ are vertical asymptotes for the graph of $f(x) = \tan x$.

18.5.3 Oblique Asymptotes

Let f be a function defined in an unbounded interval, for example $(b, +\infty)$, and P$(x,$ $f(x))$ a point of the graph of f. The line t of equation $y = mx + n$ is called an *oblique asymptote* for the graph of f if, called Q the point of t having the same abscissa x of P,

$$\lim_{x \to +\infty} |PQ| = 0, \tag{18.17}$$

i.e., the distance of the line t and the graph of f approaches zero, as x approaches $+\infty$ (Fig. 18.20).

The limit (18.17) allows to identify, when it exists, an oblique asymptote for the graph of f. Indeed, since the length of the segment PQ is

$$|PQ| = |f(x) - (mx + n)| = |f(x) - mx - n|$$

the value of the limit

$$\lim_{x \to +\infty} |f(x) - mx - n| = 0 \tag{18.18}$$

is obtained. Equation (18.18) implies

Fig. 18.20 The line t is an oblique asymptote for the graph of f

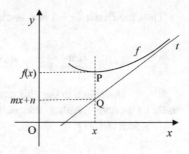

$$\lim_{x \to +\infty} (f(x) - mx - n) = 0 \tag{18.19}$$

which yields

$$\lim_{x \to +\infty} \frac{f(x) - mx - n}{x} = 0$$

and, by the properties of the limits (Sect. 18.4.1):

$$\lim_{x \to +\infty} \frac{f(x)}{x} - m - \lim_{x \to +\infty} \frac{n}{x} = 0$$

As $\lim_{x \to +\infty} \frac{n}{x} = 0$, we obtain

$$m = \lim_{x \to +\infty} \frac{f(x)}{x} \tag{18.20}$$

and, by (18.19),

$$n = \lim_{x \to +\infty} (f(x) - mx) \tag{18.21}$$

The Eqs. (18.20) and (18.21) allow to determine also possible horizontal asymptotes. The graph of f has an oblique or horizontal asymptote if and only if the limits in (18.20) and (18.21) exist and are finite.

Example 18.20 Consider the graph of the function

$$f(x) = \frac{x^2 - 4}{x + 1}$$

The function has domain $\mathbf{R} - \{-1\}$. Let us compute the left and right limits at -1:

$$\lim_{x \to -1-} \frac{x^2 - 4}{x + 1} = +\infty \quad \text{and} \quad \lim_{x \to -1+} \frac{x^2 - 4}{x + 1} = -\infty$$

Thus the line $x = -1$ is a vertical asymptote for the graph of f. Furthermore,

$$\lim_{x \to -\infty} \frac{x^2 - 4}{x + 1} = -\infty \quad \text{and} \quad \lim_{x \to +\infty} \frac{x^2 - 4}{x + 1} = +\infty$$

Since f is defined in the unbounded intervals $(-\infty, -1)$ and $(-1, +\infty)$, it makes sense to wonder whether oblique asymptotes exist. To this aim, apply the formulas (18.20) and (18.21):

$$m = \lim_{x \to -\infty} \frac{x^2 - 4}{x(x + 1)} = \lim_{x \to +\infty} \frac{x^2 - 4}{x(x + 1)} = 1$$

$$n = \lim_{x \to -\pm\infty} \left(\frac{x^2 - 4}{x + 1} - x \right) = \lim_{x \to -\pm\infty} \left(\frac{-x - 4}{x + 1} \right) = -1$$

The line $y = x - 1$ is an asymptote for the graph of f. Let us give a sketch of the graph of f in Fig. 18.21.

Example 18.21 Consider the function

$$f(x) = \frac{x^3 - 4x}{x^2 - 1}$$

The domain of f is $\mathbf{R} - \{-1, 1\}$, vertical asymptotes are $x = 1$ and $x = -1$, the oblique asymptote is $y = x$. Observe that the function f is odd, i.e., $-f(x) = f(-x)$ (see Sect. 17.5) (Fig. 18.22).

Fig. 18.21 The graph of $f(x) = \frac{x^2 - 4}{x + 1}$ with asymptotes $x = -1$ and $y = x - 1$

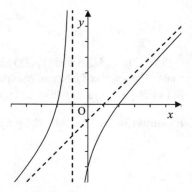

Fig. 18.22 The graph of
$f(x) = \frac{x^3 - 4x}{x^2 - 1}$

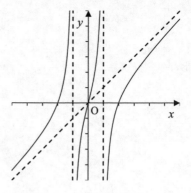

Bibliography

Adams, R.A., Essex, C.: Calculus. Single variable, Pearson Education Canada, Toronto (2006)

Anton, H.: Calculus. John Wiley & Sons Inc., New York (1980)

Lax, P., Burnstein, S., Lax, A.: Calculus with Applications and Computing. Springer-Verlag, New York (1976)

Royden, H.L., Fitzpatrick, P.M.: Real Analysis. Pearson, Toronto (2010)

Spivak, M.: Calculus. Cambridge University Press (2006)

Ventre, A.: Matematica. Fondamenti e calcolo, Wolters Kluwer Italia. Milano (2021)

Chapter 19
Continuity

19.1 Continuous Functions

Let f be a real-valued function of a real variable x. Let A be the domain of f and $c \in$ A an accumulation point of A. If f admits a limit as x approaches c, we have observed (Sect. 18.2) that the value $f(c)$ and the limit of f as $x \to c$ are completely independent values. The coincidence of these values gives rise to the concept of *continuous function*. We will deal with the classical theorems and topics of the mathematical analysis such as Bolzano's theorem, Darboux theorem or the first existence theorem of the intermediate values, uniform continuity, Cantor's and Weierstrass' theorems. A classification of discontinuities and the inverse functions of the circular functions are defined.

Definition 19.1 Let f be a real-valued function having domain A $\subseteq$ **R** and let $c \in$ A an accumulation point of A. If

$$\lim_{x \to c} f(x) = f(c) \tag{19.1}$$

then the function f is said to be *continuous* at c.

Remark 19.1 The Definition 19.1 implies that the following conditions must be met in order that $f(x)$ be continuous at c (Fig. 19.1):

c belongs to the domain of f

$\lim_{x \to c} f(x)$ exists and is equal to ℓ

$\ell = f(c)$

Definition 19.1 is expressed in terms of Definition 18.1 as follows:

Definition 19.2 Let f be a real-valued function whose domain is A $\subseteq$ **R** and let $c \in$ A be an accumulation point of A. The function f is said to be *continuous* at c if for every neighborhood J of $f(c)$ there exists a neighborhood I of c such that $f(x) \in$ J, whenever x belongs to I $\cap$ A. (Observe that it is no longer necessary to impose the condition $x \neq c$).

© The Author(s), under exclusive license to Springer Nature Switzerland AG 2023   321
A. G. S. Ventre, *Calculus and Linear Algebra*,
https://doi.org/10.1007/978-3-031-20549-1_19

Fig. 19.1 f continuous at c

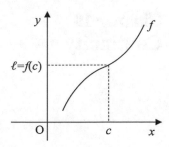

The Definition 19.2 may be reformulated in terms of Definition 18.2 as follows:

Definition 19.3 The function $f(x)$ is called *continuous* at c if for every $\varepsilon > 0$ there exists a real number $\delta > 0$ such that $|f(x) - f(c)| < \varepsilon$, whenever $0 < |x - c| < \delta$.

Obviously, the identical function $f(x) = x$ is continuous at every point $c \in \mathbf{R}$, i.e.,

$$\lim_{x \to c} x = c$$

Then equality (19.1) can be written:

$$\lim_{x \to c} f(x) = f\left(\lim_{x \to c} x\right) \tag{19.2}$$

Therefore, the fact that f is continuous at c, means that the interchange of the symbol f and the symbol $\lim_{x \to c}$ of the passage to the limit, is allowed.

The function f continuous at any point of its domain A is said to be *continuous* in A.

19.2   Properties of Continuous Functions

From the properties of limits (Sect. 18.4) we obtain some important consequences:

Theorem 19.1 If f and g are continuous in $c \in \mathbf{R}$, then the sum $f + g$, the difference $f - g$, the product $f \cdot g$ and the ratio $\frac{f}{g}$ (if $g(c) \neq 0$) are continuous functions at c. If f is continuous at c, the absolute value $|f|$ is continuous at c.

The *composite* of two continuous functions is continuous. In fact, according to Theorem 18.7, the following theorem holds.

Theorem 19.2 [Continuity of the composite function]. If function f has domain A and g has domain B, if $c \in$ A and $f(c) \in$ B, and if f is continuous at c and g is continuous at $f(c)$, then the function $g(f(x))$ is continuous at c.

Formula (19.2) is generalized as follows: if the function $g(f(x))$ is composite of two continuous functions, then

$$\lim_{x \to c} g(f(x)) = g\left(\lim_{x \to c} f(x)\right) \qquad (19.3)$$

It is shown that the elementary functions are continuous in their respective domains. Hence, from Theorem 19.2, the following is obtained.

Theorem 19.3 The functions power, exponential, logarithm, sine, cosine, tangent and their composite functions are continuous in the respective domains.

Examples

$$\lim_{x \to \pi} \sin x = \sin \pi = 0,$$

$$\lim_{x \to 0} \ln(1-x^2) = \ln 1 = 0,$$

$$\lim_{x \to 1} \frac{1}{\sqrt{1+e^x}} = \frac{1}{\sqrt{1+e}}$$

$$\lim_{x \to +\infty} \ln \frac{2x-1}{x+1} = \ln \lim_{x \to +\infty} \frac{2x-1}{x+1} = \ln 2,$$

$$\lim_{x \to +\infty} e^{\frac{\sqrt{x}}{x-1}} = e^{\lim_{x \to +\infty} \frac{\sqrt{x}}{x-1}} = e^0 = 1$$

$$\lim_{x \to 1} \frac{2}{1+e^{\frac{-1}{x}}} = \frac{2}{1+\lim_{x \to 1} e^{\frac{-1}{x}}} = \frac{2}{1+\frac{1}{e}} = \frac{2e}{1+e}$$

The interchange (19.3) of the symbol of function f and the symbol of limit is explicitly applied in the examples.

Let us state the following theorem already (18.4.2) proved in a wider context.

Theorem 19.4 [Theorem of the permanence of the sign]. If the function f defined in an open interval I_0 is continuous and positive (negative) at the point c of I_0, then there exists a neighborhood I of c such that $f(x) > 0$ ($f(x) < 0$), for every $x \in I$. (Fig. 19.2).

Fig. 19.2 The theorem of the permanence of the sign

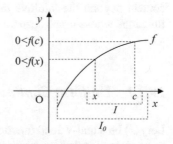

Fig. 19.3 Bolzano's
theorem

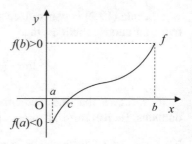

It is immediate to realize the following:

Theorem 19.5 Let the function f be defined in the interval I; if it is continuous at a point c such that in any neighborhood of c the function f assumes both non-positive and non-negative values, then $f(c) = 0$.

Proof. By Theorem 19.4 it cannot turn out neither $f(c) > 0$, nor $f(c) < 0$; whence it necessarily follows $f(c) = 0$.   □

Let us mention the following statements concerning continuous functions.

Theorem 19.6 [Bolzano's theorem, or theorem of the existence of the zeros]. If the function f is continuous in a closed and bounded interval $[a, b]$ and if $f(a) < 0$ and $f(b) > 0$, then there exists a point c in the open interval (a, b), such that $f(c) = 0$. (Fig. 19.3)

Theorem 19.7 [Darboux's theorem, or the first existence theorem of the intermediate values]. If the function f is continuous in a closed and bounded interval $[a, b]$, then it takes all the values between $f(a)$ and $f(b)$.

Theorem 19.8 If the function f is defined in a closed and bounded interval $[a, b]$, if f is monotone in $[a, b]$ and takes all the values between $f(a)$ and $f(b)$, then f is continuous in $[a, b]$.

Remark 19.2 Theorem 19.8 offers the opportunity for expressing an intuitive consideration. The attribute "continuous", given to a function, suggests the idea, when drawing the graph, that the pen need never leave the paper, i. e., it can be drawn without "interruptions". However, the definition 19.1 of continuity has a wider content beyond the intuitive meaning of traceability; in fact, there are continuous functions whose graphs cannot be drawn on a sheet of paper.

19.2.1 *Uniform Continuity*

Let $f(x)$ be a real-valued function of a real variable defined in an interval I. However two points x' and x'' are given in the interval, consider the number

$$|f(x') - f(x'')|$$

If $f(x)$ is continuous in I, then (see Definition 19.3) for every $\varepsilon > 0$ and fixed the point x'', there exists a number $\delta > 0$ such that $|f(x') - f(x'')| < \varepsilon$ whenever $0 < |x' - x''| < \delta$. The value δ in general depends on both ε and the particular point x''. If the number δ depends *only* on ε, and *not* on x'', then the function f is called *uniformly continuous* in the interval I.

In other words:

Definition 19.4 A function $f(x)$ whose domain is the interval I is said to be *uniformly continuous* in I if for every $\varepsilon > 0$ there exists a number $\delta > 0$ such that $|f(x') - f(x'')| < \varepsilon$ for any pair x', x'' of elements in I such that $|x' - x''| < \delta$.

It is obvious that if $f(x)$ is uniformly continuous in the interval I, then $f(x)$ is continuous at any point x of I. Therefore:

Theorem 19.9 A function uniformly continuous in the interval I is also continuous in I.

We limit ourselves to stating the following theorem:

Theorem 19.10 [Cantor's theorem]. A function $f(x)$ continuous in a bounded and closed interval I is uniformly continuous in I.

Let us prove the following:

Theorem 19.11 If $f(x)$ is continuous in a bounded and closed interval $[a, b]$, then $f(x)$ is bounded in the interval.

Proof. By Cantor's theorem the function f is uniformly continuous in $[a, b]$; hence, for every $\varepsilon > 0$ we can find a natural number n such that

$$|f(x') - f(x'')| < \varepsilon$$

whenever

$$|x' - x''| \le \frac{b - a}{n}$$

where x' and x'' are in $[a, b]$.

Let us subdivide the interval into n congruent (Sect. 7.1) intervals by means of the points

$$x_0 = a, x_1, x_2, \ldots, x_n = b$$

If x is a point belonging to $[a, b]$ such that $x_i \le x \le x_{i+1}$, then

$$|x - x_i| \le \frac{b - a}{n} \text{ implies} |f(x) - f(x_i)| \le \varepsilon$$

and

$$f(x_i) - \varepsilon \le f(x) \le f(x_i) + \varepsilon$$

If h denotes the minimum of the numbers $f(x_i)-\varepsilon$ and k the maximum of the numbers $f(x_i) + \varepsilon$, then

$$h \le f(x) \le k$$

what suffices to prove that the range of f is bounded.   □

Theorem 19.12 [Weierstrass' theorem]. Let $f(x)$ be a continuous function in a bounded and closed interval $[a, b]$. Then $f(x)$ assumes the maximum value M and the minimum value m in $[a, b]$.

Proof. By Theorem 19.11 f is bounded in $[a, b]$. Then there exists the supremum M of f (Sect. 7.1.2). We have to prove that there is a point $x\in[a, b]$ such that $f(x) = M$. For every $\varepsilon > 0$, by the properties of the supremum M there exists x such that $M-f(x) < \varepsilon$. Then

$$\frac{1}{M-f(x)} > \frac{1}{\varepsilon} \tag{19.4}$$

and given the arbitrariness of ε, the function $\frac{1}{M-f(x)}$ is not bounded. Let us suppose by contradiction

$$f(x) \ne M,$$

for every $x \in [a, b]$. As $M-f(x)$ is continuous and non-null in $[a, b]$, we conclude that the ratio $\frac{1}{M-f(x)}$ is continuous and bounded in $[a, b]$: but this conclusion contradicts (19.4). The contradiction arose for having supposed $f(x) \ne M$, for every $x \in [a, b]$. Hence, there exists x in $[a, b]$ such that $f(x) = M$. Therefore, M is the maximum of $f(x)$ in $[a, b]$.

Likewise, we can prove that there exists an $x' \in [a, b]$ such that $f(x') = m = inf$ $f(x) = min f(x)$. (Fig. 19.4).

□

Fig. 19.4 Weierstrass' theorem: m minimum of f, M maximum of f

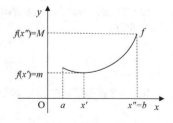

Fig. 19.5 Second existence
theorem of the intermediate
values

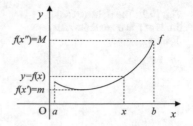

Theorem 19.13 [Second existence theorem of the intermediate values]. If the function f is continuous in a bounded and closed interval $[a, b]$, then f assumes all the values between the minimum m and the maximum M.

In other words, if f is continuous in the bounded and closed interval $[a, b]$, then, for every point y of the interval $[m, M]$, there exists x in $[a, b]$ such that $f(x) = y$ (Fig. 19.5).

19.3 Discontinuity

Let $f(x)$ be a real-valued function defined in $A \subseteq \mathbf{R}$. If $c \in A$ and f is not continuous at c, then the point c is called a *point of discontinuity*, or a *discontinuity* of f, and f is said to be *discontinuous* at c.

Let us classify the discontinuities as follows.

A. The point c is a discontinuity of f and $\lim_{x \to c} f(x)$ exists, is finite and $\lim_{x \to c} f(x) \neq$ $f(c)$ (Fig. 19.6).

Then the function f is said to admit a *removable discontinuity* at the point $c \in \mathbf{R}$. The nomenclature is due to the fact that a new function g continuous at c can be defined, such that:

$$g(x) = f(x) \quad \text{for every } \ x \neq c$$
$$g(c) = \lim_{x \to c} f(x)$$

Fig. 19.6 The limit of f at c
is $\ell \neq f(c)$, $\ell \in \mathbf{R}$

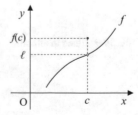

Fig. 19.7 The right and left
limits of f at c are finite and
distinct

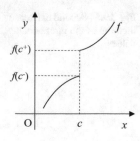

B. The limits $\lim\limits_{x \to c+} f(x)$ and $\lim\limits_{x \to c-} f(x)$ exist, are finite and $\lim\limits_{x \to c+} f(x) \neq \lim\limits_{x \to c-} f(x)$.
The difference $f(c^+) - f(c^-)$ is called the *jump* of f at c. The function f is said to
have a *discontinuity of the first kind*, or a *jump discontinuity* at c (Fig. 19.7).

Example B1 The function $\mathrm{sign}\,x$ (see Example 18.9):

$$f(x) = 1, \quad \text{if } x \geq 0$$
$$f(x) = -1, \quad \text{if } x < 0$$

is continuous at any non-null x. The point 0 is a discontinuity of the first kind
for f. (Fig. 19.8)

Example B2 The plane (Fig. 19.9) maintains an altitude $q(x)$ from the ground
and its trajectory is projected on the line xc. As long as x approaches c from the
left, the value of the altitude is $q(x)$. The left limit of q at c is $q(x)$, the right limit of
q at c is $q(x) - h$. The point c is a discontinuity of the first kind of the function q.

C. Neither case (A) nor case (B) occurs.
Then the point c is called a *discontinuity of the second kind at c*.

Fig. 19.8 $\lim\limits_{x \to 0+} \mathrm{sign}\,x = 1$,
$\lim\limits_{x \to 0-} \mathrm{sign}\,x = -1$

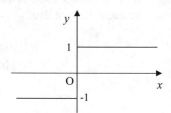

Fig. 19.9 Plane in altitude

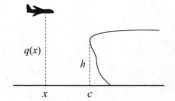

Fig. 19.10 $f(x) = 2^{\frac{1}{x}}$

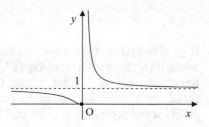

Example C1 The function $j(x)$ defined in (Sect. 7.1.3) has a discontinuity of the second kind at any $x \in [0, 1]$.

Example C2 The function

$$f(x) = \begin{cases} 2^{\frac{1}{x}} & if \ x \neq 0 \\ 0 & if \ x = 0 \end{cases}$$

has a discontinuity of second kind at $x = 0$. Observe that $2^{\frac{1}{x}} \to +\infty$ as $x \to 0^+$ and $2^{\frac{1}{x}} \to 0$ as $x \to 0^-$ (Fig. 19.10). The limits $\lim\limits_{x \to +\infty} f(x) = \lim\limits_{x \to -\infty} f(x) = 1$ indicate the existence of the horizontal asymptote $x = 1$.

19.4 Domain Convention

The domain of a real-valued function of a real variable f is the largest subset of **R** such that the expression $f(x)$ has meaning.

This statement is known as the *domain convention*.

Thus, the domain of $f(x) = x^2$ is the set of real numbers; the domain of $g(x) = \frac{1}{x}$ is the set of real numbers except 0; the set $\mathbf{R}^+ = [0, +\infty)$ is the domain of the square root function $y = \sqrt{x}$. Of course, given a function f having domain D, it is allowable to define any restriction (Sect. 5.5) of f to a proper non-empty part of D.

19.5 Curves

The concept of curve has been introduced in Sects. 7.2.2 and 7.2.3 where it was noticed that some curves are not graphs of functions and the graphs of some functions are not curves.

It seems reasonable to call *curve* the graph of a continuous function.

The equation $y = f(x)$ defines the function f and is, at the same time, the equation of the curve which is the graph of f. We say that the set A of points in the plane has an equation

$$y = f(x) \tag{19.5}$$

if (x, y) is a point of the set A, i.e., the couple (x, y) is a solution of (19.5) and, vice versa, if (x, y) is a solution of Eq. (19.5), then there exists a unique point of A which has coordinates (x, y). (Sect. 7.2 and Chap. 8). For example, the graph of the power function $f(x) = x^2$, is made of the set of couple (x, y) that are the coordinate of the points of the parabola $y = x^2$ (Sect. 17.4.1).

19.6 Continuous Functions and Inverse Functions

If f is a continuous function strictly increasing in the interval $[a, b]$, then $f(a) = m$ and $f(b) = M$, being m and M the minimum and maximum of f, respectively, whose existence is assured by Weierstrass' theorem. Besides, from the second theorem of existence of intermediate values the function f assumes any value between m and M. Then, however we fix a number y in the interval $[m, M]$, there is one and only one number x of the interval $[a, b]$ such that $f(x) = y$ (Fig. 19.11).

Furthermore, by Theorem 17.1 the function f is invertible in $[a, b]$, then the function f^{-1}, inverse of f, by Theorem 19.8, is continuous and strictly increasing in $[m, M]$.

For example, the function e^x and the function $\ln x$ are continuous and strictly increasing in the respective domains.

Likewise, if f is a continuous and strictly decreasing function in $[a, b]$, there exists the inverse f^{-1} that is continuous and strictly decreasing in $[m, M]$.

19.7 The Inverse Functions of the Circular Functions

Notice that the circular functions are periodical and consequently they do not admit inverse functions. Specific restrictions of the circular functions are invertible, the name *inverse* is reserved for these functions.

1. The inverse sine function

Fig. 19.11 The graphs of a continuous and strictly increasing function f and the inverse f^{-1}

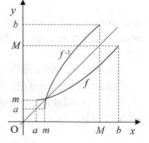

The sine function is continuous in **R**. The restriction of the sine function to the interval $\left[-\frac{\pi}{2}, \frac{\pi}{2}\right]$ is strictly increasing and takes all the values of the interval $[-1, 1]$. Therefore, restricted to the interval $\left[-\frac{\pi}{2}, \frac{\pi}{2}\right]$ the sine function is invertible and the inverse is called the *inverse sine function* or the *arcsin function* denoted arcsin$x$ or $\sin^{-1}x$. The domain of arcsin$x$ is $[-1, 1]$, where it is strictly increasing and continuous, and the range is $\left[-\frac{\pi}{2}, \frac{\pi}{2}\right]$ (Fig. 19.12). Of course, $\sin(\arcsin x) = x$.

2. The inverse cosine function

The cosine function is continuous in **R**. The restriction of the function cosine is strictly decreasing and continuous in the interval $[0, \pi]$ where it takes all the values of the interval $[-1, 1]$. Restricted to the interval $[0, \pi]$, the function cosine is invertible and its inverse is called the *inverse cosine function* or the *arccos function* denoted arccos$x$, or $\cos^{-1}x$, the domain is $[-1, 1]$ where it is strictly decreasing and continuous, and the range is $[0, \pi]$ (Fig. 19.13). Of course, $\cos(\arccos x) = x$.

3. The inverse tangent function

The restriction of the function tangent to the open interval $\left(-\frac{\pi}{2}, \frac{\pi}{2}\right)$ is continuous, strictly increasing and has range **R**. Therefore, restricted to the interval $\left(-\frac{\pi}{2}, \frac{\pi}{2}\right)$, the function tangent is invertible and its inverse is called *inverse tangent function*, denoted arctan$x$ or atan$x$ or $tg^{-1}(x)$ or arctg$x$ whose domain is **R**, where it is strictly increasing and continuous, and range $\left(-\frac{\pi}{2}, \frac{\pi}{2}\right)$ (Fig. 19.14). Of course, $tg(\arctan x) = x$.

Fig. 19.12 $y = \arcsin x$

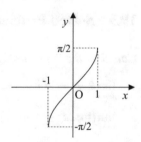

Fig. 19.13 $y = \arccos x$

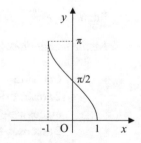

Fig. 19.14 $y = \text{atan}x$

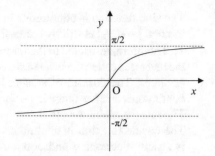

19.8 Continuity of Elementary Functions

In addition to the functions considered in Theorem 19.3, also the inverses of the circular functions belong to the class of elementary functions. In conclusion, the following extension of Theorem 19.3 to the elementary functions holds:

Theorem 19.14 The elementary functions and the composite functions of the elementary functions are continuous in their respective domains.

19.9 Solved Problems

Calculate the limits through formula (19.2).

1. $\lim\limits_{x \to 1} 2^{\frac{x}{x+1}} = 2^{\lim\limits_{x \to 1} \frac{x}{x+1}} = \sqrt{2}$

2. $\lim\limits_{x \to 0} \left(\ln\left(2x^2 + x + 1\right) \right) = \ln\left(\lim\limits_{x \to 0}\left(2x^2 + x + 1\right) \right) = \ln 1 = 0$

3. $\lim\limits_{x \to +} e^{\frac{1}{x}} = e^{\lim\limits_{x \to +} \frac{1}{x}} = e^0 = 1; \quad \lim\limits_{x \to -} e^{\frac{1}{x}} = e^{\lim\limits_{x \to -} \frac{1}{x}} = e^0 = 1$

   $\lim\limits_{x \to 0-} e^{\frac{1}{x}} = e^{\lim\limits_{x \to 0-} \frac{1}{x}} = 0; \quad \lim\limits_{x \to 0+} e^{\frac{1}{x}} = e^{\lim\limits_{x \to 0+} \frac{1}{x}} = +$

Solve the inequalities (see Sect. 8.1.3):

$$6 \arcsin x - \pi < 0 \Rightarrow \arcsin x < \tfrac{\pi}{6} \quad \left[-1 \le x < \tfrac{1}{2}\right]$$
$$2 \arcsin x + \pi > 0 \Rightarrow \arcsin x > -\tfrac{\pi}{2} \quad \left[-1 < x \le 1\right]$$
$$\arccos x > 0 \quad \left[-1 < x \le 1\right]$$
$$2 \arccos x + \pi > 0 \Rightarrow \arccos x > -\tfrac{\pi}{2} \quad \left[-1 < x \le 1\right]$$
$$4 \text{atan}x + \pi < 0 \Rightarrow \text{atan}x < -\tfrac{\pi}{4} \quad \left[x < -1\right]$$
$$\text{atan}x < 1 \quad \left[x < \tan 1\right]$$
$$\text{atan}x < 2 \quad \left[x \in \mathbf{R}\right]$$

Bibliography

Adams, R.A., Essex, C.: Calculus. Single variable, Pearson Education Canada, Toronto (2006)

Apostol, T.: Calculus. John Wiley & Sons, New York (1969)

Lax, P., Burnstein, S., Lax, A.: Calculus with Applications and Computing. Springer-Verlag, New York (1976)

Spivak, M.: Calculus. Cambridge University Press (2006)

Spiegel, M.R.: Advanced Calculus. McGraw-Hill Book Company, New York (1963)

Chapter 20
Derivative and Differential

20.1 Introduction

The concepts of derivative and differential are based on that of limit and deepen the knowledge of the real-valued functions in the analytical and geometric aspects. Differential geometry is indeed an important subject of mathematics.

Galileo (1564–1642) and his school had faced crucial questions on the subject in the unitary perspective of the scientific knowledge.

Isaac Newton (1642–1727) and Gottfried Wilhelm Leibniz (1646–1716) initiated a new vision of mathematics. Newton lays the foundations that will change the face of mathematics and physics, discovers the law of attraction of the bodies, which, along with the "calculation of fluids and pressures", marks the beginning of the "infinitesimal analysis", a priority shared with Leibniz.

Differential calculus has developed such as to induce important advancements not only in mathematics but also in natural, environmental, forecasting sciences, in psychology, medicine, architecture, engineering, linguistics. In the economic and social sciences, various situations that connect men and reality are systematically studied: the increase in satisfaction in an individual who sees an increase in his income and well-being, the attitude towards risky behaviors, the paradoxes in collective decisions. Mathematical models are defined for the study of these behaviors, and suitable functions have been introduced, such as utility functions, to interpret phenomena.

The dynamics in individual behaviors and the actions of masses of individuals, interpretable as a *continuum* of agents, are studied in mathematical terms (Aumann 1964).

A. G. S. Ventre, *Calculus and Linear Algebra*,
https://doi.org/10.1007/978-3-031-20549-1_20

20.2 Definition of Derivative

Let $f(x)$ be a real function defined in an open interval $I \subseteq \mathbf{R}$. Let x_0 be a point of I and h a non-null real number such that the point $x_0 + h$ belongs to I. The number h is named an *increment* of the variable x from x_0 and the function $f(x)$ is thereby given an *increment* $f(x_0 + h) - f(x_0)$ from $f(x_0)$.

The ratio

$$\frac{f(x_0 + h) - f(x_0)}{h}$$

is called the *difference quotient* of the function f on the interval $[x_0, x_0 + h]$.

If the limit

$$\lim_{h \to 0} \frac{f(x_0 + h) - f(x_0)}{h} \qquad (20.1)$$

exists and is finite, then it is called the *derivative* of the function f at the point x_0 and is denoted by the symbol $f'(x_0)$. A function is said to be *differentiable* at a point x_0 if it is endowed with the derivative at this point; in other words, $f(x)$ is *differentiable* at x_0 if $f'(x_0)$ exists. If f is differentiable at any point x of the interval I, it is said to be *differentiable in the interval I*. The function that associates the real number $f'(x)$ to every $x \in I$ is denoted Df or f'. The value $f'(a)$ of the derivative of f at a point a is denoted $f'(a)$ or $(f'(x))_{x=a}$. The operation to calculate the derivative of a function f is called the *differentiation* of f.

Remark 20.1 The symbols f' and $f'(x)$ have different meanings: the former denotes the function, i. e., the derivative of the function f, the latter denotes the value of the function f' at x. However, when there is no possibility of misunderstanding, we will denote by $f'(x)$ both the application and the value it assumes at x. (See Remark 5.1).

20.3 Geometric Meaning of the Derivative

Let f be a function defined in a neighborhood I of x_0. If h is a non-zero real number and $x_0 + h$ belongs to I, let us consider the points $P_0(x_0, f(x_0))$ and $P_h(x_0 + h, f(x_0 + h))$ of the graph of f. The line s passing through the points P_0 and P_h oriented as $P_0 P_h$ is called the *secant line* to the graph of f at the points P_0 and P_h. The slope of the secant line equals

$$\frac{f(x_0 + h) - f(x_0)}{h}$$

This ratio, in turn, is equal to the value of the tangent function of the angle $\gamma = \widehat{xs}$ that the x axis forms with the line s (Sect. 8.1.3d):

Fig. 20.1 Geometric
meaning of the derivative:
$f'(x_0) = \tan \alpha$

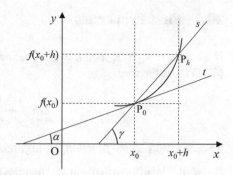

$$\tan \gamma = \tan \widehat{xs} = \frac{f(x_0 + h) - f(x_0)}{h}$$

If f is differentiable at x_0, i.e., the finite limit

$$\lim_{h \to 0} \frac{f(x_0 + h) - f(x_0)}{h} = f'(x_0)$$

exists, the line t passing through the point $P_0(x_0, f(x_0))$ is the limit position of the
secant line s as $h \to 0$ and it is named the *tangent line* to the graph of f at x_0 (or at
the point P_0) (Fig.20.1). The tangent line forms the angle $\alpha = \widehat{xt}$ with the x axis and
has slope $f'(x_0) = \tan \widehat{xt}$.

Since $f'(x_0)$ is a finite number, then $\alpha \neq \pm \frac{\pi}{2}$, i.e., the line t is not parallel to the
y axis. In other words, if f is differentiable at x_0, then the tangent line t to the graph
of f at P_0 exists and its slope is $f'(x_0)$.

The equation of the secant line s passing through P_0 and P_h is

$$y = f(x_0) + \tan \gamma (x - x_0)$$

and the equation of the tangent line t to the graph of f at x_0 is

$$y = f(x_0) + f'(x_0)(x - x_0) \tag{20.2}$$

For example, the derivative of the function $f(x) = x^2$ at the point $x = 1$ is

$$\lim_{h \to 0} \frac{f(x + h) - f(x)}{h} = \lim_{h \to 0} \frac{(1 + h)^2 - 1^2}{h} = \lim_{h \to 0} \frac{2h + h^2}{h} = \lim_{h \to 0} (2 + h) = 2$$

Thus $f'(1) = 2$. By (20.2) the tangent line t at $x = 0$ to the graph of f passes
through the point $(1, f(1)) = (1, 1)$ and has slope 2:

$$(t) \quad y = 1 + 2(x - 1)$$

20.4   First Properties

If we set $x = x_0 + h$, then $h = x - x_0$ and $h \to 0$ if only if $x \to x_0$. Then the limit (20.1) takes the form

$$\lim_{x \to x_0} \frac{f(x) - f(x_0)}{x - x_0}$$

Theorem 20.1 *If the function f is differentiable at x_0, then it is continuous there.*

Proof If f is differentiable at x_0, then the limit

$$\lim_{x \to x_0} \frac{f(x) - f(x_0)}{x - x_0}$$

exists and its value is $f'(x_0) \in \mathbf{R}$. We must prove that

$$\lim_{x \to x_0} f(x) = f(x_0)$$

or what is the same,

$$\lim_{x \to x_0} (f(x) - f(x_0)) = 0$$

From the properties of the operations with the limits (Sect. 18.4.1) we obtain

$$\lim_{x \to x_0} (f(x) - f(x_0)) = \lim_{x \to x_0} \frac{f(x) - f(x_0)}{x - x_0} \lim_{x \to x_0} (x - x_0) = f'(x_0)0 = 0 \quad \square$$

The inverse of the Theorem 20.1 does not hold.

Example 20.2 The absolute value function $f(x) = |x|$ is continuous in $\mathbf{R}$. However, we will see that the derivative of $f(x) = |x|$ does not exist at $x = 0$.

20.4.1   Derivatives of Some Elementary Functions

1. The derivative of the constant function $f(x) = c$ equals zero. Indeed, the difference quotient of the function f on the interval $[x_0, x_0 + h]$ is

$$\frac{f(x + h) - f(x)}{h} = \frac{c - c}{h}$$

and its value is zero for every $h \neq 0$. Furthermore, the limit of the quotient is 0, as $h \to 0$. In symbols, for any constant c, $Dc = 0$.

2. The derivative of $f(x) = x$ is $f'(x) = 1$, for every $x \in \mathbf{R}$. Indeed,

$$\lim_{h \to 0} \frac{f(x+h) - f(x)}{h} = \lim_{h \to 0} \frac{x+h-x}{h} = \lim_{h \to 0} \frac{h}{h} = 1$$

Then $Dx = 1$.

3. The derivative of $f(x) = x^2$ is $f'(x) = 2x$, for every $x \in \mathbf{R}$. Indeed,

$$\lim_{h \to 0} \frac{f(x+h) - f(x)}{h} = \lim_{h \to 0} \frac{(x+h)^2 - x^2}{h} = \lim_{h \to 0} (2x + h) = 2x$$

Then $Dx^2 = 2x$.

Similarly, we obtain: $Dx^3 = 3x^2$.

20.5 Operations Involving Derivatives

Let us state some basic rules that allow to perform operations with derivatives. We will consider real-valued functions defined in an open interval $I \subseteq \mathbf{R}$.

Theorem 20.2 [Sum rule] *If the functions $f(x)$ and $g(x)$ are differentiable at x_0, then the sum*

$F(x) = f(x) + g(x)$ *is differentiable at x_0 and*

$$F'(x_0) = f'(x_0) + g'(x_0)$$

Proof Let us consider the increment of F

$$F(x_0 + h) - F(x_0) = f(x_0 + h) + g(x_0 + h) - f(x_0) - g(x_0)$$

and the difference quotient

$$\frac{F(x_0 + h) - F(x_0)}{h} = \frac{f(x_0 + h) - f(x_0)}{h} + \frac{g(x_0 + h) - g(x_0)}{h}$$

As $f(x)$ and $g(x)$ are differentiable at x_0, by Theorem 18.2

$$\lim_{h \to 0} \frac{F(x_0 + h) - F(x_0)}{h} = f'(x_0) + g'(x_0)$$

as we wanted to show. □

Evidently under the hypotheses of the theorem 20.2 the *difference rule* holds:

$$D(f(x_0) - g(x_0)) = f'(x_0) - g'(x_0)$$

The theorem 20.2 clearly extends to the sum of any number n of functions:

$$D(f_1(x_0) + f_2(x_0) + \cdots + f_n(x_0)) = f_1'(x_0) + f_2'(x_0) + \cdots + f_n'(x_0)$$

For example,

$$D(x^3 - x^2 + x + 6) = Dx^3 - Dx^2 + Dx + D6 = 3x^2 - 2x + 1$$

Theorem 20.3 [Product rule]. *If the functions $f(x)$ and $g(x)$ are differentiable at x_0, then the product $F(x) = f(x)\,g(x)$ is differentiable and the following equality holds*:

$$F'(x_0) = f(x_0)g'(x_0) + f'(x_0)g(x_0) \tag{20.3}$$

Proof Let us calculate the increment of F:

$$F(x_0 + h) - F(x_0) = f(x_0 + h)g(x_0 + h) - f(x_0)g(x_0)$$
$$= f(x_0 + h)g(x_0 + h) - g(x_0)f(x_0 + h) + g(x_0)f(x_0 + h) - f(x_0)g(x_0)$$

and the difference quotient

$$\frac{F(x_0 + h) - F(x_0)}{h} =$$
$$\frac{f(x_0 + h)g(x_0 + h) - g(x_0)f(x_0 + h) + g(x_0)f(x_0 + h) - f(x_0)g(x_0)}{h} =$$
$$f(x_0 + h)\frac{g(x_0 + h) - g(x_0)}{h} + g(x_0)\frac{f(x_0 + h) - f(x_0)}{h}$$

By virtue of Theorem 18.2 the limit of the difference quotient as $h \to 0$ yields Eq. (20.3).   □

A particular case of (20.3) is $F(x) = kf(x)$, with k constant. Since $Dk = 0$, from (20.3) we obtain

$$F'(x) = kf'(x) \tag{20.4}$$

For example,

$$D\left(-8x^3 + 3x^2 + \frac{1}{2}x - 9\right) = -8Dx^3 + 3Dx^2 + \frac{1}{2}Dx - D9 = -24x^2 + 6x + \frac{1}{2}$$

Equation (20.3) extends to a product of any n differentiable functions:

$$F(x) = f_1(x)f_2(x) \ldots f_n(x)$$

Indeed,

$$F'^{(x)} = f_1'(x)f_2(x)\ldots f_n(x) + f_1(x)f_2'(x)\ldots f_n(x) + \cdots + f_1(x)f_2(x)\ldots f_n'(x)$$

A further consequence of the Theorem 20.3 is the calculation of the derivative of the function

$$(f(x))^n$$

with positive integer n, provided that the derivative of $f(x)$ exists. Since

$$(f(x))^n = f(x)f(x)\ldots f(x) \quad (n \text{ times})$$

one has

$$D(f(x))^n = f'(x)[f(x)]^{n-1} + f'(x)[f(x)]^{n-1} + \cdots + f'(x)[f(x)]^{n-1}$$

$$(n \text{ times})$$

Hence,

$$D(f(x))^n = nf'(x)[f(x)]^{n-1} \tag{20.5}$$

In particular, as $Dx = 1$, if $f(x) = x$, then we obtain

$$Dx^n = nx^{n-1} \tag{20.6}$$

Theorem 20.4 [Quotient rule]. *If the derivatives of the functions $f(x)$ and $g(x)$ at the point x_0 exist, and if $g(x_0) \neq 0$, then there exists the derivative at x_0 of the quotient $F(x) = \frac{f(x)}{g(x)}$ and*

$$F'(x) = \frac{f'(x)g(x) - f(x)g'(x)}{g^2(x)}$$

The proof is similar to that of the product rule.

20.6 Composite Functions. The Chain Rule

Theorem 20.5 [Chain rule]. *Let $y = f(x)$ be a real valued function defined in the interval I, with derivative $f'(x)$ at $x \in I$. Let $g(y)$ be a real valued function defined in an interval L, such that $f(I) \subseteq L$, with derivative $g'(y)$ at $y = f(x)$. Then the composite function $h(x) = g(f(x))$ is differentiable and the derivative at x is*

$$h'(x) = g'(f(x))f'(x) \tag{20.7}$$

Proof Consider $y_0 \in L$ and set

$$\frac{g(y)-g(y_0)}{y-y_0} - g'(y_0) = \omega(y) \tag{20.8}$$

Then

$$g(y) - g(y_0) = g'(y_0)(y - y_0) + \omega(y)(y - y_0) \tag{20.9}$$

The left-hand side of (20.8) is not defined if $y = y_0$ and then the formula (20.8) defines $\omega(y)$ only for $y \neq y_0$. However, it is evident that

$$\lim_{y \to y_0} \omega(y) = 0$$

Therefore, we can extend the function ω at y_0 by setting

$$\omega(y_0) = 0$$

So the function $\omega(y)$ is defined and continuous everywhere in L and the right-hand side of (20.9) is defined also at $y = y_0$ where equals zero. Since $y = y_0$ implies also the left-hand side of (20.9) equals zero, in conclusion the whole formula is valid when $y = y_0$.

All this stated, let x and x_0 be distinct points of I and put $y = f(x)$ and $y_0 = f(x_0)$ in the formula (20.9). Let us divide both sides of the formula so obtained by $x - x_0$,

$$\frac{g(f(x))-g(f(x_0))}{x-x_0} = g'(f(x_0))\frac{f(x)-f(x_0)}{x-x_0} + \omega(f(x))\frac{f(x)-f(x_0)}{x-x_0} \tag{20.10}$$

In virtue of the continuity of $\omega(y)$ and $f(x)$ and theorem 19.2 on the continuity of the composite function, we obtain

$$\lim_{x \to x_0} \omega(f(x)) = \lim_{y \to y_0} \omega(y) = 0$$

Observe that the right-hand side of (20.10), as x approaches x_0, has limit $g'(f(x_0))$ while the left-hand side is the *difference quotient* of $g(f(x))$. This proves formula (20.7), known as the *chain rule*.   □

Remark 20.2 The composite function $g(f(x))$ acts as follows: the function f is applied first and then g; f is the inner function and g is the outer function. By means of chain rule the derivative of the function $g(f(x))$ is the product of the derivative of the outer function g (evaluated at $f(x)$) and the derivative of the inner function (evaluated at x): $D(g(f(x))) = g'(f(x))f'(x)$. See examples below.

20.6.1 Derivatives of Some Elementary Functions

Let us state without proofs the derivatives of the following elementary functions.

$$D\log_a x = \tfrac{1}{x}\log_a e = \tfrac{1}{x\ln a} \qquad D\ln|x| = \tfrac{1}{x}$$
$$Da^x = a^x \ln a \qquad\qquad De^x = e^x$$
$$D\sin x = \cos x \qquad\qquad D\cos x = -\sin x$$
$$D\tan x = \tfrac{1}{\cos^2 x} = 1 + \tan^2 x$$

Examples (See the derivatives of the elementary functions above, (Sect. 20.5 and 20.6) and sum, product, quotient rules and Eqs. (20.4), (20.5) and (20.6).

$$D(5x^2 + 7\sin x - \cos x) = 10x + \cos x + \sin x;$$
$$D\sin^5 x = 5\sin^4 x\,D\sin x = 5\sin^4 x\cos x;$$
$$D(3x^5 + 5\sin x - 1)^3 = 3(3x^5 + 5\sin x - 1)^2 D(3x^5 + 5\sin x - 1)$$
$$\qquad\qquad = 3(3x^5 + 5\sin x - 1)^2(15x^4 + 5\cos x);$$
$$D(6x^3\cos^6 x) = 18x^2\cos^6 x - 6x^3\cos^5 x\sin x;$$
$$D(\tan x - x)^3 = 3(\tan x - x)^2\tan^2 x;$$
$$D(x^3 + \cos x)\ln x = (3x^2 - \sin x)\ln x + \tfrac{1}{x}(x^3 + \cos x).$$

20.7 Derivatives of the Inverse Functions

Let $y = f(x)$ be a real valued function, continuous and strictly monotonic in an interval I. Let $f^{-1}(y)$ be the inverse function of f. The inverse of f is necessarily strictly monotonic and continuous in the interval $f(I)$, i.e., the range of f (see Sect. 19.6).

Theorem 20.6 *If $f'(x_0)$ is the derivative of $f(x)$ at x_0 and $f'(x_0) \neq 0$, then there exists $Df^{-1}(y_0)$, i,e., the derivative at y_0 of $f^{-1}(y)$ and*

$$Df^{-1}(y_0) = \frac{1}{f'(x_0)} \qquad\qquad (20.11)$$

Proof If $y_0 = f(x_0)$, then it is also $f^{-1}(y_0) = x_0$. Therefore,

$$\frac{f^{-1}(y) - f^{-1}(y_0)}{y - y_0} = \frac{x - x_0}{f(x) - f(x_0)}$$

Since the differences $x - x_0$ and $y - y_0$ are continuous, x approaches x_0 if and only if y approaches y_0. So the previous equality implies

$$\lim_{y \to y_0} \frac{f^{-1}(y) - f^{-1}(y_0)}{y - y_0} = \lim_{x \to x_0} \frac{1}{\frac{f(x) - f(x_0)}{x - x_0}}$$

i. e., the equality (20.11).   □

20.7.1 Derivatives of the Inverses of the Circular Functions

Let us find the derivatives of the inverses of the circular functions (Sect. 18.7).

The function $y = \arcsin x$ is the inverse function of $x = \sin y$. If $-\frac{\pi}{2} < y < \frac{\pi}{2}$, then $-1 < x < 1$ and $D\sin y \neq 0$. By theorem 20.5 there exists the derivative of $\arcsin x$ in the open interval $(-1, 1)$. We have

$$D\arcsin x = \frac{1}{D\sin y} = \frac{1}{\cos y} = \frac{1}{\sqrt{1 - \sin^2 y}} = \frac{1}{\sqrt{1 - x^2}}$$

Similarly, one can prove the following equations

$$D\arccos x = \frac{1}{\sqrt{1 - x^2}}$$

$$D\arctan x = \frac{1}{1 + x^2}$$

20.8 The Derivative of the Function $(f(x))^{g(x)}$ and the Power Rule

Let $f(x)$ and $g(x)$ be differentiable functions in the interval I. Furthermore, let $f(x)$ be positive in I. Under these hypotheses, also the function

$$(f(x))^{g(x)}$$

has the derivative in I and

$$D(f(x))^{g(x)} = (f(x))^{g(x)} \left[\frac{g(x) f'(x)}{f(x)} + g'(x) \ln f(x) \right] \qquad (20.12)$$

Indeed, from the identity (see Eq. 17.5)

$$(f(x))^{g(x)} = \left(e^{g(x)\ln f(x)}\right)$$

and the theorem 20.3, the equality (20.12) follows.

Let us apply (20.12) for finding the derivative of the function x^a, $x > 0$ and any real number a:

$$Dx^a = e^{a\ln x}\left(a\frac{1}{x} + 0\ln x\right)$$

$$Dx^a = ax^{a-1}. \tag{20.13}$$

Formula (20.13) is known as the *power rule* for the derivatives.

Observe that the derivative of the power with real exponent is formally identical to the formulas (20.5) and (20.6) with natural exponent. Furthermore, if a is a rational number the equality (20.13) apply at any point x such that x^a and x^{a-1} are defined.

Exercises 20.1 Find the derivatives:

$$D3^2 = 0 \left(\text{because } 3^2 \text{ is a constant}\right)$$
$$D\sqrt{x} = Dx^{\frac{1}{2}} = \frac{1}{2}x^{\frac{1}{2}-1} = \frac{1}{2}\frac{1}{x^{\frac{1}{2}}} = \frac{1}{2\sqrt{x}}$$
$$D\frac{1}{x} = Dx^{-1} = -1x^{-1-1} = -\frac{1}{x^2}$$
$$D\sqrt[3]{x^4} = Dx^{\frac{4}{3}} = \frac{4}{3}x^{\frac{4}{3}-1} = \frac{4}{3}\sqrt[3]{x}$$
$$D\sqrt[4]{x^3} = Dx^{\frac{3}{4}} = \frac{3}{4}x^{\frac{3}{4}-1} = \frac{3}{4}x^{-\frac{1}{4}} = \frac{3}{4}\frac{1}{x^{\frac{1}{4}}} = \frac{3}{4}\frac{1}{\sqrt[4]{x}}$$
$$D\left(\sqrt{x} + \frac{1}{x} - 5\right) = D\sqrt{x} + D\frac{1}{x} - D5\frac{1}{2\sqrt{x}} - \frac{1}{x^2}$$
$$D\left(\sqrt[4]{x^3} + \sqrt[3]{x^4} - x^{10}\right) = D\sqrt[4]{x^3} + D\sqrt[3]{x^4} - Dx^{10} = \frac{3}{4}\frac{1}{\sqrt[4]{x}} + \frac{4}{3}\sqrt[3]{x} - 10x^9$$

Exercises 20.2 Find the derivatives of the functions:

$$f(x) = \frac{1}{x^2}; \quad g(x) = \frac{1}{\sqrt{x}}$$

$$\left[Ans : f'(x) = -\frac{2}{x^3} \; ; \; g'(x) = -\frac{1}{2x\sqrt{x}}\right]$$

Exercise 20.3

$$D\sin^3 x = 3\sin^2 x \cos x;$$
$$De^{\sqrt{x}} = e^{\sqrt{x}} \frac{1}{2\sqrt{x}},$$
$$D\tan^3 2x = 3\tan^2 2x \frac{1}{\cos^2 2x} 2;$$
$$D\left(\sin 2x \frac{\mathrm{tg}x + x^2}{\cos x}\right) = 2\cos 2x \frac{\mathrm{tg}x + x^2}{\cos x}$$
$$+\sin 2x \frac{(1 + \tan^2 x + 2x)\cos x + \sin x (\tan x + x^2)}{\cos^2 x};$$
$$De^{(\arcsin x)^3} = e^{(\arcsin x)^3} 3\arcsin^2 x \frac{1}{\sqrt{1 - x^2}};$$
$$D(\sin x)^{2x} = e^{2x \ln(\sin x)}\left[2\ln(\sin x) + 2x \frac{\cos x}{\sin x}\right],$$
$$D(1 + x^2)^{\cos x} = (1 + x^2)^{\cos x}\left[\frac{2x(\cos x)}{1 + x^2} - \sin x \ln(1 + x^2)\right].$$

Exercise 20.4 The equation of the tangent line at the point $x = 1$ to the graph of the function $f(x) = x^3$ is (Sect. 20.3),

$$y = f(1) + f'(1)(x - 1)$$

Since $f(1) = 1$ and $f'(1) = (3x^2)_{x=1} = 3$, we have:

$$y = 1 + 3(x - 1)$$

Hence, the tangent line has equation

$$y = 3x - 2$$

20.8.1   Summary of Formulas and Differentiation Rules

For convenience of the reader let us gather up the most important formulas and rules of differentiation.

$$Dx^a = ax^{a-1}$$
$$D\log_a x = \frac{1}{x}\log_a e = \frac{1}{x \ln a} \qquad D\ln|x| = \frac{1}{x}$$
$$Da^x = a^x \ln a \qquad\qquad\qquad De^x = e^x$$
$$D\sin x = \cos x \qquad\qquad\qquad D\cos x = -\sin x$$
$$D\tan x = \frac{1}{\cos^2 x} = 1 + \tan^2 x$$
$$D\arcsin x = \frac{1}{\sqrt{1 - x^2}} \qquad\qquad D\arccos x = -\frac{1}{\sqrt{1 - x^2}}$$
$$D\arctan x = \frac{1}{1 + x^2}$$
$$D(f(x) \pm g(x)) = Df(x) \pm Dg(x)$$
$$D(f(x)g(x)) = f(x_0)g'(x_0) + f'(x_0)g(x_0)$$
$$D\frac{f(x)}{g(x)} = \frac{f'(x)g(x) - f(x)g'(x)}{g^2(x)}$$

$$D(g(f(x))) = g'(f(x))f'(x)$$

$$D(f(x))^{g(x)} = e^{g(x)lnf(x)}D(g(x)\ln f(x)) = (f(x))^{g(x)}\left[\frac{g(x)f'(x)}{f(x)} + g'(x)\ln f(x)\right]$$

20.9 Right and Left-Hand Derivatives

Definition 20.1 If the difference quotient (Sect. 20.1).

$$\frac{f(x)-f(x_0)}{x-x_0}$$

is convergent as $x \to x_0-$ ($x \to x_0 +$), then the function f is said to be *left-hand* (*right-hand*) *differentiable* at x_0, and the limit of the difference quotient as $x \to x_0-$ ($x \to x_0 +$) is called the *left-hand* (*right-hand*) *derivative* of function f at x_0, and is denoted $f'_-(x_0)$ ($D_-f(x_0)$) or $f'_+(x_0)$ ($D_+f(x_0)$).

The function f is differentiable at x_0 if and only if the limits $f'_+(x_0)$ and $f'_-(x_0)$ exist and $f'_+(x_0) = f'_-(x_0)$ (see Sect. 18.4.5).

If $f'_+(x_0) \neq f'_-(x_0)$, then f is not differentiable at x_0 and the right and left tangent line to the graph at x_0 are distinct (Fig. 20.2).

Remark 20.3 Suppose that f is differentiable at any interior point of the closed interval $[a, b]$. If f has a right-hand derivative at the point a and a left-hand derivative at the point b, then f is defined *differentiable in the closed interval* $[a, b]$. The symbol $f'(x)$ denotes the function that is equal to the derivative of f in the open interval (a, b), with right-hand derivative of f at a and left-hand derivative at b. However, the statement "f is differentiable at x_0" will mean that the ordinary derivative exists at x_0.

Example The function $f(x) = |x|$ is not differentiable at the point 0. In fact, by definition of absolute value, we have

Fig. 20.2 The right tangent line and the left tangent line are distinct, $tg\gamma^+ = f'_+(x_0)$, $tg\gamma^- = f'_-(x_0)$

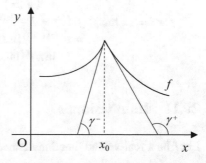

$$\lim_{x\to 0^+} \frac{|x|-0}{x-0} = \lim_{x\to 0^+} \frac{x-0}{x-0} = 1$$

$$\lim_{x\to 0^-} \frac{|x|-0}{x-0} = \lim_{x\to 0^-} \frac{-x-0}{x-0} = -1$$

20.10   Higher Order Derivatives

Let $f(x)$ be a differentiable function in the open interval (a, b). If the derivative of f is also differentiable, its derivative is denoted by $f''(x)$ or $D^2 f(x)$ and called the *derivative of the second order* of $f(x)$. Therefore, we have: $D^2 f(x) = f''(x) = D(D(f(x)))$.

The procedure of successive derivatives can be repeated n times, when the conditions exist. The nth derivative of $f(x)$ is denoted $f^{(n)}(x)$ or $D^n f(x)$, where n is called the *order* of the derivative of f, and f is said to be *n times differentiable* at x.

Examples
$$D^n e^x = e^x$$

for every natural n.

$$D^2 \sin x = D(D \sin x) = D \cos x = -\sin x$$
$$D^3 \sin x = D(D^2 \sin x) = -\cos x$$
$$D^4 \sin x = D(D^3 \sin x) = D(-\cos x) = \sin x$$
$$D^2 \tan x = D(1 + \tan^2 x) = 2 \tan x (1 + \tan^2 x)$$

Observe that the derivatives of $\sin x$ and $\cos x$ reproduce four by four.

Exercises 20.5 Calculate the derivatives:

1. $f(x) = \sin(\arccos x)$, $f'(x) = \cos(\arccos x)\frac{-1}{\sqrt{1-x^2}} = \frac{-x}{\sqrt{1-x^2}}$
   [in fact, $\cos(\arccos x) = x$]

2. $f(x) = \operatorname{atan}\sqrt{\ln \sin x}$, $f'(x) = \frac{1}{1+\ln \sin x}\frac{1}{2\sqrt{\ln \sin x}}\frac{1}{\sin x}\cos x$

3. $f(x) = \sqrt{\frac{x+1}{1-x}}$, $f'(x) = \frac{1}{2\left(\sqrt{\frac{x+1}{1-x}}\right)}\frac{1-x+x+1}{(1-x)^2} = \frac{\sqrt{\frac{1-x}{1+x}}}{(1-x)^2} = \frac{1}{\sqrt{1-x^2}}\frac{1}{1-x}$

4. $f(x) = (x \ln x)^x$ $f'(x) = D e^{x \ln(x \ln x)}$
   $= e^{x \ln(x \ln x)}\left(\ln(x \ln x) + \frac{x}{x \ln x}\left(\ln x + \frac{x}{x}\right)\right)$
   $= (x \ln x)^x\left(\ln(x \ln x) + \frac{1}{\ln x}(\ln x + 1)\right)$

20.11   Infinitesimals

Let f be a real-valued function defined in a set A of real numbers. If

$$\lim_{x \to c} f(x) = 0$$

with c finite or infinite, then f is said to be an *infinitesimal function* or *an infinitesimal* as $x \to c$.

Let us now specify what is meant by *comparison* between two infinitesimals. If f and g are real-valued functions defined in A and infinitesimals as $x \to c$, c finite or infinite, and there exists a neighborhood I of c such that $g(x) \neq 0$, for all $x \in A \cap I - \{c\}$. So the function $\frac{f(x)}{g(x)}$ is defined for every $x \in A \cap I - \{c\}$.

Furthermore, let us suppose that the limit

$$\lim_{x \to c} \frac{f(x)}{g(x)}$$

exists. The following cases occur:

$$\lim_{x \to c} \frac{f(x)}{g(x)} = 0$$
$$\lim_{x \to c} \frac{f(x)}{g(x)} = \ell \neq 0, \ \ell \text{ finite}$$
$$\lim_{x \to c} \frac{f(x)}{g(x)} = +\infty \text{ or } -\infty$$

In the first case the function f is said to be an *infinitesimal of higher order* than g at c. In the second case the functions f and g are said to be *infinitesimals of the same order* at c. In the third case the function f is said to be an *infinitesimal of lower order* than g at c.

If $\lim_{x \to c} \frac{f(x)}{g(x)}$ does not exist, then the two infinitesimals f and g as $x \to c$ are said to be *non-comparable* at c. For example, $f(x) = x \sin \frac{1}{x}$ and $g(x) = x$ are infinitesimals (Example 18.3) non-comparable at zero because $\sin \frac{1}{x}$ has no limit as $x \to 0$ (Example 18.11).

Examples 20.7

1. The function $f(x) = x - 5$ is an infinitesimal as $x \to 5$ because $\lim_{x \to 5}(x - 5) = 0$; the function $f(x) = \frac{1}{x}$ is an infinitesimal as $x \to +\infty$ because $\lim_{x \to +\infty} \frac{1}{x} = 0$.

2. The function x^2 is an infinitesimal of higher order than x as $x \to 0$ and x^3 is an infinitesimal of higher order than x^2 as $x \to 0$; indeed,

$$\lim_{x \to 0} \frac{x^2}{x} = \lim_{x \to 0} x = 0$$
$$\lim_{x \to 0} \frac{x^3}{x^2} = \lim_{x \to 0} x = 0$$

Figure 20.3 suggests an intuitive consideration. The infinitesimals x^3, x^2 and x as $x \to 0$ are different from each other: the function x^3 approaches 0 "faster" than x^2, and x^2 approaches 0 "faster" than x.

3. The functions $\sin x$ and $\tan x$ are infinitesimals of the same order as $x \to 0$. In fact,

Fig. 20.3 x^2 is an infinitesimal of higher order than x as $x \to 0$; x^3 is an infinitesimal of higher order than x^2 as $x \to 0$

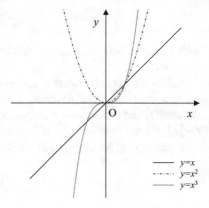

$$\lim_{x \to 0} \frac{\sin x}{\tan x} = \lim_{x \to 0} \frac{\sin x \cos x}{\sin x} = \lim_{x \to 0} \cos x = 1$$

4. The functions $x-1$ and x^2-1 are infinitesimals of the same order as $x \to 1$. In fact,

$$\lim_{x \to 1} \frac{x-1}{x^2-1} = \lim_{x \to 1} \frac{x-1}{(x-1)(x+1)} = \lim_{x \to 1} \frac{1}{x+1} = \frac{1}{2}$$

5. The function $x-1$ is an infinitesimal of *lower order* than $(x-1)^3$ as $x \to 1$. In fact,

$$\lim_{x \to 1} \frac{x-1}{(x-1)^3} = \lim_{x \to 1} \frac{1}{(x-1)^2} = +\infty$$

Definition 20.2 Let f and g be real-valued functions defined in $A \subseteq \mathbf{R}$ and infinitesimals as $x \to c$. Let a neighborhood I of c exist such that $g(x) > 0$, for all $x \in A \cap I - \{c\}$. If p is a positive real number, any power $(g(x))^p$ is an infinitesimal as $x \to c$. The function $f(x)$ is said to be an *infinitesimal of order p* with respect to the infinitesimal $g(x)$, called *principal infinitesimal*, if $f(x)$ and $(g(x))^p$ are infinitesimals of the same order at c, i.e.,

$$\lim_{x \to c} \frac{f(x)}{(g(x))^p} = \ell \neq 0, \quad \ell \text{ finite}$$

If c is finite, then $(x-c)$ is usually assumed as principal infinitesimal; if $c = +\infty$, or $c = -\infty$, $\frac{1}{|x|}$ is usually assumed as principal infinitesimal.

If q is a positive real number such that $q > p$, then $(g(x))^q$ is an infinitesimal of higher order than $(g(x))^p$. In fact, $q > p$ implies:

$$\lim_{x \to c} \frac{(g(x))^q}{(g(x))^p} = \lim_{x \to c} (g(x))^{q-p} = 0$$

20.12 Infinities

A real-valued function $f(x)$ defined in a subset $A \subseteq \mathbf{R}$ such that $f(x) \to +\infty (-\infty)$ as $x \to c$, $c \in \mathbf{R}^*$, is said to be an *infinity* at c.

The comparison of infinities is defined similarly to the comparison of infinitesimals.

Let $f(x)$ and $g(x)$ be two infinities as x approaches c and let us suppose that the limit

$$\lim_{x \to c} \frac{f(x)}{g(x)}$$

exists. The following cases occur:

$$\lim_{x \to c} \frac{f(x)}{g(x)} = 0$$
$$\lim_{x \to c} \frac{f(x)}{g(x)} = \ell \neq 0$$
$$\lim_{x \to c} \frac{f(x)}{g(x)} = +\infty \ \text{ or } \ -\infty$$

In the first case the function f is said to be an *infinity of lower order* than g as $x \to c$. In the second case the functions f and g are said to be *infinities of the same order* as $x \to c$. In the third case the function f is said to be an *infinity of higher order* than g as $x \to c$.

If $\lim_{x \to c} \frac{f(x)}{g(x)}$ does not exist, then the two infinities f and g are called *non-comparable* at c.

Definition 20.3 The function $f(x)$ is said to be an *infinity of order $p > 0$* with respect to the infinity $g(x)$ assumed positive and called *principal infinity*, if $f(x)$ and $(g(x))^p$ are infinities of the same order. For the limits as $x \to c$, $c \in \mathbf{R}$, $|x-c|^{-1}$ is assumed as principal infinity; for the limits as $x \to +\infty$ (or $x \to -\infty$), $|x|$ is assumed as principal infinity.

20.13 Differential

Let f be a function defined in an open interval (a, b) and differentiable at the point x of (a, b). We denote with Δx an increment of x such that $x + \Delta x$ belongs to the interval (a, b) and Δf the corresponding increment of f, i.e.,

$$\Delta f = \Delta f(x) = f(x + \Delta x) - f(x)$$

The *difference quotient* of f on the interval $[x, x + \Delta x]$ is

$$\frac{\Delta f}{\Delta x} = \frac{\Delta f(x)}{\Delta x} = \frac{f(x + \Delta x) - f(x)}{\Delta x}$$

Since f is differentiable at x (Sect. 20.2) the limit

$$\lim_{\Delta x \to 0} \frac{\Delta f}{\Delta x} = \lim_{\Delta x \to 0} \frac{f(x + \Delta x) - f(x)}{\Delta x} = f'(x)$$

exists and is finite; its value is the derivative of f at x; the line t passing through the point $P_0(x_0, f(x_0))$ forms the angle $\alpha = \widehat{xt}$ with the x axis and has slope $f'(x_0) = \mathrm{tg}\,\widehat{xt} = \mathrm{tg}\,\alpha$. (see Sect. 20.3).

Let us consider the points $H(x + \Delta x, f(x))$, $L(x + \Delta x, f(x + \Delta x))$ and M common to the lines t and HL (Fig. 20.4). We get (Sect. 8.1.3):

$$HM = f'(x)\Delta x$$

Therefore,

$$\Delta f = f'(x)\Delta x + ML \tag{20.14}$$

Let us now put

$$\omega(x) = ML$$

i.e., by (20.14),

$$\omega(x) = \Delta f - f'(x)\Delta x \tag{20.15}$$

Theorem 20.7 *Let f be a real-valued function defined in the open interval (a, b) and differentiable at the point x. Then the function ω is an infinitesimal of higher order than Δx, as $\Delta x \to 0$.*

Proof Since f is differentiable at x, from (20.15) we obtain:

Fig. 20.4 Δf increment of f; df differential of f

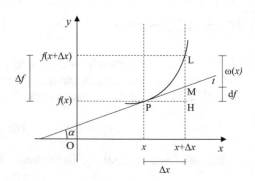

$$\lim_{\Delta x \to 0} \frac{\omega(x)}{\Delta x} = \lim_{\Delta x \to 0} \frac{\Delta f}{\Delta x} - \lim_{\Delta x \to 0} \frac{f'(x)\Delta x}{\Delta x} = f'(x) - f'(x) = 0$$

and this means that ω is an infinitesimal of higher order than Δx, as $\Delta x \to 0$. $\square$

The product $f'(x)\Delta x$ is named the *differential* of f at x and is denoted by the symbol df:

$$df = f'(x)\Delta x \tag{20.16}$$

Plugging $f(x) = x$ into (20.16), we get $dx = 1\Delta x = \Delta x$, i.e., the differential of the identical function x equals the increment of x.

Therefore, (20.16) may be rewritten as

$$df = f'(x)dx \tag{20.17}$$

that is the usual expression of the differential of f.

From (20.15) and (20.16) we obtain

$$\Delta f = f'(x)dx + \omega(x) = df + \omega(x) \tag{20.18}$$

This means that the increment of f differs from the differential of f by an infinitesimal of higher order then Δx. This justifies the approximate equality

$$\Delta f \approx df \tag{20.19}$$

(where the symbol $\approx$ means "almost equal") used in the physical and natural sciences, in economics, where approximate equality is often replaced by equality *tout court*:

$$\Delta f = df \tag{20.20}$$

From (20.17) we get

$$\frac{df}{dx} = f'(x) \tag{20.21}$$

The symbol $\frac{df}{dx}$ in (20.21) which indicates the derivative of f as the ratio of two differentials, was introduced by Leibniz and is currently used.

Once the derivative is known, the calculation of the differential is immediate. For example:

$$d(\sin x) = \cos x dx$$
$$d(4x - 1) = 4dx$$
$$d(\ln x^2) = \frac{1}{x^2} 2x dx = \frac{2}{x} dx$$

Another field of application of the "equalities" (20.18) and (20.19). In the case that the variable x is time, if $\Delta x > 0$, then Δf expresses the forecast of the increment in the value of f at the time $x + \Delta x$.

20.13.1   Differentials of Higher Order

With the same procedure followed for the derivatives, the differentials of higher order of a function are defined. The second order differential is the differential of the first differential, calculated *considering constant* the factor dx. Therefore, if we denote with d^2y or d^2f the differential of the second order, we have:

$$d^2 y = d^2 f = d(df) = D\big(f'(x)dx\big)dx = f''(x)dxdx = f''(x)dx^2$$

Just as the differentials of a function are expressed by means of the derivatives of the function, so the derivatives can be expressed by means of the differentials. Indeed, given the function $y = f(x)$, the following identities hold:

$$f'(x) = \frac{dy}{dx} = \frac{df}{dx}; \; f''(x) = \frac{d^2y}{dx^2} = \frac{d^2f}{dx^2}; \ldots; f^{(n)}(x) = \frac{d^n y}{dx^n} = \frac{d^n f}{dx^n}$$

In other words, the derivative of order n of a function is equal to the ratio of the differential of order n of the function and the n-th power of the differential of the independent variable.

20.14   Solved Problems

1. Differentiate $f(x) = x(\ln x - 1)$.
    Use the product rule: $f'(x) = \ln x - 1 + x\frac{1}{x} = \ln x - 1 + 1 = \ln x$.
2. Differentiate $f(x) = \frac{x^2}{2}\left(\ln x - \frac{1}{2}\right)$
    Use product rule: $f'(x) = \frac{2x}{2}\left(\ln x - \frac{1}{2}\right) + \frac{x^2}{2}\frac{1}{x} = x\ln x - \frac{x}{2} + \frac{x}{2} = x\ln x$.
3. Differentiate $f(x) = \ln(\tan x)$.
    Use chain rule: $f'(x) = \frac{1}{\tan x}\frac{1}{\cos^2 x} = \frac{1}{\sin x \cos x}$.
4. Differentiate $f(x) = x^{\sqrt{x}}$.
    Use rule (20.12): $f'(x) = x^{\sqrt{x}}\left[\frac{1}{2\sqrt{x}}\ln x + \sqrt{x}\frac{1}{x}\right] = \frac{x^{\sqrt{x}}}{\sqrt{x}}\left(\frac{1}{2}\ln x + 1\right)$.
5. Find the equations of the two tangent lines to the curve $f(x) = x^3 - 3x + 1$ and parallel to the line $3x - y + 1 = 0$. Calculate the distance between the two contact points.

    Solution The slope of the line is 3, the tangent line to the curve at a point x has slope $f'(x) = 3x^2 - 3$. As the given line is not vertical the necessary and sufficient condition for parallelism is the equality of the slopes: $3x^2-3 = 3$, i. e., $x = \sqrt{2}$ or $x = -\sqrt{2}$. If $x = \sqrt{2}$, then $f(x) = 1-\sqrt{2}$; if $x = -\sqrt{2}$, then $f(x) = 1 + \sqrt{2}$. Thus, the contact points are $A(\sqrt{2}, 1-\sqrt{2})$ and $B(-\sqrt{2}, 1 + \sqrt{2})$, to which correspond (see Sect. 7.2.3) the lines $y = 3(x- \sqrt{2}) + 1-\sqrt{2}$ and $y = 3(x + \sqrt{2}) + 1 + \sqrt{2}$, respectively: applying the formula of the distance of two points (Sect. 7.1.1), we obtain $|AB| = 4$.

Bibliography

Aumann R.J.: Markets with a continuum of traders. Econometrica **32**, 39–50, no.1–2 (1964)

Anton, H.: Calculus. John Wiley & Sons Inc., New York (1980)

Boyer, C.B.: A History of Mathematics. John Wiley & Sons Inc., New York (1968)

Lax, P., Burnstein, S., Lax, A.: Calculus with Applications and Computing. Springer-Verlag, New York (1976)

Royden, H.L., Fitzpatrick, P.M.: Real Analysis. Pearson, Toronto (2010)

Spivak, M.: Calculus. Cambridge University Press (2006)

Ventre, A.: Matematica. Fondamenti e calcolo, Wolters Kluwer Italia, Milano (2021)

Chapter 21
Theorems of Differential Calculus

21.1 Introduction

We deepen the study of differential calculus and present some important results. We have mentioned (Sect. 20.1) the origins of differential calculus and its spreading in the sciences.

Mechanics and astronomy, along with the advancements resulting from the studies of Galileo and Newton and Leibniz, provide a theoretical framework for all the sciences, for which the knowledge of the phenomena is translated into the construction of a mathematical and mechanical model. Started by Descartes and raised in the century of the Enlightenment, the impulse of mathematics enlivens the entire cultural process.

21.2 Extrema of a Real-Valued Function of a Single Variable

The problems which deal with finding the best way to perform a given task are called *optimization* problems. Many optimization problems consist in searching for the minimum or the maximum value of a function and determine where these values occur.

In Sect. 7.2.1 we encountered the concept of *extremum* of a real-valued function f. Infimum, supremum, minimum and maximum of f, are *extrema* of the function.

The minimum and the maximum of f in the domain A are also named *absolute minimum* and *absolute maximum*, respectively. If $f(x') = min f$, then f is said to have a *minimum point* at x', also named *absolute minimum point*; if $f(x'') = max f$, then f is said to have a *maximum point* at x'', also named *absolute maximum point*.

Examples of minimum, maximum, infimum and supremum of elementary functions (see Sects. 8.1.3, 17.4, and 19.7) are:

© The Author(s), under exclusive license to Springer Nature Switzerland AG 2023
A. G. S. Ventre, *Calculus and Linear Algebra*,
https://doi.org/10.1007/978-3-031-20549-1_21

Fig. 21.1 Relative
maximum and relative
minimum points for f

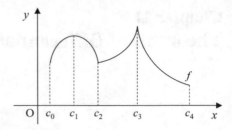

$$sup \sin x = max \sin x = 1, x \in \mathbf{R}$$
$$inf \sin x = min \sin x = -1, x \in \mathbf{R}$$
$$inf\, x^2 = minx^2 = 0, x \in \mathbf{R}$$
$$inf(x^2 - 1) = \min(x^2 - 1) = -1, x \in \mathbf{R}$$
$$sup \arcsin x = max \arcsin x = \tfrac{\pi}{2}, x \in [-1, 1]$$
$$sup \arccos x = max\arccos x = \pi, x \in [-1, 1]$$
$$sup \, \text{atan}x = \tfrac{\pi}{2}, x \in \mathbf{R}$$
$$inf \, \text{atan}x = -\tfrac{\pi}{2}, x \in \mathbf{R}$$

The functions x^3, $-x^2$, atan$x$ have no minimum; the functions x^3, x^2, atan$x$ have no maximum.

Definition 21.1 Let f be a real-valued function defined in the set $A \in \mathbf{R}$. The point c of A is said to be a *relative maximum* (*minimum*) *point* of the function f if there exists a neighborhood I of c such that

$$f(x) \le f(c) \qquad (f(x) \ge f(c)) \tag{21.1}$$

for every $x \in A \cap I$.

If c is a maximum (minimum) point of f, then (21.1) is verified at any point of A. Therefore, the maximum (minimum) point of f is also a relative minimum (relative maximum) point.

The relative maximum points or relative minimum are also called *relative extremum points* of the function. The relative maximum or relative minimum values attained by the function are also called *relative extrema* of the function; the maximum and minimum values assumed by the function are also called *absolute maximum* and *absolute minimum*, or generically *absolute extrema* of the function (Fig. 21.1).

Observe that the expressions "absolute maximum" and "absolute minimum" are often used instead of the terms "maximum" and "minimum", to avoid possible misunderstandings in a context including "relative maximum" or "relative minimum".

More specifically, if c is a point of A and an accumulation point of A, i.e., $c \in A \cap A'$, where A' is the derived set of A, the point c is said to be a *proper relative maximum* (*minimum*) *point* and $f(c)$ a *proper relative maximum* (*minimum*) if a neighborhood I of c exists such that

$$f(x) < f(c) \qquad (f(x) > f(c))$$

for every $x \in A \cap A' - \{c\}$. As reported in Fig. 21.1, we identify the following notable points:

c_0 relative minimum point of f: $f(x) \geq f(c_0)$ for every $x \in A \cap I$, $A = [c_0, c_4]$ and I a suitable neighborhood of c_0;

c_1 relative maximum point of f: $f(x) \leq f(c_1)$ in a neighborhood of c_1;

c_2 relative minimum point of f: $f(x) \geq f(c_2)$ in a neighborhood of c_2;

c_3 relative maximum point of f: (c_3 is also the maximum point of f);

c_4 relative minimum point of f: (c_4 is also the minimum point of f)

Let us observe that the maximum (minimum), if it exists, of a function f defined in the set A is unique. The same does not happen to the maximum points and minimum points: indeed, in the same set may exist several distinct points $x_1, x_2, \ldots, x_n$, all maximum points or all minimum points of f; of course, in this case it will be $f(x_1) = f(x_2) = \ldots = f(x_n) = , \ldots$. For example, the function $\cos x$, restricted to the interval $[0, 4\pi]$, has the maximum points $0, 2\pi, 4\pi$.

21.3 Fermat's and Rolle's Theorems

We state and prove some basic theorems of differential calculus.

Theorem 21.1 (Fermat's theorem) *If f is a real-valued function differentiable in the open interval (a, b), and if $c \in (a, b)$ is a relative extreme point of f, then*

$$f'(c) = 0$$

Proof We prove the theorem in the case that c is a relative maximum point for the function f. The proof, in case that c is a relative minimum point, is similar. If c is a relative maximum point, there exists a neighborhood I of the point c, $I \subseteq (a, b)$, such that

$$f(c) \geq f(x) \tag{21.2}$$

whatever the point $x \in I$ is. By (21.2), if $x < c$, then $\frac{f(x)-f(c)}{x-c} \geq 0$; if $x > c$, then $\frac{f(x)-f(c)}{x-c} \leq 0$. Therefore, in virtue of Theorem 18.4 applied to the function $\frac{f(x)-f(c)}{x-c}$, we obtain,

$$\lim_{x \to c^-} \frac{f(x)-f(c)}{x-c} \geq 0$$
$$\lim_{x \to c^+} \frac{f(x)-f(c)}{x-c} \leq 0$$

Since f is differentiable at c, the right limit equals the left limit.

Thus, by Theorem 18.8, $f'(c) = 0$. □

Fig. 21.2 Rolle's theorem:
the angle between the
tangent line t at c and the x
axis is null

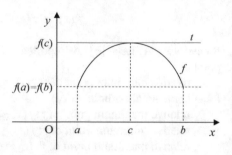

From a geometric point of view Fermat's theorem states that the tangent line to the curve $y = f(x)$ on a relative maximum or minimum point must be parallel to the x axis.

Theorem 21.2 (Rolle's theorem) *If f is a continuous function in the closed and bounded interval $[a, b]$, differentiable in the open interval (a, b) and if $f(a) = f(b)$, then there exists a point c of (a, b), such that $f'(c) = 0$.*

Proof The function f satisfies the hypothesis of Weierstrass' theorem. Then f achieves the minimum value m' and the maximum value m''. This means that there exist in $[a, b]$ two points x' and x'' such that $f(x') = m'$ and $f(x'') = m''$. One of the two cases occurs:

case 1 one of the points x' and x'' is different from the extremes of the interval $[a, b]$. Let us suppose $a < x' < b$. Then x' is a relative minimum point of f in (a, b). As f is differentiable in (a, b), by Fermat's theorem, $f'(x') = 0$. (If $a < x'' < b$, we infer $f'(x'') = 0$.)

case 2 the points x' and x'' coincide with the extremes of the interval $[a, b]$. Suppose $a = x'$ and $b = x''$. By the hypothesis $f(a) = f(b)$ we have $m' = f(x') = f(x'') = m''$, and then f is constant in $[a, b]$. Therefore, by (Sect. 20.4.1.1), $f'(x) = 0$, for every x in (a, b) (Fig. 21.2).   □

21.4 Lagrange's Theorem and Consequences

The following theorem is a generalization of Rolle's theorem, since the hypothesis $f(a) = f(b)$ is abandoned. Geometrically (Fig. 21.3), the theorem states that, if f is a continuous function in the closed and bounded interval $[a, b]$ and if it is differentiable in the open interval (a, b), there exists a point c of (a, b) such that the tangent line t to the graph of f is parallel to the line s passing through the points $(a, f(a))$, $(b, f(b))$:

$$y = f(b) + \frac{f(b) - f(a)}{b - a}(x - a) \tag{21.3}$$

Fig. 21.3 Geometrical meaning of Lagrange's theorem:
$\tan \alpha = f'(c) = \frac{f(b)-f(a)}{b-a}$

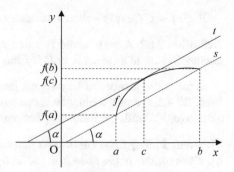

Theorem 21.3 (Lagrange's theorem) *If f is a continuous function in the closed and bounded interval [a, b] and differentiable in the open interval (a, b), then there exists a point c in the interval (a, b), such that*

$$f'(c) = \frac{f(b)-f(a)}{b-a}$$

Proof We construct an auxiliary function $g(x)$ by subtracting the right-hand side of (21.3) from $f(x)$:

$$g(x) = f(x) - f(b) - \frac{f(b) - f(a)}{b - a}(x - a)$$

It is easy to check that $g(a) = g(b)$. So, the function $g(x)$ satisfies the hypotheses of Rolle's theorem in the interval [a, b]. Hence, there is a point c in the interval (a, b) such that

$$g'(c) = f'(c) - \frac{f(b) - f(a)}{b - a} = 0$$

Thus, the equality

$$f'(c) = \frac{f(b) - f(a)}{b - a}$$

follows, as claimed. □

Theorem 21.4 (Cauchy's theorem) *Let the real-valued functions f and g be continuous in the closed interval [a, b] and differentiable in the open interval (a, b). If $g(a) \neq g(b)$ and $f'(x)$ and $g'(x)$ do not vanish simultaneously, then there exists a point c in (a, b) such that*

$$\frac{f(b)-f(a)}{g(b)-g(a)} = \frac{f'(c)}{g'(c)}$$

If $g(x) = x$, Cauchy's theorem reduces to Lagrange's.

Definition 21.2 A real-valued function f, whose domain is $D \subseteq \mathbf{R}$, is said to be *identically null* in a subset A of D if f has value 0 at any point of A.

Some consequences of Lagrange's theorem are of particular interest. We know (Sect. 20.4.1.1) that if a function is constant in (a, b) then its derivative is identically zero in (a, b). Well, the following theorem inverts the implication.

Theorem 21.5 *If f is a continuous function in the closed and bounded interval $[a, b]$, differentiable in the open interval (a, b) and if f has identically null derivative at (a, b), then f is constant in $[a, b]$.*

Proof Let x be a point in (a, b). We apply Lagrange's theorem to the restriction of f to the interval $[a, x]$. Therefore, a point c in (a, x) exists such that

$$f'(c) = \frac{f(x) - f(a)}{x - a}$$

If the derivative of f at every point of (a, b) is zero, then $f'(c) = 0$ and $f(x) = f(a)$, i.e., f takes the same value $f(a)$ at every x in (a, b).   □

Theorem 21.6 *Let f be a continuous function in the closed and bounded interval $[a, b]$. If f is differentiable in (a, b), then for every $x \in (a, b)$, $f'(x) \geq 0$ if and only if f is increasing in (a, b).*

Proof We prove that if $f'(x) \geq 0$ for every x, then f is increasing in (a, b); i.e., however two elements x' and x'' are chosen in $[a, b]$, if $x' < x''$, then $f(x') \leq f(x'')$. We apply Lagrange's theorem to the restriction of f to the interval $[x', x'']$. Then, a point c exists in (x', x''), such that

$$f'(c) = \frac{f(x'') - f(x')}{x'' - x'}$$

and since $f'(c) \geq 0$ and $x'' - x' > 0$, it follows $f(x') \leq f(x'')$. We now prove that if f is increasing in (a, b), then $f'(x) \geq 0$, for every $x \in (a, b)$. In fact, if f is increasing in (a, b), then, for every x in (a, b) and $h > 0$, such that $x + h$ is in (a, b), we have $f(x + h) \geq f(x)$. Therefore, we get

$$\frac{f(x + h) - f(x)}{h} \geq 0 \tag{21.4}$$

As the same inequality holds when $h < 0$ we get

$$\lim_{h \to 0} \frac{f(x + h) - f(x)}{h} = f'(x) \geq 0$$

as claimed.   □

Similarly, we can prove the following:

Theorem 21.7 *Let f be a continuous function in the closed and bounded interval* $[a, b]$ *and differentiable in* (a, b). *Then* $f'(x) \leq 0$ *in* (a, b) *if and only if f is decreasing in* (a, b).

Observe that the statements of Theorems 21.6 and 21.7 are double implications. The following are implications related to strictly monotonic functions.

Theorem 21.8 *Let f be a continuous function in the closed and bounded interval* $[a, b]$ *and differentiable in* (a, b). *If* $f'(x) > 0$ *in* (a, b), *then f is strictly increasing in* (a, b).

Theorem 21.9 *Let f be a continuous function in the closed and bounded interval* $[a, b]$ *and differentiable in* (a, b). *If* $f'(x) < 0$ *in* (a, b), *then f is strictly decreasing in* (a, b).

Remark 21.1 Theorems 21.8 and 21.9 cannot be inverted. For example, $f(x) = x^3$ is strictly increasing and differentiable in $\mathbf{R}$, but $f'(0) = [3x^2]_{x=0} = 0$.

Example 21.1 The function $f(x) = x^2$ is strictly decreasing in $(-\infty, 0)$ and strictly increasing in $(0, +\infty)$ because $f'(x) = 2x$.

Example 21.2 The function $f(x) = \log_2 x$ is strictly increasing in $(0, +\infty)$, in fact, for every $x > 0$, $f'(x) = \frac{1}{x} \log_2 e > 0$.

Example 21.3 The function $f(x) = \log_{\frac{1}{2}} x$ is strictly decreasing in $(0, +\infty)$, in fact,

$$f'(x) = \frac{1}{x} \log_{\frac{1}{2}} e < 0$$

for every $x > 0$.

21.5 Comments on Fermat's Theorem

If the function f continuous in $[a, b]$ and differentiable in (a, b) is endowed with a relative maximum or minimum point $c \in (a, b)$, then $f'(c) = 0$.

The inverse implication does not hold. Indeed, if f is defined in $[a, b]$, differentiable in (a, b) and if $f'(c) = 0$, for some c in (a, b), then one of the following circumstances occurs:

1. there exists a neighborhood of c in which $f'(x)$ is non-negative for $x \leq c$ and non-positive for $x \geq c$; then c is a relative maximum point;
2. there exists a neighborhood of c in which $f'(x)$ is non-positive for $x \leq c$ and non-negative for $x \geq c$; then c is a relative minimum point;

Fig. 21.4 Graph of $f(x) = 1 - x^2$

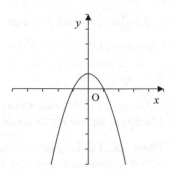

3. there exists a neighborhood of c in which $f'(x) \geq 0$; then f is increasing in the neighborhood;
4. there exists a neighborhood of c in which $f'(x) \leq 0$; then f is decreasing in the neighborhood;
5. $f'(x)$ changes sign an infinite number of times in every neighborhood of c.

Therefore, the fact that the derivative of f at c vanishes is not a sufficient condition for c to be a relative maximum or minimum point, i.e.:

$\{f'(c) = 0\}$ does not imply $\{c$ is a point of relative maximum or minimum of $f\}$

Example 21.4 Let us consider the function $f(x) = x^2$ and the derivative $f'(x) = 2x = 0$ if and only if $x = 0$. The function f has a relative minimum point in 0 (which is also an absolute minimum point). The tangent line to the graph in $(0, f(0)) = (0, 0)$ is the x axis.

Example 21.5 Let us consider the function $f(x) = 1 - x^2$ and the derivative $f'(x) = -2x = 0$ if and only if $x = 0$. The function has in 0 a relative maximum point (which is also an absolute maximum point). The tangent line to the graph at $(0, f(0)) = (0, 1)$ is the line $y = 1$ (Fig. 21.4).

Example 21.6 Let us consider the function $f(x) = x^3$; the derivative $f'(x) = 3x^2 = 0$ if and only if $x = 0$. The function f has no relative extremes in **R**; f is strictly increasing in **R** (see Sects. 18.3 and 18.4.6). The tangent line to the graph of f at the point $(0, f(0)) = (0, 0)$ is the x axis (Fig. 17.7).

21.5.1   Searching for Relative Maximum and Minimum Points

If f is differentiable in the interval (a, b), a point $c \in (a, b)$ such that $f'(c) = 0$ is called a *critical point* of f. From the classification above (Sect. 21.5) which lists five cases which proceed from the hypothesis $f'(c) = 0$, we deduce the following sufficient condition that c be a relative maximum (minimum) point.

Second derivative test. Let $f'(c) = 0$ and the second derivative of f at c exist. If $f''(c) < 0$ ($f''(c) > 0$), then f has a proper relative maximum (relative minimum) at c.

Exercise 21.1 Find the relative maxima and minima of the function $f(x) = x^4 - 2x^2$. The function is endowed with first and second derivative:

$$f'(x) = 4x^3 - 4x = 4x(x - 1)(x + 1)$$
$$f''(x) = 12x^2 - 4$$

We have $f'(x) = 0$ if and only if $x = 0, x = 1, x = -1$. As

$$f''(0) = -4 < 0$$
$$f''(1) = 8 > 0$$
$$f''(-1) = 8 > 0$$

$x = 0$ is a proper relative maximum point of f, $x = 1$ and $x = -1$ are proper relative minimum points of f. The function f attains a relative maximum $f(0) = 0$, and relative minima $f(1) = -1$ and $f(-1) = -1$.

21.5.2 Searching for the Absolute Maximum and Minimum of a Function

Let f be a continuous function in the closed and bounded interval $[a, b]$. By Weierstrass' theorem the function f has an absolute minimum m and an absolute maximum M; this means that two points, x' and x'', exist in the interval such that $f(x') = m$ and $f(x'') = M$. The points x' and x'' are absolute minimum and maximum points, respectively.

If x' and x'' are in the open interval (a, b), they are a relative minimum point and a relative maximum point, respectively. Hence, if f is differentiable in (a, b), then $f'(x') = 0$ and $f'(x'') = 0$.

However, it may happen that f is not differentiable in x' or x'', or at least one of these points is an endpoint of the interval $[a, b]$.

Therefore, let us describe a procedure for finding the absolute minimum and maximum points of a continuous function f in the closed and bounded interval $[a, b]$. The absolute minimum and maximum points of f must be searched in

a. the set of the points at which the first derivative of f is null;
b. the set of the points at which f is not differentiable;
c. the endpoints of the interval of definition of f.

If x' is the point belonging to the union of the sets (a), (b), (c) at which f attains the smallest value, then $f(x') = $ minimum of f. If x'' is the point belonging to the union of the sets (a), (b), (c) at which f attains the largest value, then $f(x'') =$ maximum of f.

If the domain of f is not a closed and bounded interval, or the function has points of discontinuity, then it may happen that the function has no maximum or minimum or neither maximum nor minimum. In cases like these, the study of the graph of f will come in handy.

Exercise 21.2 Find the relative and the absolute extrema of the function

$$f(x) = \sqrt{8x - x^2}$$

The domain of the function is the set of the points x such that $8x - x^2 \geq 0$, i.e., the closed interval $[0, 8]$. where f is continuous. By Theorem 19.12 the function f is endowed with absolute minimum and absolute maximum. The derivative of f is

$$f'(x) = \frac{4 - x}{\sqrt{8x - x^2}}$$

and $f'(x) = 0$ if and only if $x = 4$. Calculate the second derivative

$$f''(x) = \frac{-(8x - x^2) - (4 - x)^2}{\left(\sqrt{8x - x^2}\right)(8x - x^2)}$$

and apply the test

$$f''(4) = -\frac{1}{4} < 0$$

Therefore, $x = 4$ is a relative maximum point of f. As $f(4) = 4$ and $f(0) = f(8) = 0$, the point $x = 4$ is the absolute maximum of f and the points $x = 0$ and $x = 8$ endpoints of the domain, are absolute minimum points at which the value of f is 0.

Exercise 21.3 Inquire $f(x) = \frac{1}{x-3}$ concerning relative extrema and find the intervals on which the function is increasing or decreasing.

The domain of f is $\mathbf{R} - \{3\}$. The derivative of f is

$$f'(x) = -\frac{1}{(x-3)^2}$$

that is non-null for every $x \neq 3$. Then f has no relative extrema; as $f'(x) < 0$ on both sides of 3, f is decreasing for $x < 3$ and $x > 3$.

21.6 de l'Hospital's Rule

Let us state a theorem useful when the evaluation of $\lim\limits_{x \to c} \frac{f(x)}{g(x)}$, that has one of *indeterminate forms* (see (18.9)) $\frac{0}{0}$ or $\frac{\infty}{\infty}$, is examined. The theorem, attributed to the marquis Guillaume de l'Hospital (1661–1704), is not a panacea: there are indeterminate forms resistant to de l'Hospital's rule.

Theorem 21.10 (de l'Hospital's rule) *Let $f(x)$ and $g(x)$ be differentiable functions in a neighborhood of the point c, possibly except the point c. If the $\lim\limits_{x \to c} \frac{f(x)}{g(x)}$ is an indeterminate form $\frac{0}{0}$ or $\frac{\infty}{\infty}$, if $g(x) \neq 0$ in a neighborhood of c and $g'(x) \neq 0$, for every $x \neq c$, if the limit*

$$\lim_{x \to c} \frac{f'(x)}{g'(x)}$$

exists, then also the limit $\lim\limits_{x \to c} \frac{f(x)}{g(x)}$ *exists and the two limits are equal*:

$$\lim_{x \to c} \frac{f(x)}{g(x)} = \lim_{x \to c} \frac{f'(x)}{g'(x)}$$

L'Hospital's rule is extended to the cases $x \to +\infty$, $x \to -\infty$, and to one sided limits such as $x \to c+$, $x \to c-$.

Example 21.7 The limit

$$\lim_{x \to 0} \frac{\sin x}{x} \tag{21.5}$$

is an indeterminate form $\frac{0}{0}$. We may apply de l'Hospital's rule and construct the limit of the ratio of the derivatives:

$$\lim_{x \to 0} \frac{\cos x}{1} = 1$$

Hence, the limit (21.5) exists and equals 1

$$\lim_{x \to 0} \frac{\sin x}{x} = 1$$

Example 21.8 The limit

$$\lim_{x \to 0} \frac{e^x - 1 + \sin x}{\ln(x + 1)}$$

is an indeterminate form $\frac{0}{0}$. We may apply de l'Hospital's rule and construct the limit of the ratio of the derivatives:

$$\lim_{x \to 0} \frac{e^x + \cos x}{\frac{1}{x+1}} = 2$$

Hence, the given limit exists and equals 2:

$$\lim_{x \to 0} \frac{e^x - 1 + \sin x}{\ln(x+1)} = 2$$

Example 21.9 The $\lim\limits_{x \to +\infty} \frac{\ln x}{x^2}$ is an indeterminate form $\frac{\infty}{\infty}$.

We may apply de l'Hospital's rule. Let us calculate the limit of the ratio of the derivatives,

$$\lim_{x \to +\infty} \frac{\frac{1}{x}}{2x} = \lim_{x \to +\infty} \frac{1}{2x^2} = 0$$

Hence, the given limit also exists and equals 0.

Remark 21.2 If, after applying de l'Hospital's rule, the $\lim\limits_{x \to c} \frac{f'(x)}{g'(x)}$ is of type $\frac{0}{0}$ or $\frac{\infty}{\infty}$, in case $f'(x)$ and $g'(x)$ satisfy the due conditions, the procedure can be repeated and the limit $\lim\limits_{x \to c} \frac{f''(x)}{g''(x)}$ calculated. If the last limit exists, finite or infinite, then also $\lim\limits_{x \to c} \frac{f(x)}{g(x)}$ and $\lim\limits_{x \to c} \frac{f'(x)}{g'(x)}$ exist and the three limits are equal.

Example 21.10 Let us now consider an indeterminate form not resolved by de l'Hospital's rule.

Evaluate

$$\lim_{x \to +\infty} \frac{\sqrt{x^2 + 1}}{x}$$

The limit is of type $\frac{\infty}{\infty}$. Let us apply de l'Hospital's rule and calculate the limit of the ratio of the derivatives,

$$\lim_{x \to +\infty} \frac{x}{\sqrt{x^2 + 1}}$$

The $\lim\limits_{x \to +\infty} \frac{x}{\sqrt{x^2+1}}$ is the limit of the inverse of the given function and it is also an indeterminate form of type $\frac{\infty}{\infty}$. So it is useless to reapply de l'Hospital's rule. Then we proceed directly:

$$\lim_{x \to +\infty} \frac{\sqrt{x^2 + 1}}{x} = \lim_{x \to +\infty} \frac{\sqrt{x^2\left(1 + \frac{1}{x^2}\right)}}{x}$$

$$= \lim_{x \to +\infty} |x| \frac{\sqrt{1 + \frac{1}{x^2}}}{x} = \lim_{x \to +\infty} x \frac{\sqrt{1 + \frac{1}{x^2}}}{x} = 1$$

21.7 More on the Indeterminate Forms

We introduced (Sect. 18.4.1) some indeterminate forms

$$0 \cdot \infty, \quad +\infty - \infty, \quad \frac{\infty}{\infty}, \frac{0}{0} \tag{21.6}$$

and we also verified that it is possible, in some cases, to eliminate with simple transformations, the indeterminacy that does not allow to calculating the limit, or ascertain its existence. The rule of de l'Hospital offers the tools to cope with the indeterminate forms $0 \cdot \infty, +\infty -\infty$, that can be traced back to ratios.

The case $0 \cdot \infty$ occurs when we want to study the limit at the point c, finite or infinite, of a product $f(x) g(x)$ whose factors are an infinitesimal and an infinity. As

$$f(x) \, g(x) = \frac{f(x)}{\frac{1}{g(x)}}$$

the limit of $f(x) g(x)$ can be transformed into the ratio of two infinitesimals or two infinities to which de l'Hospital's rule is applicable.

In order to deal with the indeterminate form $+\infty -\infty$, let us observe that the identity holds

$$f(x) - g(x) = \frac{\frac{1}{g(x)} - \frac{1}{f(x)}}{\frac{1}{g(x)f(x)}}$$

Therefore, the limit of the difference $f(x) - g(x)$ is brought back to limits of infinitesimals.

Other indeterminate forms besides (21.6) can occur when examining the limit:

$$\lim_{x \to c} f(x)^{g(x)} \tag{21.7}$$

with $f(x) > 0$. By the equality (see Sect. 17.9)

$$f(x)^{g(x)} = e^{g(x) \ln f(x)}$$

and from the continuity of the exponential function, we get

$$\lim_{x \to c} f(x)^{g(x)} = e^{\lim_{x \to c} g(x) \ln f(x)}$$

The limit of the exponent, $\lim_{x \to c} g(x) \ln f(x)$, is of type $0 \cdot \infty$, in one of the four cases:

1. $\lim_{x \to c} g(x) = 0$ and $\lim_{x \to c} f(x) = 0$
2. $\lim_{x \to c} g(x) = 0$ and $\lim_{x \to c} f(x) = +\infty$
3. $\lim_{x \to c} g(x) = +\infty$ and $\lim_{x \to c} f(x) = 1$
4. $\lim_{x \to c} g(x) = -\infty$ and $\lim_{x \to c} f(x) = 1$

Thus, in these cases the limit (21.7) appears in one of the forms:

$$0^0, +\infty^0, 1^{+\infty}, 1^{-\infty}$$

that consequently are *indeterminate forms*.

Example 21.11 The limit

$$\lim_{x \to 0} x \ln x$$

is an indeterminate form of type $0 \cdot \infty$. Let us transform the given limit into

$$\lim_{x \to 0} x \ln x = \lim_{x \to 0} \frac{\ln x}{\frac{1}{x}}$$

then apply de l'Hospital's rule to the indeterminate form $\frac{\infty}{\infty}$:

$$\lim_{x \to 0} \frac{\frac{1}{x}}{-\frac{1}{x^2}} = -\lim_{x \to 0} x = 0$$

Therefore, $\lim_{x \to 0} x \ln x = 0$.

Example 21.12 $\lim_{x \to 0} x^x$ is an indeterminate form 0^0. From the previous example we obtain: $\lim_{x \to 0} x^x = \lim_{x \to 0} e^{x \ln x} = e^0 = 1$.

Example 21.13 Evaluate $\lim_{x \to 0} \left(\frac{1}{e^x - 1} - \frac{1}{x} \right)$. This is an indeterminate form of type $+\infty$ $-\infty$. Reduce to a unique fraction and apply de l'Hospital's rule

$$\lim_{x \to 0} \frac{x - e^x + 1}{(e^x - 1)x} = \lim_{x \to 0} \frac{-e^x + 1}{xe^x + e^x - 1}$$

The right side is an indeterminate form of type $\frac{0}{0}$. Apply de l'Hospital's rule:

$$\lim_{x \to 0} \frac{-e^x}{xe^x + 2e^x} = -\frac{1}{2}$$

Remark 21.3 It is worth stressing that the following symbols are not indeterminate forms. The right side of each equality is the value of the symbol on the left

$$(+\infty)^{-\infty} = 0; \quad 0^{+\infty} = 0; \quad 0^{-\infty} = +\infty; \quad (+\infty)^{+\infty} = +\infty$$

21.8 Parabola with Vertical Axis

The real-valued function

$$y = px^2 + q + u \tag{21.8}$$

with p, q, u real numbers, has domain $\mathbf{R}$. In the plane xy, Eq. (21.8) defines a curve, which is a parabola with vertical symmetry axis. The roots x_1, x_2 of the equation $px^2 + qx + u = 0$ (Sect. 12.1) are

$$x_1 = \frac{-q - \sqrt{q^2 - 4pu}}{2p} \qquad x_2 = \frac{-q + \sqrt{q^2 - 4pu}}{2p}$$

If $q^2 - 4pu \geq 0$, then x_1 and x_2 are real numbers, if $q^2 - 4pu < 0$, then x_1 and x_2 are imaginary numbers and the parabola does not intersect the x axis.

The axis of symmetry of the parabola $y = x^2$ is the y axis. The axis of symmetry of the parabola (21.8) is the vertical line passing through the points of abscissa $\frac{x_1 + x_2}{2} = \frac{1}{2}\frac{-q}{p}$, whatever the sign of $q^2 - 4pu$ is.

Observe that the midpoint of the segment $x_1 x_2$ has a real abscissa $\frac{1}{2}\frac{-q}{p}$, even if the points x_1, x_2 are conjugate complexes numbers (see Sect. 12.1). Then, the equation of the symmetry axis of parabola is $x = \frac{-q}{2p}$.

Example 21.14 The axis of symmetry of parabola $y = x^2 - 2x$ is the line $x = 1$. The parabola has the points $x_1 = 0$ and $x_2 = 2$ in common with x axis. The point of the parabola (21.8) which has abscissa $\frac{-q}{2p}$ is called the *vertex* of the parabola. The vertex of the parabola $y = x^2 - 2x$ has coordinates $(1, -1)$ (Fig. 21.5).

Fig. 21.5 Parabola $y = x^2 - 2x$

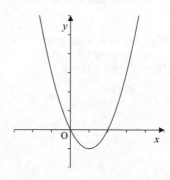

21.9   Approximation

A real-valued function P is called a *polynomial* in the real variable x if it has the form

$$P(x) = a_0 + a_1 x + a_2 x^2 + \cdots + a_n x^n$$

where n is a non-negative integer, $a_0, a_1, a_2, \ldots, a_n$ are real numbers called *coefficients* of the polynomial and, if $n > 0$ then $a_n \neq 0$. The number n, the highest degree of the powers of x, is called the *degree* of the polynomial. (The degree of the null polynomial, $0 = 0$, is not defined.) The domain of a polynomial is **R**.

The polynomial has derivatives of order however high.

We dealt with polynomials in Chap. 12. Examples of polynomials are the linear function and the power function with natural exponent.

Among the topics studied in the analysis since the origins, we find the calculations of the numerical values of functions such as $\sin x$, $\ln x$, e^x. The goal is to approximate the given function by means of polynomials so that the error belongs to a certain tolerance range.

The concept of differential introduces those of *approximation* and *error*.

We have found that the tangent line at c to the graph of f differentiable at c represents the behavior of the graph in a neighborhood of point c, and the y ordinate of a point of the tangent line provides an approximate value of $f(x)$, when x is near the point c.

The equation of the tangent line at c to the graph of f is $y = f(c) + f'(c)(x - c)$. The expression $f(x) \approx f(c) + f'(c)(x - c) = y$ (see Sect. 20.13) denotes that $f(c) + f'(c)(x - c)$ provides a linear approximation of the value $f(x)$ in a neighborhood of c (Fig. 21.6).

The linear function $g_1(x) = f(c) + f'(c)(x - c)$, i.e., the polynomial of first degree $g_1(x)$ describes the behavior of $f(x)$ near $x = c$ more precisely than any other polynomial of first degree since both $g_1(x)$ and $f(x)$ have the same value and the same derivative at $x = c$.

By Theorem 20.6 we know that the distance $\Delta f = f(x) - f(c)$, i.e., the measure of the error made when replacing y for $f(x)$,

Fig. 21.6 $\omega(x)$ is the error of the approximation $f(x) \approx f(c) + f'(c)(x - c) = y$

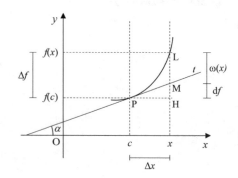

$$\Delta f - d(f) = \omega(x) = f(x) - \big(f(c) + f'(c)(x-c)\big)$$

is an infinitesimal of higher order than $x - c$, as $x \to c$.

21.9.1 Quadratic Approximation

Definition 21.3 If n is a non-negative integer number, the *factorial* of n, denoted $n!$, is defined by:

$$n! = n(n-1)(n-2)(n-3)\cdots 3\cdot 2\cdot 1, \text{ if } n \text{ is positive and } 0! = 1.$$

The symbol $n!$ reads "factorial of n". For example, $1! = 1, 2! = 2\cdot 1 = 2, 3! = 3\cdot 2\cdot 1 = 6, 4! = 4\cdot 3! = 24, 5! = 120, 6! = 720$.

We know that the linear function

$$y = f(c) + f'(c)(x-c)$$

provides the simplest approximation of $f(x)$ near the point c.

A more precise approximation is the *quadratic approximation*, obtained by means of the polynomial

$$g(x) = px^2 + qx + u$$

whose graph is a parabola with the vertical axis.

Given the function f, endowed with first and second derivative in a neighborhood of c, we will determine the coefficients p, q and u, in order to identify the parabola with vertical axis that approximates the graph of f. It is convenient to express $g(x)$ as a quadratic polynomial in the variable $x - c$:

$$g(x) = p(x-c)^2 + q(x-c) + u \tag{21.9}$$

Let us state the conditions that g must fulfil in order to be the best approximation of f.

First condition: if g approximates f in a neighborhood of c, then it must verify the equality $g(c) = f(c)$. Therefore, by (21.9),

$$f(c) = g(c) = u \tag{21.10}$$

Second condition: the derivatives of f and g at c are equal. Since $g'(x) = 2p(x-c) + q$, then

$$f'(c) = g'(c) = q \tag{21.11}$$

Third condition: the second derivatives of f and g at c are equal:

$$f''(c) = g''(c) = 2p \tag{21.12}$$

Hence, the conditions (21.10)–(21.12) determine the coefficients p, q and u. In fact, $u = f(c)$, $q = f'(c)$, $p = \frac{f''(c)}{2}$ and the polynomial $g(x)$ has the form

$$g(x) = f(c) + f'(c)(x - c) + \frac{f''(c)}{2}(x-c)^2 \tag{21.13}$$

The Eq. (21.13) is the required *quadratic approximation* of f at c. To denote that $g(x)$ approximates $f(x)$ let us write:

$$f(x) \approx f(c) + f'(c)(x - c) + \frac{f''(c)}{2}(x-c)^2$$

The polynomial (21.13) is called the *Taylor's polynomial of the second degree* for f near the point c; it describes the behavior of $f(x)$ near $x = c$ more precisely than any other polynomial of second degree because both $g(x)$ and $f(x)$ have the same value, the same first derivative and the same second derivative at $x = c$.

21.10   Taylor's Formula

The following proposition generalizes formula (21.13).

Proposition 21.1 *If $f(x)$ is a polynomial of degree n, then the equality holds*

$$f(x) \approx g_n(x) = f(c) + f'(c)(x - c) + \frac{f''(c)}{2!}(x - c)^2 + \cdots + \frac{f^{(n)}(c)}{n!}(x - c)^n \tag{21.14}$$

The formula (21.14), called the Taylor's formula for the polynomial $g_n(x)$ states that a polynomial of degree n is determined if the value of the polynomial is known along with the values of the derivatives up to the order n at any point c.

Let us recall (Sect. 20.10) that a function is said to be *n times differentiable* if the derivatives up to and including order n exist. In particular, a function twice differentiable is endowed with first and second derivative.

Formula (21.14) is a particular case of a formula valid for any real-valued function defined in an open interval (a, b) and n times differentiable in (a, b). Let $f(x)$ be such a function and x and c two points in (a, b). Let us put

$$r_n(f, c, x) = f(x) - \left[f(c) + f'(c)(x - c) + \frac{f''(c)}{2!}(x - c)^2 + \cdots + \frac{f^{(n-1)}(c)}{(n-1)!}(x-c)^{n-1} \right] \tag{21.15}$$

If $f(x)$ is a polynomial of degree n, then by (21.14)

$$r_n(f, c, x) = \frac{f^{(n)}(c)}{n!}(x-c)^n$$

The following theorem gives an expression of $r_n(f, c, x)$ for any function defined and n times differentiable in an open interval.

Theorem 21.11 *Let $f(x)$ be a function defined in the open interval (a, b) and c any point in (a, b). If f is n times differentiable in the interval (a, b), then the following equality holds:*

$$f(x) = f(c) + f'(c)(x - c) + \frac{f''(c)}{2!}(x - c)^2 + \cdots + \frac{f^{(n-1)}(c)}{(n-1)!}(x-c)^{n-1} \\ + \frac{f^{(n)}(d)}{n!}(x-c)^n \qquad (21.16)$$

where d is a point between c and x and distinct from both.

We observe that the number d which appears in (21.16) is unknown in general: it depends on the points c and x and on the order n of the derivative of f.

The formula (21.16) is called *Taylor's formula* or *Taylor's expansion* of *order n* and *initial point c* relative to the function f. The quantity

$$r_n(f, c, x) = \frac{(x-c)^n}{n!} f^{(n)}(d)$$

called the *remainder* in the *Lagrange's form*, is the *error term*, $r_n(f, c, x) = f(x) - g_n(x)$ of the approximation $f(x) \approx g_n(x)$.

As n increases, less and less inaccurate approximations of $f(x)$ are obtained.

Remark 21.4 The statement of Theorem 21.11 asserts that d is between c and x. This means that d belongs to the interval (c, x) if $c < x$, or d belongs to the interval (x, c) if $x < c$, or $d = c = x$ if $c = x$.

Taylor expansion of order n and initial point 0 is also called *MacLaurin expansion* of order n.

Example 21.15 Let us find MacLaurin expansion of order 3 relative to the function $f(x) = e^x$.

Let us carry out the derivatives:

$$f'(x) = f''(x) = f'''(x) = e^x$$

and the values at $x = 0$

$$f'(0) = f''(0) = f'''(0) = e^0 = 1$$

Therefore,

$$e^x = 1 + x + \frac{x^2}{2!} + \frac{e^d}{3!}x^3$$

with d lying between 0 and x. The formula is valid for any real x because the hypotheses of the theorem are satisfied in $\mathbf{R}$. The value of the error of the approximation is

$$r_3(f, 0, x) = \frac{e^d}{3!}x^3$$

Exercise 21.4 Find the Taylor's second degree polynomial $g_2(x)$ of initial point $c = 0$ for the function

$$f(x) = e^{\sin x}$$

The Taylor's second degree polynomial of initial point 0 is

$$g_2(x) = f(0) + f'(0)x + \frac{f''(0)}{2}x^2$$

where

$$\begin{aligned}
f(x) &= e^{\sin x}, &\Rightarrow\quad f(0) &= 1 \\
f'(x) &= e^{\sin x}\cos x, &\Rightarrow\quad f'(0) &= 1 \\
f''(x) &= e^{\sin x}\cos^2 x - e^{\sin x}\sin x, &\Rightarrow\quad f''(0) &= 1
\end{aligned}$$

Hence,

$$g_2(x) = 1 + x + \frac{x^2}{2}$$

21.11   Convexity, Concavity, Points of Inflection

The ways of bending of a curve or a surface of the space are important characters for the understanding the shape of the figure. A rock-cut site is a hollow, a cavern. Even the summit of a mountain can give the idea of concavity or convexity. So, also a rope in the throat of a pulley resembles a concave or convex curve.

We have studied (Sect. 4.2.1) the concepts of concavity and convexity from a geometric point of view. We now consider the analytical aspects, related with the real-valued functions.

21.11.1 Convexity and Concavity

Let f be a differentiable function in the interval I, bounded or not, of $\mathbf{R}$ and let c be a point in I. The line t tangent to the graph of f in c has equation (Sect. 20.3)

$$y = f'(c)(x - c) + f(c)$$

The function

$$F(x) = f'(c)(x - c) + f(c)$$

is defined in $\mathbf{R}$ and its graph is the line t.

Definition 21.4 We say that the line t is *below* the graph of f if $F(x) \leq f(x)$, for every x in I; we say that the line t is *above* the graph of f if $F(x) \geq f(x)$, for every $x \in$ I.

Definition 21.5 The function f is said to be a *convex function* in the interval I if, for each x in I, the tangent line to the graph of f in x is below the graph of f (Fig. 21.7). The function f is said to be a *concave function* in the interval I if, for each x in I, the tangent line to the graph of f in x is above the graph of f (Fig. 21.8).

The graph of a convex (concave) function is also called a convex (concave) curve. The graph of x^2 is a convex curve in $\mathbf{R}$.

Fig. 21.7 Convex function f

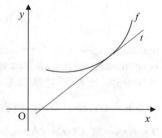

Fig. 21.8 Concave function f

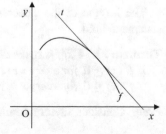

Theorem 21.12 *Let f be twice differentiable in the interval* I. *If f is convex in* I, *then* $f''(x) \geq 0$, *for every x in* I.

Proof Let x_1, x_2 be two distinct points in I and $x_1 < x_2$. Let

$$F_1(x) = f'(x_1)(x - x_1) + f(x_1)$$

be the equation of the tangent line to the graph of f at the point x_1. If f is convex in I, then

$$F_1(x) \leq f(x),$$

whatever $x \in$ I. In particular, if $x = x_2$, we obtain

$$f'(x_1)(x_2 - x_1) + f(x_1) \leq f(x_2)$$

Symmetrically, if $x = x_1$,

$$f'(x_2)(x_1 - x_2) + f(x_2) \leq f(x_1)$$

Let us sum the two inequalities to obtain

$$f'(x_1)(x_2 - x_1) + f(x_1) + f'(x_2)(x_1 - x_2) + f(x_2) \leq f(x_1) + f(x_2)$$

Let us cancel $f(x_2) + f(x_1)$ from each of the two sides,

$$f'(x_1)(x_2 - x_1) + f'(x_2)(x_1 - x_2) \leq 0$$

and since $x_2 - x_1 > 0$, we get $f'(x_1) \leq f'(x_2)$. As a consequence, the derivative f' is increasing over I and, in virtue of Theorem 21.6, the second derivative of f is non-negative, $f''(x) \geq 0$, in I. $\qquad\square$

Similarly, one can prove the following

Theorem 21.13 *Let f be twice differentiable in the interval* I. *If f is concave in* I, *then* $f''(x) \leq 0$, *for every* $x \in$ I.

The inverse of the Theorems 21.13 and 21.14, which we collect in a single statement, hold.

Theorem 21.14 *If f is twice differentiable in the interval* I, *then:*
if $f''(x) \geq 0$, *for every x in* I, *then f is convex in* I;
if $f''(x) \leq 0$, *for every x in* I, *then f is concave in* I.

Proof Consider Taylor's formula for $n = 2$ at initial point $c \in$ I:

$$f(x) = f(c) + f'(c)(x - c) + \frac{f''(d)}{2}(x - c)^2 \tag{21.17}$$

with x in I and d between x and c. The equation of the tangent line to the graph of f at c is

$$F(x) = f'(c)(x - c) + f(c) \tag{21.18}$$

For every $x \in$ I, plugging (21.18) into (21.17) we get

$$F(x) - f(x) = -\frac{f''(d)}{2}(x-c)^2$$

Then, if $f''(x) \geq 0$, for every x in I, we have $F(x) - f(x) \leq 0$, for every x in I. Therefore, f is convex in I. If $f''(x) \leq 0$, for every x in I, we have $F(x) - f(x) \geq 0$. Therefore, f is concave in I. $\qquad\qquad\qquad\qquad\qquad\qquad\qquad\qquad\qquad\qquad\square$

21.11.2   Points of Inflection

Definition 21.6 Let f be twice differentiable in the open interval I. We say that $x_0 \in$ I is a *point of inflection* of f if $f''(x_0) = 0$ and the function f is convex on one side and concave on the other side of the point x_0.

The inflection points of a twice differentiable function f are the roots of the equation $f''(x) = 0$ at which the second derivative f'' changes the sign.

Let us show the convexity and concavity properties of some elementary functions.

Example 21.16 The function $f(x) = x^2$ is convex in **R**, in fact $f''(x) = 2 \geq 0$, for every x belonging to **R**.

Example 21.17 The function $f(x) = x^3$ is convex in $(0, +\infty)$ and is concave in $(-\infty, 0)$, in fact $f''(x) = 6x$; and the second derivative of f has the sign of x.

Example 21.18 The exponential function $f(x) = a^x$ is convex in **R**, for every positive $a \neq 1$.

Example 21.19 The function $\log_a x$ is concave in $(0, +\infty)$ for every $a > 1$, convex in $(0, +\infty)$ for $a \in (0, 1)$.

Example 21.20 The point $x = 0$ is an inflection point for $f(x) = x^3$, in fact $f''(0) = (6x)_{x=0} = 0$ and (see Example 21.14) near $x = 0$, f'' changes the sign.

Example 21.21 Let consider the restriction $f(x)$ of $\tan x$ to the open interval $\left(-\frac{\pi}{2}, \frac{\pi}{2}\right)$.

The function $f(x)$ is concave in $\left(-\frac{\pi}{2}, 0\right)$ and convex in $\left(0, \frac{\pi}{2}\right)$ and has a unique inflection point at $x = 0$. In fact, (see Sect. 19.10), $D^2 \tan x = 2\tan x(1 + \text{tg}^2 x)$, and for $x \in \left(-\frac{\pi}{2}, \frac{\pi}{2}\right)$ it is $D^2 \tan x = 0$ if and only if $x = 0$. Furthermore, the sign of f'' coincides with the sign of x.

Fig. 21.9 Little man on the
logarithm

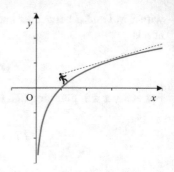

Example 21.22 The function $f(x) = \operatorname{atan} x$ is defined in **R**, convex in $(-\infty, 0)$, concave in $(0, +\infty)$ and has a single inflection point in $x = 0$. Indeed,

$$f''(x) = D^2\operatorname{atan} x = \frac{-2x}{\left(1 + x^2\right)^2}$$

Thus the second derivative of $\operatorname{atan} x$ equals zero at the unique point $x = 0$; besides, f is convex in $(-\infty, 0)$ and concave in $(0, +\infty)$.

21.11.3 Defiladed Objects

We illustrate some intuitive and applicative aspects of the concepts of concavity and convexity.

A concave curve lies under each tangent line in one of its points: then a two-dimensional little man walking on the curve sees the curve up to a certain point and on the remaining part, out of sight, ignores existence and form.

Even though he reaches points of the curve having ordinate no matter how high, as well as the tangent line at any point of the curve, a possibly not bounded section of the curve remains unknown to the observer: this is the case of the logarithmic curve to base greater than 1 (Fig. 21.9).

More practically, it happens that a walker on a mountain path looks into high from time to time and has the illusion of being close to the top, but he will know what is missing from the summit when he has reached it, when he dominates, he will know the shape of the mountain, when the eye will be able to send straight lines to the points of the whole surface: the optical cone, formed by the lines that pass through the points of the apparent contour of mountain.

Due to G. Monge is a geometric construction to hide, defilade architectural artefacts, especially fortifications. The fences must be raised so that the fortification is protected by observation from the outside and grazing fire. If the ground is flat, the problem is not difficult; if the terrain is a cavea, then complications arise.

The geometric project consists in determining a suitable plane on which a wall for concealment is built.

Historical background. Gaspard Monge (1746–1818) was an enthusiastic patriot of the French Revolution. He held important positions: he was assigned a role by the Constituent Assembly in the reform of weights and measures, he was minister of the navy, founded, administered and taught at the École Polytechnique, where he had Napoleon as an instructor. It was he, according to some historians, who signed the death sentence of Louis XVI. In twenty days he managed to recruit 900,000 men who formed the army (September 20, 1792) that saved France from the central empires. Right arm of Napoleon in Egypt. A great scholar, adored by his students, creator of descriptive geometry for military purposes, he had just as much weight in the development of differential geometry.

Gaspard Monge

21.12 Drawing the Graph of a Function

In order to study a real-valued function f and drawing its graph, one can proceed by taking into account some indications.

1. Determine the domain of the function.
2. Calculate the limits at the points of accumulation that do not belong to the domain, at the extremes of the domain, to infinity.
3. Determine any asymptotes.
4. Calculate the derivative f': this allows to know the parts of the domain where f is increasing or decreasing, relative maximum points and relative minimum points (Sect. 20.4).

Fig. 21.10 Graph of $f(x) =$ $e^{\frac{1}{x-1}}$

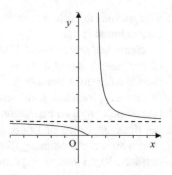

5. Calculate the second derivative of f: this allows to know the parts of the domain where f is convex or concave, possible inflection points (Sect. 21.11).
6. Sometimes the function is even or odd (Sect. 17.5). It is worth noting it.

It is inadvisable to tackle the study of a function by searching for the parts of the domain where the function assumes positive or negative values (the so called study of the positive part of f).

It is a sometimes tiring and not very illuminating calculation, which students sometimes face at the beginning of the study, consuming in brute calculations the first period, the freshest, of their commitment. In general, the trend of the graph is well understood by applying the first four steps, and one has a fairly clear idea after the first three.

Example 21.23 Draw the graph of the function

$$f(x) = e^{\frac{1}{x-1}}$$

1. The domain of the function is the set $\mathbf{R} - \{1\}$.
2. Calculate the limits of the function f as x approaches $-\infty$, $+\infty$ and the point 1.

$$\lim_{x \to -\infty} f(x) = 1 \quad \lim_{x \to +\infty} f(x) = 1$$
$$\lim_{x \to 1-} f(x) = 0 \quad \lim_{x \to 1+} f(x) = +\infty$$

3. The point $x = 1$ is a discontinuity of the second kind. The line $x = 1$ is a vertical asymptote. The line $y = 1$ is a horizontal asymptote (Fig. 21.10).
4. The derivative of $f(x)$ is $f'(x) = e^{\frac{1}{x-1}} \frac{-1}{(x-1)^2}$. The derivative $f'(x)$ is negative in the domain of f, so f is *decreasing* in the open interval $(-\infty, 1)$ and in in the open interval $(1, +\infty)$.

Note that $f(0) = e^{-1} = \frac{1}{e}$ and $2 < e < 3$ implies $\frac{1}{3} < \frac{1}{e} < \frac{1}{2}$.

Example 21.24 Consider the function

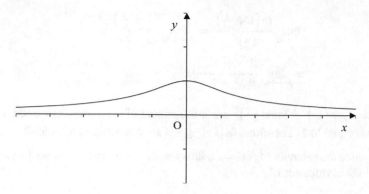

Fig. 21.11 Graph of $f(x) = \frac{1}{\sqrt{1+x^2}}$

$$f(x) = \frac{1}{\sqrt{1+x^2}}$$

1. The domain of the function is the set **R**.
2. The limits at the extremes of the domain are

$$\lim_{x \to -\infty} \frac{1}{\sqrt{1+x^2}} = 0 \quad \lim_{x \to +\infty} \frac{1}{\sqrt{1+x^2}} = 0$$

3. Then the line $y = 0$ is a horizontal asymptote.
4. The derivative of f is

$$f'(x) = D\frac{1}{\sqrt{1+x^2}} = \frac{\frac{-2x}{2\sqrt{1+x^2}}}{1+x^2} = \frac{-x}{(1+x^2)\sqrt{1+x^2}}$$

This means that $f'(x)$ has the opposite sign to x; i.e., the derivative is positive if and only if x is negative, the derivative is negative if and only if x is positive. Hence f is increasing in the interval $(-\infty, 0]$ and decreasing in $[0, +\infty)$. Furthermore, $x = 0$ is a point of relative and absolute maximum where $f(0) = 1$.

The graph of f is sketched in Fig. 21.11.

21.13 Solved Problems

1. Calculate the order of infinitesimal of the function $ln\left(1 + \frac{1}{x^2}\right)$ as $x \to +\infty$ with respect to the infinitesimal $\frac{1}{|x|}$ assumed as principal infinitesimal (see Definition 20.2).

Solution. Apply de L'Hospital's rule to the limit:

$$\lim_{x\to+\infty}\frac{ln\left(1+\frac{1}{x^2}\right)}{\left(\frac{1}{x}\right)^p}=\lim_{x\to+\infty}\frac{\left(1+\frac{1}{x^2}\right)\frac{-2}{x^3}}{-p\left(\frac{1}{x}\right)^{p+1}}$$

$$=\lim_{x\to+\infty}\frac{2}{p}\frac{x^2}{x^2+1}\frac{1}{x^3}x^{p+1}=\lim_{x\to+\infty}\frac{2}{p}\frac{x^p}{x^2+1} \tag{21.19}$$

If $p=2$, $ln\left(1+\frac{1}{x^2}\right)$ and $\left(\frac{1}{x}\right)^p$ are infinitesimal of the same order since the limit (21.19) is equal to 1. Therefore, $ln\left(1+\frac{1}{x^2}\right)$ is an infinitesimal of order 2 w. r. to $\frac{1}{|x|}$.

2. Examine the behavior of $f(x)=x(atanx)$ as $x\to-\infty$ and $x\to+\infty$. In particular, find the asymptotes of f.

Solution.

$$\lim_{x\to-\infty}x(atanx)=(-\infty)\left(-\frac{\pi}{2}\right)=+\infty$$
$$\lim_{x\to+\infty}x(atanx)=(+\infty)\frac{\pi}{2}=+\infty$$

Let us search if f has oblique asymptotes.

$$\lim_{x\to-\infty}\frac{f(x)}{x}=\lim_{x\to-\infty}atanx=-\frac{\pi}{2}$$
$$\lim_{x\to-\infty}\left(x(atanx)+\frac{\pi}{2}x\right)=\lim_{x\to-\infty}x\left(atanx+\frac{\pi}{2}\right)$$

The second limit takes the indeterminate form $\infty\bullet 0$. Let the form be altered so as to give the indeterminate form $\frac{0}{0}$:

$$\lim_{x\to-\infty}\frac{atanx+\frac{\pi}{2}}{\frac{1}{x}}$$

to which to apply de L'Hospital's rule

$$\lim_{x\to-\infty}\frac{\frac{1}{1+x^2}}{-\frac{1}{x^2}}=\lim_{x\to-\infty}\frac{-x^2}{x^2+1}=-1$$

Hence le line $y=-\frac{\pi}{2}x-1$ is an oblique asymptote as $x\to-\infty$.
Likewise, we find that the line $y=-\frac{\pi}{2}x-1$ is an oblique asymptote as $x\to+\infty$.
The function f can not have vertical asymptotes because it is continuous in **R**.

3. Given the function

$$f(x)=xe^{-\frac{1}{|x|}}$$

determine domain, symmetries, limits as $x\to0$, $x\to+\infty$ and $x\to-\infty$, asymptotes, first and second derivatives, monotonicity and convexity and the tangent at the origin.

Solution.

The domain of f is $D = \mathbf{R} - \{0\}$. As $e^{-\frac{1}{|x|}}$ is positive in D, the sign of $xe^{-\frac{1}{|x|}}$ coincide with the sign of x. The function f is odd (see Sect. 17.5); indeed,

$$f(-x) = (-x)e^{-\frac{1}{|x|}} = -xe^{-\frac{1}{|x|}} = -f(x)$$

Then the graph of the function is symmetric with respect to the origin. Therefore, we study the function for $x > 0$. If $x < 0$, the graph will be constructed by the symmetry. As $x > 0$ implies $|x| = x$, the function may be rewritten

$$f(x) = xe^{-\frac{1}{x}}$$

for $x > 0$.

Let us calculate the limits of $f(x) = xe^{-\frac{1}{x}}$ as $x \to 0^+$ and $x \to +\infty$:

$$\lim_{x \to 0+} xe^{-\frac{1}{x}} = 0 \cdot 0 = 0$$

$$\lim_{x \to +\infty} xe^{-\frac{1}{x}} = +\infty e^0 = +\infty$$

The derivative of $f(x) = xe^{-\frac{1}{x}}$ is

$$f'(x) = e^{-\frac{1}{x}} + xe^{-\frac{1}{x}}\frac{1}{x^2} = \left(1 + \frac{1}{x}\right)e^{-\frac{1}{x}} > 0$$

for every $x > 0$. Hence, f is strictly increasing in $(0, +\infty)$ (see Sect. 21.4). The second derivative

$$f''(x) = -\frac{1}{x^2}e^{-\frac{1}{x}} + \left(1 + \frac{1}{x}\right)\frac{1}{x^2}e^{-\frac{1}{x}} = \frac{1}{x^3}e^{-\frac{1}{x}}$$

is positive. Then $f(x)$ is convex in $(0, +\infty)$ (see Sect. 21.11).

Let us seek for oblique asymptotes as $x \to +\infty$ and $x \to -\infty$:

$$\lim_{x \to +\infty} \frac{f(x)}{x} = \lim_{x \to +\infty} e^{-\frac{1}{x}} = e^0 = 1$$

$$\lim_{x \to +\infty} (f(x) - x) = \lim_{x \to +\infty} x\left(-1 + e^{-\frac{1}{x}}\right) \quad \text{[inderminate form } \infty \bullet 0]$$

Let us set the limit in the form $\frac{\infty}{\infty}$ and apply de l'Hospital rule:

$$\lim_{x \to +\infty} \frac{-1 + e^{-\frac{1}{x}}}{\frac{1}{x}} = \lim_{x \to +\infty} \frac{\frac{1}{x^2}e^{-\frac{1}{x}}}{-\frac{1}{x^2}} = -1$$

Fig. 21.12 Graph of
$f(x) = xe^{-\frac{1}{|x|}}$

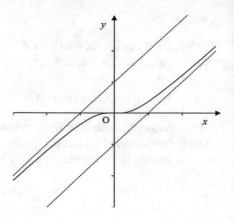

Hence, the line $y = x - 1$ is an asymptote as $x \to +\infty$. Due to the symmetry with respect to the origin, the line $-y = -x - 1$, symmetrical of $y = x - 1$ is an asymptote as $x \to -\infty$.

Let calculate the limit of the derivative as $x \to 0^+$. As $\lim\limits_{x \to 0+} e^{-\frac{1}{x}} = 0$, we get

$$\lim_{x \to 0+} \left(1 + \frac{1}{x}\right) e^{-\frac{1}{x}} = 0$$

Hence the right tangent to the graph of f at the origin is the x axis. Even because of the symmetry w. r. to the origin also the left tangent is the x axis. The point $x = 0$ is an inflection point.

Figure 21.12 gives an idea of the graph of f.

4. Determine the Taylor's polynomials of the second and the third degree, $P_2(x)$ and $P_3(x)$, for the function

$$f(x) = \frac{1}{2} x e^{3x}$$

near the point $c = 0$.

Solution. The expression of $P_2(x)$ and $P_3(x)$ for a generic function $f(x)$ (see Sect. 21.10) is

$$P_2(x) = f(0) + x \, f'(0) + \frac{x^2}{2} f''(0)$$
$$P_3(x) = f(0) + x \, f'(0) + \frac{x^2}{2} f''(0) + \frac{x^3}{3!} f'''(0)$$

Let us calculate the following derivatives of f:

$$f(x) = \tfrac{1}{2}xe^{3x} \qquad\qquad\qquad\qquad f(0) = 0$$
$$f'(x) = \tfrac{1}{2}e^{3x} + \tfrac{3}{2}xe^{3x} \qquad\qquad\qquad f'(0) = \tfrac{1}{2}$$
$$f''(x) = \tfrac{3}{2}e^{3x} + \tfrac{3}{2}e^{3x} + \tfrac{9}{2}xe^{3x} = 3e^{3x} + \tfrac{9}{2}xe^{3x} \quad f''(0) = 3$$
$$f'''(x) = 9e^{3x} + \tfrac{9}{2}e^{3x} + \tfrac{27}{2}xe^{3x} \qquad\qquad f'''(0) = 9 + \tfrac{9}{2} = \tfrac{27}{2}$$

Therefore, we obtain:

$$P_2(x) = \tfrac{1}{2}x + \tfrac{3}{2}x^2$$
$$P_3(x) = \tfrac{1}{2}x + \tfrac{3}{2}x^2 + \tfrac{9}{4}x^3$$

5. Determine the Taylor's formula of order 5 and initial point $c = \tfrac{\pi}{2}$ for the function

$$f(x) = \cos(2x)$$

and compute the remainder in the Lagrange's form.

Solution.

The required formula for a generic $f(x)$ is

$$f(x) = f(c) + (x-c)f'(c) + \tfrac{(x-c)^2}{2!}f''(c) + \tfrac{(x-c)^3}{3!}f'''(c) + \tfrac{(x-c)^4}{4!}f^{(4)}(c)$$
$$+ \tfrac{(x-c)^5}{5!}f^{(5)}(d)$$

where d is a point between c and x, distinct from both.
    The necessary values of $f(x)$ and the derivatives are:

$$\begin{aligned}
f(x) &= \cos(2x) & f\left(\tfrac{\pi}{2}\right) &= \cos\pi = -1 \\
f'(x) &= -2\sin(2x) & f'\left(\tfrac{\pi}{2}\right) &= -2\sin\pi = 0 \\
f''(x) &= -4\cos(2x) & f''\left(\tfrac{\pi}{2}\right) &= -4\cos\pi = 4 \\
f'''(x) &= 8\sin(2x) & f'''\left(\tfrac{\pi}{2}\right) &= 8\sin\pi = 0 \\
f^{(4)}(x) &= 16\cos(2x) & f^{(4)}\left(\tfrac{\pi}{2}\right) &= 16\cos\pi = -16 \\
f^{(5)}(x) &= -32\sin(2x) & f^{(5)}(d) &= -32\sin(2d)
\end{aligned}$$

Then the required Taylor's formula:

$$\cos(2x) = -1 + \frac{\left(x-\tfrac{\pi}{2}\right)^2}{2!}4 + \frac{\left(x-\tfrac{\pi}{2}\right)^4}{4!}(-16) + \frac{\left(x-\tfrac{\pi}{2}\right)^5}{5!}(-32\sin(2d))$$

Bibliography

Anton, H.: Calculus. Wiley, New York (1980)
Lax, P., Burnstein, S., Lax, A.: Calculus with Applications and Computing. Springer-Verlag, New York (1976)
Miranda, C.: Lezioni di Analisi matematica I, II. Liguori, Napoli (1978)
Royden, H.L., Fitzpatrick, P.M.: Real Analysis. Pearson, Toronto (2010)
Spivak, M.: Calculus. Cambridge University Press (2006)
Stoka, M.: Corso di matematica. Cedam, Padova (1988)
Ventre, A.: Matematica. Fondamenti e calcolo. Wolters Kluwer Italia, Milano (2021)

Chapter 22
Integration

22.1 Introduction

Integration is a topic of central importance in mathematics, rich in applications. It is a powerful means of investigation in every field of science. Historically, the germs of the theory are attributed to Archimedes and numerous scholars have developed the subject until today. Our approach is the one due to Riemann.

22.2 The Definite Integral

Let us consider a natural number n, the closed and bounded interval $[a, b]$ and $n + 1$ points in it, ordered as follows:

$$a = x_0 < x_1 < x_2 < \ldots < x_{i-1} < x_i < x_{i+1} < \ldots < x_{n-1} < x_n = b \qquad (22.1)$$

The points (22.1) define a *decomposition* D of the interval $[a, b]$ into consecutive intervals, the generic of which we denote $[x_i, x_{i+1}]$. $i = 0, 1, \ldots, n - 1$.

The *size of a decomposition* D is defined as the maximum d_D of the lengths $x_{i+1} - x_i$. Let us fix a point c_i in each interval $[x_i, x_{i+1}]$.

Let us now consider a real-valued function f defined and bounded in $[a, b]$. The sum

$$S_D(f) = \sum_{i=0}^{n-1} f(c_i)(x_{i+1} - x_i)$$

is named the *Riemann sum* relative to the decomposition D, the function f and the choice of the points c_i in $[x_i, x_{i+1}]$.

An example of Riemann sum is described in Fig. 22.1, where f *is* a positive function and the decomposition of $[a, b]$ is made of congruent intervals $[x_i, x_{i+1}]$.

© The Author(s), under exclusive license to Springer Nature Switzerland AG 2023
A. G. S. Ventre, *Calculus and Linear Algebra*,
https://doi.org/10.1007/978-3-031-20549-1_22

Fig. 22.1 The sum of the areas of the rectangles is a Riemann sum relative to a positive function f

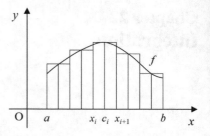

Definition 22.1 The bounded real-valued function f is said to be *integrable* on $[a, b]$ if there exists a real number I fulfilling the property:

(P) for every real number $\varepsilon > 0$ there exists a real number $\delta > 0$, such that for every subdivision D of size $d_{\mathrm{D}} < \delta$, it follows $|S_D(f) - I| < \varepsilon$.

The number I is called the *definite integral* of f on the interval $[a, b]$ and is denoted by the symbol

$$I = \int_a^b f(x)dx$$

The extremes of the interval $[a, b]$ are called the *extremes* or the *limits* of integration and f is said to be the *integrand function* or the *integrand*.

The following properties, some of which we will state without proof, hold.

Let f be *integrable* on $[a, b]$.

1.   We put, by definition

$$\int_a^b f(x)dx = -\int_b^a f(x)dx$$

i.e., interchanging the extremes of integration alter the sign of the integral.

2.   The equality applies

$$\int_a^a f(x)dx = 0$$

3.   If f is the constant function $f(x) = c$, then

$$\int_a^b cdx = c(b-a)$$

4. If f is monotone and bounded on $[a, b]$, then f is integrable on $[a, b]$.
5. If f is continuous in $[a, b]$, then f is integrable on $[a, b]$.
6. If the functions $f(x)$ and $g(x)$ are bounded and integrable on $[a, b]$, whatever the constants h and k are, the equality holds

$$\int_a^b (hf(x) + kg(x))dx = h \int_a^b f(x)dx + k \int_a^b g(x)dx$$

7. If f is non-negative, bounded and integrable on $[a, b]$, then

$$\int_a^b f(x)dx \geq 0$$

8. Let the functions $f(x)$ and $g(x)$ be bounded and integrable on $[a, b]$. If $f(x) \leq g(x)$, for every x in $[a, b]$, then

$$\int_a^b f(x)dx \leq \int_a^b g(x)dx$$

9. Let the functions $f(x)$ be bounded and integrable on $[a, b]$. If e' and e'' are the infimum and the supremum of f, respectively, then

$$e'(b-a) \leq \int_a^b f(x)dx \leq e''(b-a) \qquad (22.2)$$

10. (Mean value theorem) If f is continuous in a bounded interval $[a, b]$, then there exists a point c in $[a, b]$ such that

$$f(c)(b-a) = \int_a^b f(x)dx \qquad (22.3)$$

Proof From property 5, the function f is integrable. Then divide the inequalities (22.2) by $(b - a)$:

$$e' \leq \frac{1}{b-a} \int_a^b f(x)\, dx \leq e''$$

where e' and e'' are, in virtue of Weierstrass' theorem, the minimum and the maximum of f in $[a, b]$, respectively. By the Theorem 19.13, the function f assumes all the values in the closed interval $[e', e'']$ and hence also the value $\frac{1}{b-a}\int_a^b f(x)dx$.

Thus, a point c in $[a, b]$ exists such that

$$f(c) = \frac{1}{b-a}\int\limits_a^b f(x)dx$$

Hence, the equality (22.3) follows.   □

11. If the function f is integrable on a closed interval, then whatever the points a, b, c are in the interval, the function f is integrable on each of the intervals $[a, b]$, $[b, c]$ and $[c, a]$ and the equality holds:

$$\int\limits_a^c f(x)dx = \int\limits_a^b f(x)dx + \int\limits_b^c f(x)dx$$

12. If $f(x)$ is bounded and integrable on $[a, b]$, then the function $|f(x)|$ is integrable on $[a, b]$ and the inequality holds:

$$\left| \int\limits_a^b f(x)dx \right| \leq \int\limits_a^b |f(x)|dx$$

22.3 Area of a Plane Set

We define the concept of area for sets of points in the plane. We know (Chap. 8) the equation of the *circumference* with center C and radius $a(> 0)$. The *open circle* with center C and radius a is defined as the set of points $P(x, y)$ of the plane that have a distance from C less than a,

$$\{P : |CP| < a\}$$

The *circle* with center C and radius a is defined as the set of points of the plane that have a distance from C less than or equal to a,

$$\{P : |CP| \leq a\}$$

Let A be a set of points on the plane. The set A is said to be *a bounded set* if there exists a circle that contains it. A point P of A is said to be an *interior* point of A if P is

Fig. 22.2 Area $A = \frac{1}{2}bh$

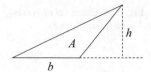

the center of an open circle entirely made up of points of A; a point Q not belonging to the set A is said to be an *exterior point* of A if Q is the center of a circle entirely made up of points not belonging to A.

A point in the plane is called a *boundary point* of A if it is neither interior nor exterior of A, i. e., in each circle having the center in that point there are both points belonging to A and points not belonging to A.

The set of points of the plane which are boundary points of A is called the *boundary* of A.

Let us express the concept of *polygon* in terms of the introduced nomenclature.

Definition 22.2 Any bounded set of points in the plane whose boundary consists of a finite number of line segments is called a *polygon*.

Definition 22.3 The area of the triangle A with base b and height h is defined by the number area $A = \frac{1}{2}bh$ (Fig. 22.2).

From the area of the triangle we are able to calculate the area of any polygon. In fact, each polygon can be decomposed into a finite number of triangles two-by-two without interior points in common, so that the union of all these triangles is congruent with the polygon. In general, given a polygon there are several decompositions of the polygon into triangles without interior points in common and such that the union of all these triangles is congruent with the polygon, however the sum of the areas of the triangles of any of such decompositions of the polygon is a number that does not vary with the decomposition: this number is called the *area of the polygon*.

The area of the polygon in Fig. 22.3 equals the sum of the areas of the triangles, with no interior points in common, in which the polygon is decomposed by means of the diagonals (dashed segments).

Remark 22.1 In any polygon, which is not a triangle, each of the segments connecting two non-consecutive vertices is called *diagonal*. If n is the number of vertices of a polygon, then $n - 3$ diagonals pass through any vertex. The points of

Fig. 22.3 Pentagon divided
into triangles by two
diagonals

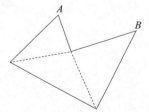

Fig. 22.4 Plane set of points
A with interior points

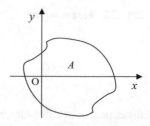

each diagonal of a convex polygon (Sect. 4.5.1) are interior points of the polygon, except the extremes. A convex polygon is decomposed by the diagonals sent by any vertex into $n - 2$ triangles, two by two with no interior points in common. In each concave polygon there is at least one diagonal, whose points are not interior points of the polygon (see the segment with endpoints A and B, Fig. 22.3).

We assume the following property:

M. If the polygon P is contained in the polygon Q, then $area P \leq area Q$.

Let A be a bounded plane set endowed with interior points (Fig. 22.4).

Let us consider the (infinite) numerical set **A** of the areas of the polygons contained in A and not coincident with A and the (infinite) numerical set **A**′ of the areas of the polygons containing A and not coincident with A. For the property M the numerical sets **A** and **A**′ are separate (Sect. 6.9). Moreover, if the two sets **A** and **A**′ are contiguous (Sect. 6.9), then the set of points A is said to be *Jordan measurable*, or simply, *measurable*, and the separating element a of the sets **A** and **A**′ is called the *area* of the plane set A, $a = $ area A, or the *(Jordan) measure* of A.

Furthermore, a set A which has no interior points is said to be *measurable* and has *measure zero* if the set of the areas of polygons containing A has infimum zero.

Thus, for example, the area of a segment is zero since the segment has no internal points and can be contained in a rectangle with area no matter how small.

The following properties apply:

13. If the sets A and B of points of the plane are measurable and A is contained in B, then

$$area A \leq area B$$

14. If A_1 and A_2 are two measurable sets and A_2 is contained in A_1, then

$$area(A_1 - A_2) = area A_1 - area A_2$$

Fig. 22.5 The region below
the curve represents the set
$A(f)$

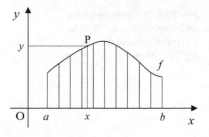

22.4 The Definite Integral and the Areas

Consider a function f continuous in the bounded and closed interval $[a, b]$, which
assumes non-negative values. By property 5, f is integrable in $[a, b]$. Let $A(f(x))$,
or $A(f)$, denote the set of the points in the plane (Fig. 22.5) whose coordinates (x, y)
satisfy the inequalities

$$a \leq x \leq b, 0 \leq y \leq f(x)$$

The following property holds:

15.   The set $A(f)$ is a measurable set and its area is given by

$$area\,A(f) = \int_a^b f(x)dx$$

22.5 The Integral Function

Let x be a point in the interval $[a, b]$ and f a bounded and integrable function on $[a,
b]$. From property 11, the function f is integrable in the interval $[a, x]$, for every x in
$[a, b]$. As x varies in $[a, b]$, the function

$$F(x) = \int_a^x f(t)dt$$

called the *integral function* of f, is a function of the extreme x, defined in the domain
$[a, b]$.

Remark 22.2 The function $f(t)$ is the restriction of $f(x)$ to the interval $[a, x]$. We
will still be referring this restriction to as f (Fig. 22.6)

Remark 22.3 Referring to Fig. 22.6, if x varies in $[a, b]$, the set $A(f(t))$, $t \in [a, x]$,
scrolls like a curtain.

Fig. 22.6 If f is nonnegative
and integrable on $[a, b]$, then
$F(x)$ is the area of the set
$A(f(t)), t \in [a, x]$

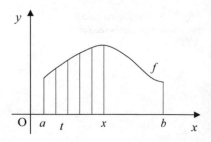

Let us show some important properties of the integral function F.

Theorem 22.1 *If the function f is bounded and integrable in the interval $[a, b]$, then the integral function F is continuous in $[a, b]$.*

Proof Let e'' be the least upper bound of f. Furthermore, let c be any point in $[a, b]$. We have:

$$|F(x) - F(c)| = \left| \int_a^x f(t)dt - \int_a^c f(t)dt \right|$$

property 1 implies: $= \left| \int_c^a f(t)dt + \int_a^x f(t)dt \right|$

property 11 implies: $= \left| \int_c^x f(t)dt \right|$

property 12 implies: $\leq \int_c^x |f(t)|dt$.

property 8 implies: $\leq \int_c^x |e''|dt$.

property 3 implies: $\leq |e''||x - c|$

Since the limit as $x \to c$ of the product $|e''||x - c|$ equals zero, we obtain

$$\lim_{x \to c} F(x) = F(c)$$

as we wanted to show.   □

Theorem 22.2 (Torricelli-Barrow's theorem) *If f is a continuous function in $[a, b]$, then the integral function $F(x)$ is differentiable in the open interval (a, b) and*

$$F'(x) = f(x) \tag{22.4}$$

Proof Let us consider the difference quotient of the function F at the point x in (a, b):

$$\frac{F(x+h) - F(x)}{h} = \frac{\int_a^{x+h} f(t)dt - \int_a^x f(t)dt}{h} = \frac{\int_x^a f(t)dt + \int_a^{x+h} f(t)dt}{h}$$

$$= \frac{\int_x^{x+h} f(t)dt}{h}$$

As f is continuous in $[a, b]$, in virtue of mean value theorem, a point c exists between x and $x + h$ such that

$$F(x + h) - F(x) = \int_x^{x+h} f(t)dt = hf(c)$$

Since c is between x and $x + h$, if h approaches 0, c approaches x.
Thus, by continuity of f in $[a, b]$, we obtain:

$$\lim_{h \to 0} \frac{F(x + h) - F(x)}{h} = \lim_{x \to c} f(c) = f(x)$$

i.e., the equality (22.4).   □

Historical background, Isaac Newton (Woolsthorpe-by-Colsterworth, 1642–London, 1727) and Gottfried Wilhelm Leibniz (Leipzig, 1646–Hannover, 1716). Since the time of Archimedes a major topic of interest was the problem of the calculation of the areas. Since then, volumes and volumes have been written to solve particular cases of the measurement of plane areas. Leibniz and Newton formulated a general method that enabled to calculate a vast set of areas through integration. Often Leibniz and Newton are named together, as if they were discovering and solving problems in collaboration. In reality, the Leibniz and Newton pair are the subject of one of the most complex priority disputes in the history of science.

22.6 Primitive Functions

Definition 22.4 Let f be a bounded and integrable function on $[a, b]$. The function f is said to admit a *primitive function*, or a *primitive*, in (a, b) if there exists a differentiable function $G(x)$ in (a, b) such that

$$G'(x) = f(x)$$

The function $G(x)$ is named a *primitive (function)* of f in (a, b). It is clear that if G is a primitive of f, then also the function $H(x) = G(x) + k$, being k a constant, is a primitive of f. This means that if f admits a primitive, then f admits infinitely many primitives.

Let us consider some examples.

Example 22.1 The function x^2 is a primitive of the function $2x$ because $Dx^2 = 2x$; $x^2 + 1$ is also a primitive of $2x$.

Example 22.2 The function $\frac{x^3}{3}$ is a primitive of x^2 because $D\frac{x^3}{3} = x^2$.

Example 22.3 The function $\sin^2 x$ is a primitive of $2\sin x \cos x$, since $D(\sin^2 x) = 2\sin x \cos x$; the function $\sin^2 x + k$, for every constant k, is also a primitive of $2\sin x \cos x$.

In terms of primitive, Torricelli-Barrow's theorem is stated as follows:

Theorem 22.3 *If f is a continuous function in $[a, b]$, then the integral function of f is a primitive of f in (a, b).*

Let us prove the following

Theorem 22.4 *If the function $f(x)$ admits primitives in the open interval (a, b), then any two primitives differ by a constant.*

Proof Let $G(x)$ and $H(x)$ be primitives of f in (a, b); i.e.,

$$G'(x) = f(x) \text{ and } H'(x) = f(x)$$

Therefore, $G'(x) = H'(x)$. This equality implies that the derivative of the function $H(x) - G(x)$ is identically null in (a, b) and, by Theorem 21.5, the function $H(x) - G(x)$ is constant in (a, b), i.e., $H(x) - G(x) = c$, being c any constant. As a conclusion, if the function f admits a primitive G in (a, b), all the primitives of f in (a, b) are obtained from the formula $H(x) = G(x) + c$, for every $c \in \mathbf{R}$, and only the primitives of G have the form $G(x) + c$, as c varies in $\mathbf{R}$. $\square$

We are now able to prove the *Fundamental Theorem of calculus.*

Theorem 22.5 (Leibniz-Newton's theorem) *If the function f is continuous in the interval $[c, d]$, and a and b are distinct points in the open interval (c, d), then*

$$\int_a^b f(x)dx = G(b) - G(a) \qquad (22.5)$$

where $G(x)$ is any primitive of f.

Proof Since f is continuous in the interval $[c, d]$, by properties 11 and 5, the function f is integrable in $[a, b]$ and we may consider the integral function of f in $[a, b]$

$$F(x) = \int_a^x f(t)dt$$

As F and G are primitives of f by Theorem 22.4 we have

$$G(x) = \int_a^x f(t)dt + c \tag{22.6}$$

where c is a constant. If we set $x = a$ in (22.6), we obtain by property 2

$$G(a) = \int_a^a f(t)dt + c = c \tag{22.7}$$

Equation (22.7) tells us that the constant c equals $G(a)$, $c = G(a)$. Let us put $x = b$ in (22.6)

$$G(b) = \int_a^b f(t)dt + c$$

and replace c with $G(a)$ to get

$$G(b) = \int_a^b f(t)dt + G(a)$$

Therefore, (22.5) follows. □

The difference $G(b) - G(a)$ is often indicated by one of the symbols $[G(x)]_a^b$ or $[G(x)]_{x=a}^{x=b}$.

Exercise 22.1 Find the area of the plane region under the parabola $y = x^2$ with $x \in [0, 1]$. In other words, the area of the region $A(f(x))$ (Sect. 22.4) has to be evaluated, where $f(x) = x^2$ and $A(f(x))$ is the set of the points (Fig. 22.7) having coordinates (x, y) such that

$$0 \le x \le 1 \text{ and } 0 \le y \le x^2$$

Fig. 22.7 Area of the region under the parabola

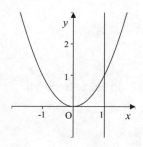

Fig. 22.8 Area under the
curve $y = \sin x$

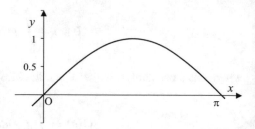

As $G(x) = \frac{x^3}{3}$ is a primitive of x^2 (see Example 22.2) *areaA* equals

$$\int_0^1 x^2 dx = G(1)-G(0) = \left(\frac{x^3}{3}\right)_{x=1} - \left(\frac{x^3}{3}\right)_{x=0} = \frac{1}{3}$$

Exercise 22.2 Find the area of the plane region A under the curve $y = f(x) = \sin x$, with $x \in [0, \pi]$ (Fig. 22.8).

The region A is the set of the points (x, y) such that

$$0 \le x \le \pi \text{ and } 0 \le y \le \sin x$$

Since a primitive of $\sin x$ is $G(x) = -\cos x$, we obtain

$$areaA = \int_0^\pi \sin x dx = G(\pi)-G(0) = -\cos \pi -(-\cos 0) = -(-1)-(-1) = 2.$$

Exercise 22.3 Examine the intervals of monotonicity of the function

$$f(x) = \frac{d}{dx} \int_0^x \left(e^{-x^2} - \frac{1}{2}\right) dt$$

Solution From Leibniz-Newton's theorem we know that

$$\frac{d}{dx} \int_0^x f(t)dt = f(x)$$

In our case we get

$$\frac{d}{dx} \int_0^x \left(e^{-x^2} - \frac{1}{2}\right) dt = e^{-x^2} - \frac{1}{2}$$

and the problem is brought back to the study of the monotonicity of $f(x) = e^{-x^2} - \frac{1}{2}$.

The derivative is $f'(x) = -2xe^{-x^2}$, whose sign coincides with that of $-x$. Therefore, f is increasing for $x < 0$ and decreasing for $x > 0$.

22.7 The Indefinite Integral

The theorem of Leibniz-Newton states that, in order to calculate the definite integral of a continuous function f, defined in the interval $[a, b]$, it is sufficient to know a primitive G of f in (a, b) and then apply (22.5).

Definition 22.5 The set of the primitives of f in (a, b) is called the *indefinite integral* of f and is denoted by the symbol

$$\int f(x)dx$$

By Theorem 22.4, if G is a primitive of the continuous function f in $[a, b]$, then the indefinite integral of f is the set of functions $G(x) + c$, with c any real constant. Therefore, the indefinite integral of f is characterized by equality:

$$\int f(x)dx = G(x) + c$$

where f is the integrand function and c the arbitrary constant.

22.7.1 Indefinite Integral Calculation

From the formulas of the derivatives of the elementary functions we deduce as many formulas of indefinite integration.

To realize the equalities, it is sufficient to verify that the derivatives of the right-hand side are equal to the integrand functions. For example, from the formula:

$$Dx^a = ax^{a-1}$$

it follows that the function $\frac{1}{a}x^a$ is a primitive of x^{a-1} and hence,

$$\int x^{a-1}dx = \frac{1}{a}x^a + c$$

Replacing a with $a + 1$ we obtain

$$\int x^a dx = \frac{x^{a+1}}{a+1} + c$$

which holds for any $a \neq -1$. If $a = -1$, we use the formula

$$D\ln|x| = \frac{1}{x}$$

which yields:

$$\int \frac{1}{x} dx = \ln|x| + c$$

From the other formulas of differentiation we similarly obtain the indefinite integrals of elementary functions. These indefinite integrals, usually named *immediate indefinite integrals*, are the elements for finding integrals of composite elementary functions.

22.7.2 Some Immediate Indefinite Integrals

1. $\int x^a dx = \frac{x^{a+1}}{a+1} + c, a \neq -1$
2. $\int \frac{1}{x} dx = \ln|x| + c$
3. $\int a^x dx = a^x \log_a e + c \Rightarrow \int e^x dx = e^x + c$
4. $\int \sin x dx = -\cos x + c$
5. $\int \cos x dx = \sin x + c$
6. $\int \frac{1}{\cos^2 x} dx = \tan x + c$
7. $\int \frac{1}{\sqrt{1-x^2}} dx = \arcsin x + c$
8. $\int \frac{-1}{\sqrt{1-x^2}} dx = \arccos x + c$
9. $\int \frac{1}{1+x^2} dx = \operatorname{atan} x + c$

Let $G(x)$ be a primitive of $f(x)$, i.e.,

$$\int f(x) dx = G(x) + c \qquad (22.8)$$

which means, by definition,

$$G'(x) = f(x) \qquad (22.9)$$

Consider the differential dx of the independent variable. From (20.17) we get

$$f(x)dx = G'(x)dx = dG \qquad (22.10)$$

Thus, equalities (22.8)–(22.10) yield

$$\int dG = G(x) + c \tag{22.11}$$

The formula (22.11) is a target that enable us to compute immediately the indefinite integral.

22.7.3 A Generalization of Indefinite Integration Formulas

The property of linearity also applies to indefinite integrals:

$$\int (hf(x) + kg(x))dx = h \int f(x)dx + k \int g(x)dx \tag{22.12}$$

whatever the constants h and k are. In particular,

$$h \int f(x)dx = \int hf(x)dx \tag{22.13}$$

The equalities (22.12) and (22.13) are widely used when computing the integrals. Furthermore, when calculating integrals the application of the composite functions differentiation rule

$$D(g(f(x))) = g'(f(x))f'(x)$$

is essential.

Let us proceed to a generalization of the formulas 1–9 (see Sect. 22.7.2).

1. If $a \neq -1$, from the differentiation rule of composite functions we have:

$$Df(x)^{a+1} = (a+1)f(x)^a f'(x)$$

that implies

$$D\frac{f(x)^{a+1}}{a+1} = f(x)^a f'(x)$$

and, therefore, by definition

$$\int f(x)^a f'(x)dx = \frac{f(x)^{a+1}}{a+1} + c \tag{22.14}$$

or

$$\int f(x)^a df(x) = \frac{f(x)^{a+1}}{a+1} + c$$

Exercise 22.4   Find the indefinite integral

$$\int \sqrt{x^4-2}(4x^3)dx = \int (x^4-2)^{\frac{1}{2}} D(x^4-2)dx = [\text{by formula (22.14)}]$$

$$= \int (x^4-2)^{\frac{1}{2}} d(x^4-2) = \frac{1}{\frac{3}{2}}(x^4-2)^{\frac{3}{2}} + c$$

Exercise 22.5   Find the indefinite integral

$$\int x^3\sqrt{x^4-2}dx = \frac{1}{4}\int x^3\sqrt{x^4-2}(4x^3)dx = \frac{1}{4}\cdot\frac{2}{3}(x^4-2)^{\frac{3}{2}} + c$$

2.   From differentiation rule

$$D\ln f(x) = \frac{1}{f(x)}f'(x)$$

and the definition of indefinite integral we obtain

$$\int \left(\frac{1}{f(x)}\right)df(x) = \int \left(\frac{1}{f(x)}\right)f'(x)dx = \ln|f(x)| + c$$

Exercise 22.6   Find the indefinite integral

$$\int \frac{1}{2x+1}dx = \frac{1}{2}\int \frac{2}{2x+1}dx = \frac{1}{2}\int \frac{d(2x+1)}{2x+1} = \frac{1}{2}\ln|2x+1| + c$$

Exercise 22.7   Find the indefinite integral

$$\int \tan x\, dx = \int \frac{\sin x}{\cos x}dx = -\int -\sin x\frac{dx}{\cos x} = -\ln|\cos x| + c$$

3.   The differentiation formula

$$De^{f(x)} = e^{f(x)}f'(x)$$

implies

$$\int e^{f(x)}df(x) = \int e^{f(x)}f'(x)dx = e^{f(x)} + c$$

Exercise 22.8 Find the indefinite integral

$$\int e^{-x}dx = -\int -e^{-x}dx = -\int e^{-x}d(-x) = -e^{-x} + c$$

Exercise 22.9 Find the indefinite integral

$$\int e^{\frac{1}{x}}\frac{1}{x^2}dx = -\int -\frac{1}{x^2}e^{\frac{1}{x}}dx = -\int e^{\frac{1}{x}}d\frac{1}{x} = -e^{\frac{1}{x}} + c$$

Exercise 22.10 Find the indefinite integral

$$\int \frac{dx}{e^x + 1} = [\text{divide numerator and denominator by } e^x]$$

$$= \int e^{-x}\frac{dx}{1 + e^{-x}} = -\int -e^{-x}\frac{dx}{1 + e^{-x}} = -ln(1 + e^{-x}) + c$$

4. The differentiation formula

$$D \sin f(x) = (\cos f(x))f'(x)$$

implies the indefinite integration rule

$$\int \cos f(x)df(x) = \sin f(x) + c$$

Exercise 22.11 Find the indefinite integral:

$$\int \frac{\cos \sqrt{x}}{\sqrt{x}}dx = 2\int \frac{\cos \sqrt{x}}{2\sqrt{x}}dx = 2\int \cos \sqrt{x}d\sqrt{x} = 2\sin \sqrt{x} + c$$

5. The formula

$$D \cos f(x) = (-\sin f(x))f'(x)$$

implies

$$\int \sin f(x)df(x) = -\cos f(x) + c$$

Exercise 22.12 Find the indefinite integral

$$\int x\sin(x^2 + 5)dx = \frac{1}{2}\int \sin(x^2 + 5)d(x^2 + 5) = -\frac{1}{2}\cos(x^2 + 5) + c$$

6. The formula

$$Dtan(f(x)) = \frac{1}{\cos^2 f(x)} f'(x)$$

implies

$$\int \frac{1}{\cos^2 f(x)} f'(x)dx = \int \frac{1}{\cos^2 f(x)} df(x) = \tan f(x) + c$$

Exercise 22.13   Find the indefinite integrals

$$\int \frac{1}{\cos^2 x^2} 2xdx = \int \frac{1}{\cos^2 x^2} dx^2 = \tan x^2 + c$$

$$\int \frac{\sin x}{\cos^3 x} dx = \int \frac{\sin x}{\cos x} \frac{1}{\cos^2 x} dx = \int \tan x d(\tan x) = \frac{1}{2}\left(\tan^2 x\right) + c$$

7. The formula

$$Darcsin f(x) = \frac{1}{\sqrt{1-f^2(x)}} f'(x)$$

yields

$$\int \frac{1}{\sqrt{1-f^2(x)}} df(x) = \arcsin f(x) + c$$

Exercise 22.14   Evaluate the indefinite integral

$$\int \frac{1}{\sqrt{4-x^2}} dx = \int \frac{1}{2\sqrt{1-\left(\frac{x}{2}\right)^2}} dx = \int \frac{1}{\sqrt{1-\left(\frac{x}{2}\right)^2}} d\frac{x}{2} = \arcsin\frac{x}{2} + c$$

8. From the formula

$$\int \frac{-1}{\sqrt{1-f^2(x)}} df(x) = \arccos f(x) + c$$

we obtain

$$\int \frac{-1}{\sqrt{4-x^2}} dx = \arccos\frac{x}{2} + c$$

9. From the formula

$$\int \frac{1}{1 + f^2(x)} df(x) = \operatorname{atan} f(x) + c$$

we obtain

$$\int \frac{1}{a^2 + x^2} dx = \int \frac{1}{a^2 \left[1 + \left(\frac{x}{a}\right)^2\right]} dx = \frac{1}{a} \int \frac{1}{1 + \left(\frac{x}{a}\right)^2} d\frac{x}{a} = \frac{1}{a} \operatorname{atan} \frac{x}{a} + c$$

22.8 Integration by Parts

Sometimes to calculate the integral, definite or indefinite, some manipulation is used, to bring the integral back to a known form. Not that there is a strict rule, but it can take some experience, trial and error.

22.8.1 Indefinite Integration Rule by Parts

A procedure which can come in handy is the *integration by parts*, based on the product differentiation rule (Sect. 20.5). Let the functions $u(x)$ and $v(x)$ be differentiable in an interval. Consider the derivative

$$D(u(x)v(x)) = u(x)v'(x) + u'(x)v(x)$$

which, passing to the indefinite integrals of both sides, leads to:

$$\int D(u(x)v(x))dx = \int u(x)v'(x)dx + \int u'(x)v(x)dx$$

The product $u(x)v(x)$ is a primitive of the derivative $D(u(x)v(x))$. Therefore,

$$\int D(u(x)v(x))dx = u(x)v(x) + c$$

Then we have

$$u(x)v(x) = \int u(x)v'(x)dx + \int u'(x)v(x)dx \tag{22.15}$$

(the constant c is absorbed by the indefinite integrals of the right-hand side). From (22.15) we obtain:

$$\int u(x)v'(x)dx = u(x)v(x) - \int u'(x)v(x)dx \qquad (22.16)$$

known as the *indefinite integration rule by parts.*

If we want to calculate $\int u(x)v'(x)dx$ and it is easier for us to calculate $\int u'(x)v(x)dx$, we can use the formula (22.16). The indefinite integration formula by parts can be rewritten as

$$\int u(x)dv(x) = u(x)v(x) - \int v(x)du(x) \qquad (22.17)$$

and summarized in the form

$$\int udv = uv - \int vdu \qquad (22.18)$$

Definition 22.6 The function u is called the *finite factor* and the function dv is called the *differential factor.*

Exercise 22.15 Find $\int xe^x dx$.

Apply formula (22.17), where $u(x) = x$ is the finite factor and $d(e^x) = e^x dx$ the differential factor. Therefore, $v(x) = e^x$ and

$$\int xe^x dx = \int xd(e^x) = xe^x - \int e^x dx = xe^x - e^x + c$$

Exercise 22.16 Find

$$\int x\sin x dx$$

Let us apply (22.18). Put $u = x$ and $dv = \sin x dx = d(-\cos x)$. Therefore, $v = -\cos x$:

$$\int x\sin x dx = -x\cos x - \int -\cos x dx = -x\cos x + \sin x + c$$

Exercise 22.17 Find

$$\int x^2 \ln x dx$$

Let $u = \ln x$ and $dv = x^2 dx = d\left(\frac{x^3}{3}\right)$. Therefore, $v = \frac{x^3}{3}$:

$$\int x^2 \ln x dx = \frac{x^3}{3}\ln x - \int \frac{x^3}{3}\frac{1}{x}dx = \frac{1}{3}\left(x^3 \ln x - \int x^2 dx\right) = \frac{1}{3}\left(x^3 \ln x - \frac{x^3}{3}\right) + c$$

Exercise 22.18 Find

$$\int \cos^2 x \, dx$$

Let us integrate by parts:

$$\int \cos^2 x \, dx = \int \cos x \cos x \, dx = \int \cos x \, d(\sin x) =$$

$$= \cos x \sin x - \int -\sin x \sin x \, dx$$

$$= \cos x \, \sin x + \int \sin^2 x \, dx$$

$$= \cos x \sin x + \int (1 - \cos^2 x) \, dx$$

$$= \cos x \sin x + x - \int \cos^2 x \, dx$$

So, the first side equals the last:

$$\int \cos^2 x \, dx = \cos x \, \sin x + x - \int \cos^2 x \, dx$$

and moving the integral to the left-hand side:

$$\int \cos^2 x \, dx = \frac{1}{2}(\cos x \, \sin x + x) + c$$

Exercise 22.19 Find $\int \frac{\ln x}{x} dx$.
   Integrating by parts:

$$\int \frac{\ln x}{x} dx = (\ln x)^2 - \int \frac{\ln x}{x} dx$$

and moving the integral to the left-hand side: $\int \frac{\ln x}{x} dx = \frac{1}{2}(\ln x)^2 + c$.

22.8.2 Definite Integration Rule by Parts

By Theorem 22.5 and product rule (Sect. 20.6) we obtain:

$$\int_a^b (u(x)v'(x) + u'(x)v(x)) \, dx = [u(x)v(x)]_a^b$$

Therefore,

$$\int_a^b u(x)v'(x)dx = [u(x)v(x)]_a^b - \int_a^b u'(x)v(x))$$

Exercise 22.20 Evaluate $\int_0^{\frac{\pi}{2}} x\sin x dx$

$$\int_0^{\frac{\pi}{2}} x\sin x dx = -[x\cos x]_0^{\frac{\pi}{2}} + \int_0^{\frac{\pi}{2}} \cos x dx = \sin\frac{\pi}{2} = 1$$

Exercise 22.21 Evaluate $\int_1^e \frac{lnx}{x} dx$

$$\int_1^e \frac{lnx}{x} dx = [(lnx)^2]_1^e - \int_1^e \frac{lnx}{x} dx$$

Therefore,

$$\int_1^e \frac{lnx}{x} dx = \frac{1}{2}$$

22.9   Area of a Normal Domain

Let $f(x)$ be a continuous function in the closed interval $[a, b]$, such that $f(x) \geq 0$ for every x in $[a, b]$. We already know how to find the area of the plane region $A(f)$ (Sect. 22.4), determined by the graph of f and the interval $[a, b]$. Let us now generalize the result.

Assume that f and g are continuous functions on $[a, b]$, such that, for every $x \in [a, b]$, $f(x) \leq g(x)$ and $f(x) < g(x)$, for every x in (a, b). The set B of the points $P(x, y)$, such that

$$a \leq x \leq b \text{ and } f(x) \leq y \leq g(x)$$

is called a *normal domain* with respect to the x axis.

Theorem 22.6 *The area of the region B is given by the formula*

$$area B = \int_a^b (g(x) - f(x))dx \tag{22.19}$$

Fig. 22.9 B normal domain
with respect to the x axis
(case $0 \leq f(x) \leq g(x)$)

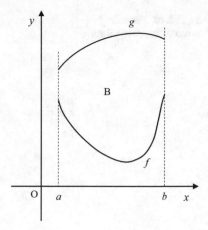

Proof If $0 \leq f(x) \leq g(x)$ (Fig. 22.9) the regions $A(f)$ and $A(g)$ are above the x axis
and under the curves $y = g(x)$ and $y = f(x)$, respectively. Then the region B is the
difference between the region $A(g)$ and the region $A(f)$. Clearly $A(f)$ is included in
$A(g)$ and $B = A(g) - A(f)$. By Property 14,

$$area\,B = area\,A(g) - area\,A(f)$$

and by Property 15,

$$area\,B = \int_a^b g(x)dx - \int_a^b f(x)dx.$$

Thus, by Property 6, the equality (22.19) follows.

In the case that one or both functions $y = f(x)$ and $y = g(x)$ assume negative values
(Fig. 22.10), the equality (22.19) still holds.

Indeed, let $m < 0$ be the minimum of f in $[a, b]$, which exists in virtue of
Weierstrass' theorem. Shift the region B along with both the curves vertically upwards
by a number not greater than $|m|$. Now the upper curve is the graph of $y = g(x) + |m|$
and the lower curve is the graph of $y = f(x) + |m|$ (Fig. 22.11). So we go back to
the case $0 \leq f(x) + |m| \leq g(x) + |m|$ Thus,

$$area\,B = \int_a^b ((g(x) + |m|) - (f(x) + |m|))dx = \int_a^b (g(x) - f(x))dx$$

$\square$

Fig. 22.10 Function f
assumes negative values

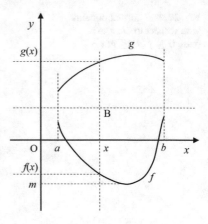

Fig. 22.11 General case of
a normal domain

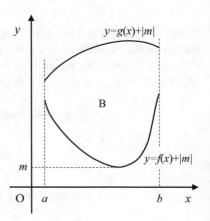

Exercise 22.22 Let us find the area of the normal domain B w. r. to the x axis defined
by the functions $f(x) = x^2$, $g(x) = x$, on the interval $[0,1]$ (Fig. 22.12). The functions
are continuous in **R** and $x^2 < x$ in $(0, 1)$. So, being $\frac{x^3}{3}$ a primitive of x^2, and $\frac{x^2}{2}$ a
primitive of x, we have

$$area B = \int_0^1 (g(x) - f(x))dx = \int_0^1 (x - x^2)dx = \left(\frac{x^2}{2} - \frac{x^3}{3}\right)_{x=1} - \left(\frac{x^2}{2} - \frac{x^3}{3}\right)_{x=0}$$

$$= \frac{1}{2} - \frac{1}{3} = \frac{1}{6}$$

Exercise 22.23 Evaluate the area of the normal domain B defined by the functions
$g(x) = x + 5$ and $f(x) = x^2 - 3x$ in the interval $[a, b]$, where a, b are the abscissas
of the common points to the two graphs.

Solving the equations simultaneously we find a and b:

Fig. 22.12 Normal domain
w.r. to x axis defined by
$g(x) = x$ and $f(x) = x^2$ and x
in $[0, 1]$

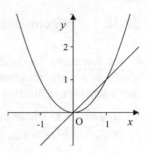

$$x^2 - 3x = x + 5 \text{ implies } x^2 - 4x - 5 = 0$$

that has roots -1 and 5. So $[a, b] = [-1, 5]$ and $f(x) < g(x)$ in the open interval $(-1, 5)$:

$$area\,B = \int_{-1}^{5} (x + 5 - (x^2 - 3x))dx = \int_{-1}^{5} (-x^2 - 4x + 5)dx =$$

$$= \left[-\frac{x^3}{3} + 2x^2 + 5x \right]_{x=-1}^{x=5} = 36$$

Remark 22.4 The notion of *normal domain* with respect to y axis is defined similarly. If $h(y)$ and $k(y)$ are two continuous functions in the interval $[c, d]$, such that $h(y) < k(y)$ in the open interval (c, d), the set of points $P(x, y)$ such that: $c \le y \le d$, $h(y) \le x \le k(y)$ is called a *normal domain* w. r. to the y axis.

22.10 Trigonometric Integrals

Some trigonometric identities (Sect. 8.1.4) are useful for the evaluation of some integrals.

Example 22.4 Find $\int \sin^2 x dx$.

$$\int \sin^2 x dx = \frac{1}{2} \int (1 - \cos 2x)dx = \frac{1}{2}x - \frac{1}{4}\sin 2x + c$$

Example 22.5 Find $\int_{-\frac{\pi}{2}}^{\frac{\pi}{2}} \cos^2 x dx$.

$$\int_{-\frac{\pi}{2}}^{\frac{\pi}{2}} \cos^2 x dx = \int_{-\frac{\pi}{2}}^{\frac{\pi}{2}} (1 - \sin^2 x)\,dx = \left[x - \frac{1}{2}x + \frac{1}{4}\sin 2x \right]_{-\frac{\pi}{2}}^{\frac{\pi}{2}} = \frac{\pi}{2}$$

22.10.1 Trigonometric Substitutions

If the integrand function contains one of the irrational forms $\sqrt{a^2-b^2x^2}$ or $\sqrt{a^2+b^2x^2}$, and not other irrational forms, then the integrand can be transformed into a function, containing trigonometric functions of a new variable, as follows:

1. if $\sqrt{a^2-b^2x^2}$ occurs in the integrand use the substitution $x = \frac{a}{b}\sin t$
2. if $\sqrt{a^2 + b^2x^2}$ occurs in the integrand use the substitution $x = \frac{a}{b}\tan t$

Example 22.6 Calculate the definite integral

$$\int\limits_{-2}^{2} (2-x)\sqrt{4-x^2}dx$$

Let us apply the substitution (1): $x = 2\sin t$. Hence, $dx = 2\cos t\, dt$ and

$$\sqrt{4-x^2} = \sqrt{4-4\sin^2 t} = 2\cos t$$

Let us replace also the extremes of integration: if $x = -2$, then $2\sin t = -2$, $\sin t = -1$ and $t = -\frac{\pi}{2}$; similarly, if $x = 2$, then $t = \frac{\pi}{2}$. By Example 22.5, we get:

$$\int\limits_{-2}^{2} (2-x)\sqrt{4-x^2}dx = \int\limits_{-\frac{\pi}{2}}^{\frac{\pi}{2}} (2-2\sin t)(2\cos t)(2\cos t)dt$$

$$= 8\int\limits_{-\frac{\pi}{2}}^{\frac{\pi}{2}} \cos^2 t\, dt - 8\int\limits_{-\frac{\pi}{2}}^{\frac{\pi}{2}} \cos^2 t\, (\sin t)dt$$

$$= 8\left[\frac{1}{2}t + \frac{1}{4}\sin t\right]_{-\frac{\pi}{2}}^{\frac{\pi}{2}} - \frac{8}{3}[\cos^3 t]_{-\frac{\pi}{2}}^{\frac{\pi}{2}} = 4\pi$$

22.11 Improper Integrals

The concept of definite integral $\int_a^b f(x)dx$ requires that the following two conditions be met:

1. the function $f(x)$ is bounded and
2. the interval $[a, b]$ is closed and bounded.

We will generalize the concept of definite integral to the case that one of the two conditions is not satisfied.

22.11.1   *Improper Integrals Over Bounded Intervals*

A.   Let us consider a function $f(x)$ continuous in the right-open interval $[a, b)$. Since the function is continuous in any closed interval $[a, x]$, $a \leq x < b$, the integral

$$\int_a^h f(x)dx$$

exists and is finite for any h such that $a \leq h < b$. If the limit

$$\lim_{h \to b} \int_a^h f(x)dx \qquad\qquad (22.20)$$

exists and is finite the function $f(x)$ is said to have *convergent improper integral* over the interval $[a, b)$ and, by definition, we set:

$$\int_a^b f(x)dx = \lim_{h \to b} \int_a^h f(x)dx$$

If the limit (22.20) does not exist or its value is $+\infty$ or $-\infty$, we say that the improper integral of $f(x)$ does *not exist* or it is *divergent*, respectively.

B.   Similarly, consider the case that $f(x)$ be continuous in the left-open interval $(a, b]$. Since the function is continuous in any interval $[x, b]$, $a < x \leq b$, the integral

$$\int_h^b f(x)dx$$

exists and is finite for any h such that $a < h \leq b$. If the limit

$$\lim_{h \to b} \int_a^h f(x)dx$$

exists and is finite the function $f(x)$ is said to have *convergent improper integral* and, by definition, we set:

$$\int\limits_{a}^{b} f(x)dx = \lim_{h\to a} \int\limits_{h}^{b} f(x)dx \qquad (22.21)$$

If the limit does not exist or its value is $+\infty$ or $-\infty$, the improper integral. does not exist or is divergent.

C. If $f(x)$ has a discontinuity in an interior point c of the interval $[a, b]$, then $f(x)$ is said to have *convergent improper integral* if both the following limits exist and are finite:

$$\lim_{b\to c-} \int\limits_{a}^{b} f(x)dx \qquad \lim_{d\to c+} \int\limits_{d}^{b} f(x)dx$$

By definition we set:

$$\int\limits_{a}^{b} f(x)dx = \lim_{b\to c-} \int\limits_{a}^{b} f(x)dx + \lim_{d\to c+} \int\limits_{d}^{b} f(x)dx \qquad (22.22)$$

Example 22.7 (related to the case (C)). The function

$$f(x) = \frac{1}{\sqrt[3]{(x-1)^2}} \qquad (22.23)$$

is an *infinity* (Sect. 20.12) at the point $c = 1$. Let us examine the improper integral:

$$\int\limits_{0}^{2} \frac{1}{\sqrt[3]{(x-1)^2}} dx$$

Let us start calculating the definite integral of $f(x)$ over $[0, b]$, $0 \le b < 1$, and the definite integral of $f(x)$ over $[d, 2]$, $1 < d \le 2$, i.e.,

$$\int\limits_{0}^{b} \frac{1}{\sqrt[3]{(x-1)^2}} dx \qquad \int\limits_{d}^{2} \frac{1}{\sqrt[3]{(x-1)^2}} dx \qquad (22.24)$$

Observe that the integrals (22.24) exist because f is continuous in $[0, 2] - \{1\}$. We obtain:

$$\int\limits_{0}^{b} \frac{1}{\sqrt[3]{(x-1)^2}} dx = \int\limits_{0}^{b} (x-1)^{-\frac{2}{3}} dx = \frac{1}{1-\frac{2}{3}}(b-1)^{1-\frac{2}{3}} - \frac{1}{1-\frac{2}{3}}(0-1)^{1-\frac{2}{3}} = 3\sqrt[3]{b-1} + 3$$

$$(22.25)$$

$$\int_{d}^{2} \frac{1}{\sqrt[3]{(x-1)^2}} dx = \int_{d}^{2} (x-1)^{-\frac{2}{3}} dx = 3 - 3\sqrt[3]{d-1} \qquad (22.26)$$

By (22.25) and (22.26) let us evaluate the limits:

$$\lim_{b \to 1-} \int_{0}^{b} \frac{1}{\sqrt[3]{(x-1)^2}} dx = \lim_{b \to 1-} \left(3\sqrt[3]{b-1} + 3\right) = 3$$

$$\lim_{d \to 1+} \int_{d}^{2} \frac{1}{\sqrt[3]{(x-1)^2}} dx = \lim_{d \to 1+} \left(3 - 3\sqrt[3]{d-1}\right) = 3$$

that are both finite.

Therefore, from (22.22) the value of the improper integral (22.23) is

$$\int_{0}^{2} \frac{1}{\sqrt[3]{(x-1)^2}} dx = 6$$

Example 22.8 Examine the improper integral

$$\int_{0}^{1} \frac{1}{x} dx$$

The unbounded function $f(x) = \frac{1}{x}$ is an infinity as x tends to 0. Let a be a point such that $0 < a < 1$. Let us calculate the integral

$$\int_{a}^{1} \frac{1}{x} dx = ln1 - lna = -lna$$

and its limit as $a \to 0+$:

$$\lim_{a \to 0+} \int_{a}^{1} \frac{1}{x} dx = \lim_{a \to 0+} (-lna) = -\lim_{a \to 0+} lna = +\infty$$

(Observe that $lna < 0$ since $0 < a < 1$.)

Thus, $f(x) = \frac{1}{x}$ is not integrable over $[0,1]$ (Fig. 22.13).

Fig. 22.13 The restriction of $\frac{1}{x}$ to $(0, 1]$

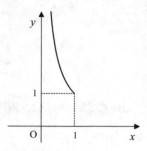

Let us state the following *existence theorem*, which refers to Sect. 20.11, for the improper integrals of unbounded functions.

Theorem 22.7 *Let a closed interval* $[a, b]$ *and a point c in it be given. Let the function* $f(x)$ *be continuous in the set* $[a, b] - \{c\}$ *and integrable over any closed interval not including c. Assuming* $\left|\frac{1}{x-c}\right|$ *as principal infinity as* $x \rightarrow c$, *if* $f(x)$ *is an infinity of order p,* $0 < p < 1$, *as* $x \rightarrow c$, *then the function* $f(x)$ *has a convergent improper integral over* $[a, b]$.

Let us observe that the Examples 22.7 and 22.8 may be solved or discussed as applications of Theorem 22.7.

22.11.2   *Improper Integrals Over Unbounded Intervals*

Definition 22.7 Let $f(x)$ be a real-valued function defined on the interval $[a, +\infty)$. If $f(x)$ is continuous in $[a, b]$, for every real number $b > a$, and if the limit

$$\lim_{b \to +\infty} \int_a^b f(x)dx \tag{22.27}$$

exists and is finite, the function $f(x)$ is said to have a *convergent improper integral*, or a *convergent integral in the improper sense*, over the interval $[a, +\infty)$ and, by definition, we set

$$\int_a^{+\infty} f(x)dx = \lim_{b \to +\infty} \int_a^b f(x)dx$$

If the limit (22.27) exists, but it is infinite, then the improper integral is said to be *divergent*.

If the limit (22.27) does not exist, or it is infinite, then we say that f is *not integrable* in the improper sense over $[a, +\infty)$.

An analogous definition holds concerning the function $f(x)$ defined on the interval $(-\infty, b]$. We set, by definition,

$$\int_{-\infty}^{b} f(x)dx = \lim_{a \to -\infty} \int_{a}^{b} f(x)dx$$

provided that the limit exists and is finite.

If the function $f(x)$ is defined on $\mathbf{R}$ we set, by definition,

$$\int_{-\infty}^{+\infty} f(x)dx = \lim_{a \to -\infty, b \to +\infty} \int_{a}^{b} f(x)dx$$

provided that the limit exists and is finite.

Example 22.9 Let us evaluate

$$\int_{0}^{+\infty} e^{-x}dx \tag{22.28}$$

Let first calculate

$$\int_{0}^{b} e^{-x}dx = -\int_{0}^{b} e^{-x}d(-x) = -\left(e^{-b}-e^{0}\right) = 1-e^{-b}$$

and then

$$\lim_{b \to +\infty} \int_{0}^{b} e^{-x}dx = \lim_{b \to +\infty} \left(1-e^{-b}\right) = 1$$

and

$$\int_{0}^{+\infty} e^{-x}dx = 1$$

Hence, the improper integral (22.28) is convergent and its value is 1 (Fig. 22.14).

Example 22.10 Examine the integrability of $f(x) = \frac{2x}{1+x^2}$ over the interval $[0, +\infty)$. If $b > 0$, let us calculate

Fig. 22.14 Graph of e^{-x}

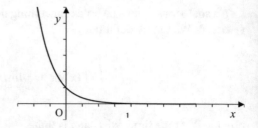

$$\int\limits_{0}^{b} \frac{2x}{1+x^2}dx = \int\limits_{0}^{b} \frac{d(1+x^2)}{1+x^2}dx = ln(1+b^2)-ln1 = ln(1+b^2)$$

Since $\lim\limits_{b\to+\infty} ln(1+b^2) = +\infty$, the function $\frac{2x}{1+x^2}$ is not integrable in the improper sense and the integral $\int_0^{+\infty} \frac{2x}{1+x^2}dx$ is divergent.

Example 22.11 Let us study the integral over the interval $[1, +\infty)$:

$$\int\limits_{1}^{+\infty} \frac{dx}{x^p}$$

The integrability of the function depends on the exponent p. In fact, let us describe the cases $p = 1$ and $p = 2$ (Fig. 22.15). The x axis is an asymptote for the graphs of $f(x) = \frac{1}{x}$ and $g(x) = \frac{1}{x^2}$,

$$\lim_{x\to\infty} \frac{1}{x} = \lim_{x\to\infty} \frac{1}{x^2} = 0$$

If $p = 1$ and $b > 1$, we obtain:

Fig. 22.15 The graphs of $\frac{1}{x}$ and $\frac{1}{x^2}$

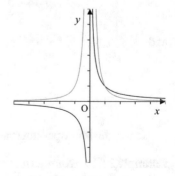

$$\int\limits_{1}^{b} \frac{1}{x} dx = lnb$$

Then

$$\lim_{b \to +\infty} \int\limits_{1}^{b} \frac{1}{x} dx = \lim_{b \to +\infty} lnb = +\infty$$

If $p = 2$ and $b > 1$, we obtain:

$$\int\limits_{1}^{b} \frac{dx}{x^2} = -\frac{1}{b} - (-1) = 1 - \frac{1}{b}$$

Then

$$\lim_{b \to +\infty} \int\limits_{1}^{b} \frac{dx}{x^2} = 1$$

Let us state the following *existence theorem*, which refers to Sect. 20.12, for the improper integral over unbounded intervals.

Theorem 22.8 *Let $f(x)$ be defined in the interval $[a, +\infty)$, integrable in $[a, b]$, for every real number $b > a$. Let us assume $g(x) = \frac{1}{x}$ as principal infinitesimal (Sect. 20.11) as $x \to +\infty$, if $f(x)$ is an infinitesimal of higher order than or equal order to p, $p > 1$, then the function $f(x)$ is integrable, in the improper sense, over $[a, +\infty)$. If $f(x)$ is an infinitesimal, as $x \to +\infty$, of order less than or equal to 1, and there exists a real number $h_0 \geq h$ such that, for every $x \geq h_0$, $f(x)$ maintains the sign, then the function f is not integrable in the improper sense, over $[a, +\infty)$.*

Let us observe that the Examples 22.10 and 22.11 may be solved or discussed as applications of Theorem 22.8.

22.12 Problems Solved. Indefinite and Improper Integrals

1. Find the following indefinite integrals:

  a. $\int 5dx = 5x + c$

  b. $\int x^2 dx = \frac{x^3}{3} + c$

  c. $\int \sqrt[3]{x} dx = \frac{x^{\frac{1}{3}+1}}{\frac{1}{3}+1} + c = \frac{3}{4} \sqrt[3]{x^4} + c$

d. $\int \sqrt[5]{x^6}dx = 5\frac{x^{\frac{11}{5}}}{11} + c$

e. $\int \frac{dx}{x^3} = \int x^{-3}dx = \frac{x^{-2}}{-2} + c = -\frac{1}{2x^2} + c$

f. $\int \frac{dx}{2\sqrt{x}} = \sqrt{x} + c$

g. $\int \frac{dx}{\sqrt[3]{x^3}} = \int x^{-\frac{1}{3}}dx = -\frac{3}{2}\sqrt[3]{x^2} + c$

2. Find the following indefinite integrals:

h.
$$\int \frac{x}{x+1}dx = \int \frac{(x+1)-1}{x+1}dx = \int \left(1 - \frac{1}{x+1}\right)dx$$
$$= \int dx - \int \frac{1}{x+1}dx = x - \ln|1+x| + c$$

i. $\int \frac{x-1}{x+1}dx = \int \frac{(x+1)-2}{x+1}dx = \int dx - 2\int \frac{dx}{x+1} = x - 2\ln|x+1| + c$

k.
$$\int \frac{x}{(x-1)^2}dx = \int \frac{x-1+1}{(x-1)^2}dx = \int \frac{x-1}{(x-1)^2}dx + \int \frac{dx}{(x-1)^2}$$
$$= \int \frac{dx}{x-1} + \int (x-1)^{-2}dx = \ln|x-1| - \frac{1}{x-1} + c$$

3. Examine the improper integral $\int_1^{+\infty} \ln\left(1 + \frac{1}{x^2}\right)dx$.

We found (Sect. 21.14, problem 1) that the integrand function $\ln\left(1 + \frac{1}{x^2}\right)$ is an infinitesimal of order 2 as $x \to +\infty$ with respect to $\frac{1}{|x|}$. As a consequence, by Theorem 22.8, $\ln\left(1 + \frac{1}{x^2}\right)$ is integrable in the improper sense on $[1, +\infty)$. In order to calculate the integral $\int_1^{+\infty} \ln\left(1 + \frac{1}{x^2}\right)dx$, we start by considering the indefinite integral

$$I = \int \ln\left(1 + \frac{1}{x^2}\right)dx$$

we will integrate by parts using the formula:

$$\int u dv = vu - \int v du \qquad (22.29)$$

where $u = \ln\left(1 + \frac{1}{x^2}\right), dv = dx, v = x, du = \frac{x^2+1}{x^2}\frac{-2}{x^3}dx$.
Therefore, by (22.29) we obtain:

$$I = x\ln\left(1 + \frac{1}{x^2}\right) - \int x \frac{-2}{x(x^2+1)}dx = x\ln\left(1 + \frac{1}{x^2}\right) + 2\text{atan}x + c.$$

If $b > 1$,

$$\int_1^b \ln\left(1 + \frac{1}{x^2}\right)dx = \left[x\ln\left(1 + \frac{1}{x^2}\right) + 2\text{atan}x\right]_1^b$$

$$= b\,ln\left(1 + \frac{1}{b^2}\right) + 2\mathrm{atan}b - ln2 - 2\frac{\pi}{4}$$

Since

$$\lim_{b\to+\infty} ln\left(1 + \frac{1}{b^2}\right) = \lim_{b\to+\infty} \frac{ln\left(1 + \frac{1}{b^2}\right)}{\frac{1}{b}} = \lim_{b\to+\infty} \frac{\frac{b^2}{b^2+1}\frac{-2}{b^3}}{\frac{-1}{b^2}} = \lim_{b\to+\infty} \frac{2b}{b^2+1} = 0$$

we conclude:

$$\int_1^{+\infty} ln\left(1 + \frac{1}{x^2}\right)dx = \lim_{b\to+\infty} 2\mathrm{atan}b - ln2 - \frac{\pi}{2} = 2\frac{\pi}{2} - ln2 - \frac{\pi}{2} = \frac{\pi}{2} - ln2.$$

Bibliography

Adams, R.A., Essex, C.: Calculus. Single Variable. Pearsons Canada, Toronto (2010)

Anton, H.: Calculus. Wiley, New York (1980)

Ayres, F., Jr., Mendelson, E.: Calculus. McGraw-Hill, New York (1999)

Fedrizzi, M., Oieni, A., Ventre, A.: Prove scritte di matematica generale. Cedam, Padova (1996)

Lax, P., Burnstein, S., Lax, A.: Calculus with Applications and Computing. Springer-Verlag, New York (1976)

Miranda, C.: Lezioni di Analisi matematica. Liguori Editore, Napoli (1978)

Stoka, M.: Corso di Matematica. Cedam, Padova (1988)

Chapter 23
Functions of Several Variables

23.1 Introduction

Let us recall the formula of the distance of two points $P(x_1, y_1)$, $Q(x_2, y_2)$ in the plane (Sect. 7.1.1):

$$d(P, Q) = |P\,Q| = \sqrt{(x_2 - x_1)^2 + (y_2 - y_1)^2}$$

and in the space (Sect. 9.8):

$$d(P, Q) = |P\,Q| = \sqrt{(x_2 - x_1)^2 + (y_2 - y_1)^2 + (z_2 - z_1)^2}$$

where $P(x_1, y_1, z_1)$ and $Q(x_2, y_2, z_2)$ are points in the space.

The cartesian product (Sect. 3.2) of the intervals $[a, b]$ and $[c, d]$ is the set of couples (x, y) in $\mathbf{R}^2$ defined by

$$[a, b] \times [c, d] = \left\{ (x, y) \in \mathbf{R}^2 : a \leq x \leq b, c \leq y \leq d \right\}$$

The cartesian product $[a, b] \times [c, d]$, also called a *closed interval* of $\mathbf{R}^2$, defines the set of the points (x, y) in the rectangle $l = [a, b] \times [c, d]$ (Fig. 23.1).

Unless otherwise indicated, the intervals $[a, b]$ and $[c, d]$ are non degenerate (Sect. 6.5), i.e., they do not reduce to a single point.

The cartesian product of the open intervals (a, b) and (c, d) is the set of the couples (x, y) in $\mathbf{R}^2$ defined by

$$(a, b) \times (c, d) = \left\{ (x, y) \in \mathbf{R}^2 : a < x < b, c < y < d \right\}$$

The cartesian product $(a, b) \times (c, d)$ is an open set, called *open interval* of $\mathbf{R}^2$ or *open rectangle*.

Given the point $(x_0, y_0) \in \mathbf{R}^2$ and the positive real numbers h and k, the product

© The Author(s), under exclusive license to Springer Nature Switzerland AG 2023   425
A. G. S. Ventre, *Calculus and Linear Algebra*,
https://doi.org/10.1007/978-3-031-20549-1_23

Fig. 23.1 I is an interval of
$\mathbf{R}^2$, the point $P(x, y) \in$
I; $a \leq x \leq b, c \leq y \leq d$

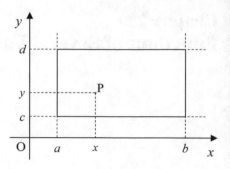

$$(x_0 - h, x_0 + h) \times (y_0 - k, y_0 + k)$$

is an open interval, called open rectangle or the rectangular neighborhood with center (x_0, y_0) and half-widths h, k.

Similarly, an *interval J* of $\mathbf{R}^3$ is, by definition, a *rectangular parallelepiped*, i.e., the set of triples $(x, y, z) \in \mathbf{R}^3$ defined by

$$J = [a, b] \times [c, d] \times [e, f] = \{(x, y, z) \in \mathbf{R}^3 : a \leq x \leq b, c \leq y \leq d, e \leq y \leq f\}$$

Let h be a positive real number and C a point in the plane $\mathbf{R}^2$. The set $S_2(C, h)$ of the points P in the plane having distance from C less than h is named *the open circle* (or *the open disk*) with *center* C and *radius h*, or the *circular neighborhood* of the point C with radius h. In symbols:

$$S_2(C, h) = \{P\mathbf{R}^2 : d(P, C) < h\}$$

A *closed circle* or *closed disk* with center C and radius $h > 0$ is defined as the set $\overline{S}_2(C, h)$ of the points P having distance from C less than or equal to h:

$$\overline{S_2}(C, h) = \{P\mathbf{R}^2 : d(P, C) \leq h\}$$

Let h be is a positive real number and C a point in the space $\mathbf{R}^3$. The set $S_3(C, h)$ of the points P in the space $\mathbf{R}^3$ having distance from C less than h is called the *open ball* or *spherical neighborhood* with center C and radius h:

$$S_3(C, h) = \{P\mathbf{R}^3 : d(P, C) < h\}$$

A *closed ball* with center C and radius h is defined as the set $S_3(C, h)$ of the points P in the space which have a distance from C less than or equal to h:

$$\overline{S_3}(C, h) = \{P \in \mathbf{R}^3 : d(P, C) \leq h\}$$

Remark 23.1 It is shown that each circle in $\mathbf{R}^2$ contains an interval of $\mathbf{R}^2$ and is contained in an interval of $\mathbf{R}^2$; and each sphere contains an interval of $\mathbf{R}^3$ and is contained in an interval of $\mathbf{R}^3$.

23.2 The Real *n*-Dimensional Space

We have stated (Chap. 12) that the set $\mathbf{R}^n$, $n \geq 1$,, of the *n*-tuples of real numbers, along with the operations of addition of two *n*-tuples and multiplication of a real number by an *n*-tuple defines a real vector space named *n-coordinate real space*, or *numerical vector space* of dimension *n*. We are familiar with the plane and the 3-coordinate real space. The intuition can falter when we have to think of parts of $\mathbf{R}^n$, with $n > 3$.

Whatever the natural number *n* is, we call *points* of $\mathbf{R}^n$ the *n*-tuples of real numbers. Then a point of $\mathbf{R}^n$ is an *n*-tuple $(x_1, x_2, \ldots, x_n)$ of real coordinates x_i, $i = 1, 2, \ldots, n$

The concept of *distance* introduced in $\mathbf{R}^2$ and $\mathbf{R}^3$ extends to $\mathbf{R}^n$, for every natural number *n*.

Indeed, the *distance*, or *Euclidean distance* $d(P, Q)$ of two points $P(x_1, x_2, \ldots, x_n)$ and $Q(y_1, y_2, \ldots, y_n)$ of the space $\mathbf{R}^n$ is defined by the equality

$$d(P, \ Q) = \sqrt{(y_1 - x_1)^2 + (y_2 - x_2)^2 + \ldots + (y_n - x_n)^2}$$

Therefore, the distance *d* is a real-valued function, which assigns the number $d(P, Q)$ to each couple (P, Q) of points of $\mathbf{R}^n$. The number $d(P, Q)$ is the value of the distance of the two points P and Q. The symbols d(P, Q) and |PQ| are equivalent.

The following properties of the distance apply, whatever the points P, P_1, P_2, P_3 of $\mathbf{R}^n$ are:

1. $d(P, P) = 0$
2. $d(P_1, P_2) \geq 0$
3. if $d(P_1, P_2) = 0$, then $P_1 = P_2$
4. $d(P_1, P_2) = d(P_2, P_1)$
5. $d(P_1, P_2) + d(P_2, P_3) \geq d(P_1, P_3)$

Property 1 says that the distance of a point from itself is 0. Property 2 states that distance assumes only non-negative values. Property 3 means that if two points have zero distance, then they coincide. Equation 4 expresses the symmetry of the distance. Inequality 5 is called the *triangular property* of distance: the sum of the lengths of two sides of a triangle is not less than the length of the third side.

The Cartesian product of *n* intervals $I_1, I_2, \ldots, I_n$ of $\mathbf{R}$,

$$I = I_1 \times I_2 \times \ldots \times I_n$$

is called *interval* I of $\mathbf{R}^n$ or *n-dimensional interval* I.

The product $I = I_1 \times I_2 \times \ldots \times I_n$ is the rectangular neighborhood of the point P whose coordinates are the midpoints of the intervals $I_i, i = 1, 2, \ldots, n$

If C is a point of $\mathbf{R}^n$ and $h > 0$, an *open n-dimensional sphere*, or an *open sphere* of $\mathbf{R}^n$, with *center* C and *radius* h, is, by definition, the set $S_n(C, h)$ of the points P of $\mathbf{R}^n$ whose distance from C is lower than h:

$$S_n(C, h) = \{P \in \mathbf{R}^n : d(P, \ C) < h\}$$

An open sphere $S_n(P, h)$ with center $P \in \mathbf{R}^n$ and radius $h > 0$ is called a *spherical neighborhood* of the point P.

A *closed n-dimensional sphere*, or a *closed sphere* of $\mathbf{R}^n$ with *center* C and *radius* $h > 0$ is, by definition, the set $\overline{S}_n(C, h)$ of the points P of $\mathbf{R}^n$, whose distance from C is less than or equal to h:

$$\overline{S_n}(C, h) = \left\{P \in \mathbf{R}^n : d(P, \ C) \le h\right\}$$

For every natural number n, the properties hold:

- each sphere in $\mathbf{R}^n$ contains an interval of $\mathbf{R}^n$ and is contained in an interval of $\mathbf{R}^n$;
- given any two distinct points P, Q of $\mathbf{R}^n$, a neighborhood of P and a neighborhood of Q exist and are disjoint.

Unless otherwise indicated, by "neighborhood" we mean "spherical neighborhood", or in particular, "circular neighborhood" in the plane $\mathbf{R}^2$.

Let us stress that a spherical neighborhood of a point is an open set.

A point P in the subset A of $\mathbf{R}^n$ is said to be an *interior point* of A if there exists a neighborhood of P contained in A. The set, possibly empty, of the interior points to A is called *the interior* of A and is denoted by $\mathring{A}$.

A subset A of $\mathbf{R}^n$ is said to be an *open set*, or simply *an open*, in $\mathbf{R}^n$ if each point of A is an interior point of A, i.e., each point of A has a neighborhood consisting entirely of points of A; in particular, A is an open in $\mathbf{R}^2$ if each of its points is the center of an open circle entirely contained in A. The set A is open if and only if A coincides with its interior, $A = \mathring{A}$.

For example, the set of points P in the plane $\mathbf{R}^2$ such that

$$S_2(O, 1) = \{P \in \mathbf{R}^2 : d(P, O) < 1\}$$

is an open set: the open circle having center at the origin $O(0, 0)$ of coordinates and radius 1. Any point P belongs to $S_2(O, 1)$ if and only if the coordinates (x, y) of P satisfy the inequality

$$x^2 + y^2 < 1$$

A point P of $\mathbf{R}^n$ is called an *accumulation point*, or *limit point*, of the subset A of $\mathbf{R}^n$ if each neighborhood of P contains a point of A distinct from P. Notice that P is not required to belong to A.

The set of accumulation points of A is called the *derived set* of A and denoted A'. A point P in A is, by definition, an *isolated point* of A if it is not an accumulation point of A; this means that there exists a neighborhood I of P such that P is the only point of A belonging to I.

A point P of $\mathbf{R}^n$ is called a *closure point*, or *adherent point*, of a set A in $\mathbf{R}^n$ if P $\in A$ or P is an accumulation point for A. The set of the adherent points of A is called the *closure* of A and it is denoted by $\overline{A}$. Obviously, $\overline{A} = A \cup A'$, i.e., the *closure* of A is the union of A and the derived set A'.

A set of points A of $\mathbf{R}^n$ is called a *closed set*, or simply *a closed*, in $\mathbf{R}^n$ if $\overline{A} = A$, i.e., A contains the accumulation points of A in $\mathbf{R}^n$. The empty set is assumed to be closed. From the definition it follows that the entire space $\mathbf{R}^n$ is closed in itself.

For example, the set of points (x, y) in the plane, such that $x^2 + y^2 \leq 1$, i.e., the closed circle with the origin at the center and radius 1, is a closed set.

Definition 23.1 A *separation* of the subset A of $\mathbf{R}^n$ is a pair A_1, A_2 of non-empty subsets of A such that $A_1 \cup A_2 = A$, $A_1 \cap A_2 = \emptyset$, and both A_1 and A_2 are closed in $\mathbf{R}^n$. A subset which has no separation is said to be *connected*.

If the intervals $I_1, I_2, \ldots, I_n$ of $\mathbf{R}$ are open, then the product

$$I_1 \times I_2 \times \ldots \times I_n$$

is an open set of $\mathbf{R}^n$. If the intervals $I_1, I_2, \ldots, I_n$ of $\mathbf{R}$ are closed, then the product $I_1 \times I_2 \times \ldots \times I_n$ is a closed subset of $\mathbf{R}^n$.

Recall (Chap. 1) that the *complement* of the subset A in $\mathbf{R}^n$ is the set $A^c = \mathbf{R}^n - A$.

A point is said to be *exterior* to the subset A of $\mathbf{R}^n$ if it is interior to the complement A^c.

Let us go back to the concept of *boundary* (see Sect. 22.3). A point P of $\mathbf{R}^n$ is called a *boundary point* of A if P is neither interior nor exterior to A. It is shown that in every neighborhood of a boundary point of A there are both points of A and points of the complement A^c: the property is characteristic of the boundary points. The set ∂A of the boundary points of A is called the *boundary* of A.

Let us mention the following properties:

a. whatever the subset A of $\mathbf{R}^n$ is, we have: $\partial A = \partial A^c$ and $A \cup \partial A = A \cup A'$;
b. the set $A \cup \partial A = A \cup A'$ is closed;
c. a set is closed if and only if it contains its boundary;
d. the boundary of $\mathbf{R}^n$ is the empty set.

23.3 Examples of Functions of Several Variables

There are relations (Chap. 3) that involve a certain quantity, or number, z variable along with other quantities, x, y, ... For example, if we deposit a capital C in the bank at the annual interest rate i, after one year we are able to withdraw an amount or capitalized value M. We can think of a function f of the capital C and the interest rate i, which produces in one year a new capital M:

$$M = f(C, i) \tag{23.1}$$

Let us give some details about the function f, in order to have it in a specific form. In the absence of risks, we assume that the capital C available at time 0 produces an amount $M > C$ at the time $t > 0$. The positive difference $I = M - C$ is called the *interest* (the yield) that is obtained by using the capital C in the time interval $[0, t]$. The quotient

$$i = \frac{M-C}{C} = \frac{I}{C}$$

is defined as the *interest rate*, or *rate of return*, of the investment of capital C in the time interval $[0, t]$. From the previous equalities we get: $iC = M - C$. Therefore,

$$M = C(1 + i) \tag{23.2}$$

is a specific expression of (23.1). The factor $(1 + i)$ is called the *capitalization factor*.

One more example of a variable magnitude depending on two others is the volume V of the solid cylinder, which is calculated by multiplying the area A of the base circle, which has radius r, by the height h,

$$V = Ah = \pi r^2 h \tag{23.3}$$

Therefore, the volume V of the cylinder is the value of a function f of the radius r and the height h. In symbols,

$$V = f(r, h) \tag{23.4}$$

This function assumes the specific form (23.3). The formulas (23.2) and (23.3) define functions of two variables.

23.4 Real-Valued Functions of Two Real Variables

A function of two real variables is a function whose domain is a subset A of $\mathbf{R}^2$; a real-valued function is, as it is known, a function that assumes real values, i.e., a function whose range is a subset of **R**. Therefore, a real-valued function of two real variables is a function that associates one and only one real number with a couple (x, y) of numbers belonging to the domain A; x and y are called the (*independent*) *variables*. The functions (23.1)–(23.4) are real-valued functions of two real variables.

Similarly, real-valued functions of $n \geq 3$ real variables are defined.

It is well-known that a couple of numbers (x, y) is identified with a point P of $\mathbf{R}^2$, then the function f of the couple (x, y) is a function of the point P. Therefore, we write $f(\mathrm{P}) = f(x, y)$ and use equivalently the symbols $f(\mathrm{P})$ and $f(x, y)$. In other words a real-valued function of two real variables is a function $f: \mathbf{R}^2 \to \mathbf{R}$ which associates the real number $f(\mathrm{P}) = f(x, y)$ with the point $\mathrm{P}(x, y)$.

In order to express the identification of the points P of the space and the triples of the coordinates (x, y, z), we write $f(\mathrm{P}) = f(x, y, z)$ and use equivalently the symbols $f(\mathrm{P})$ and $f(x, y, z)$.

Example 23.1 The function f that associates the product xy with the couple (x, y) of real numbers and the function g that associates the difference $x - y$ with (x, y) are real-valued functions of two real variables. In symbols,

$$f(x, y) = xy$$
$$g(x, y) = x - y$$

We just spend a few more words on the concept of *graph* (Sect. 7.1.2) of a real-valued function of two real variables.

If $f(x, y)$ is defined in $A \subseteq \mathbf{R}^2$, the graph of f is the subset G of $\mathbf{R}^3$ consisting of the points $Q(x, y, z)$ of the space $\mathbf{R}^3$ such that the point (x, y) belongs to A and $z = f(x, y)$. In symbols:

$$G = \left\{ Q(x, y, z) \in \mathbf{R}^3 : (x, y) \in A \text{ and } z = f(x, y) \right\}$$

Example 23.2 Let us go back to the functions $f(x, y) = xy$ and $g(x, y) = x - y$.. The function f associates the number $z = xy$ to the point $\mathrm{P}(x, y)$ in the domain A; then the point $Q(x, y, xy)$ is a point of the graph of f. The function g associates the number $z = x - y$ to the point $\mathrm{P}(x, y)$ of the plane; then the point $Q(x, y, x - y)$ belongs to the graph of g.

Clearly, the domain of a real-valued function of two real variables is a subset of $\mathbf{R}^2$. Of course, the domain convention (Sect. 19.4) is valid: indeed, the domain of a real-valued function of two real variables f is the largest subset of $\mathbf{R}^2$ such that the expression $f(x, y)$ has meaning.

The identification of the domain of f is a basic step in the study of the function.

Example 23.3 Consider the function, and observe that any couple (x, y) in $\mathbf{R}^2$ may be involved $f(x, y) = x^2 - xy + 6$, by the arithmetic operations used to build $x^2 - xy + 6$. Thus, the domain of $f(x, y)$ is $\mathbf{R}^2$; in fact, f associates the number $x^2 - xy + 6$ to each point (x, y). The graph of f is the subset of $\mathbf{R}^3$ consisting of the triples $(x, y, x^2 - xy + 6)$ which are determined as (x, y) varies in the plane $\mathbf{R}^2$. For example, the graph includes the following points of $\mathbf{R}^3$:

(0, 0, 6), obtained by setting $x = 0$, $y = 0$ in the expression $x^2 - xy + 6$,
(1, 0, 7), obtained by setting $x = 1$, $y = 0$.
and, again, the points
(3, −1, 18), (3, 3, 6), (4, 4, 6), and so on.

In the cases we will consider, it happens, as in the previous example, that the points of the graph form a surface of the space, giving for now the word "surface" the meaning that our intuition suggests, such as the covering of a station, or the surface of a mountain, or a veil. The points of the surface are projected orthogonally on the points of the domain of f, in the case of Example 23.3, on the whole plane. Furthermore, every line perpendicular to the xy plane, conducted through a point of the domain of f, meets the graph of f, that is the surface, in one and only one point.

Example 23.4 In order to find the domain of the function

$$f(x, y) = \sqrt{9 - x^2 - y^2}$$

we must identify the set of points (x, y) such that $9 - x^2 - y^2 \geq 0$, i.e.,

$$x^2 + y^2 \leq 9$$

The equation $x^2 + y^2 = 9$ defines the circumference with center $(0, 0)$ and radius 3. A point (x, y) belongs to the domain of f if and only if it verifies the inequality $x^2 + y^2 \leq 9$. The points of the domain of f are the points of the xy plane whose distance from $(0, 0)$ is less than or equal to 3.
The semi-circumference $y = \sqrt{9 - x^2}$ lies on the half-plane $y \geq 0$, the semi-circumference $y = -\sqrt{9 - x^2}$ lies on the half-plane $y \leq 0$.
The interior points of the circle $x^2 + y^2 \leq 9$ satisfy the condition $9 - x^2 - y^2 > 0$. The exterior points of the circle satisfy the condition $9 - x^2 - y^2 < 0$ (Fig. 23.2).

23.5 More About the Domain of $f(x, y)$

Finding the domain of the real-valued function of two real variables $f(x, y)$ means to identify the set of points of $\mathbf{R}^2$, whose coordinates satisfy the appropriate conditions for the definition, i.e., the existence of f. These conditions lead to a system of inequalities and equations whose solutions form the set of the points that constitute the domain of f.
For example, the line r of equation $y = x + 1$ in the plane $\mathbf{R}^2$ identifies two regions in the plane, which are half-planes (Fig. 23.3) that have in common the line. The point $(0, 0)$ is in the half-plane below the line, the point $(0, 2)$ is above it. If we want to know the domain of the function $z = \ln(x - y + 1)$, we need to find the region that contains the points (x, y) such that

Fig. 23.2 Circumference
$x^2 + y^2 = 9$,
semi-circumferences
$y = \sqrt{9-x^2}$ and
$y = -\sqrt{9-x^2}$

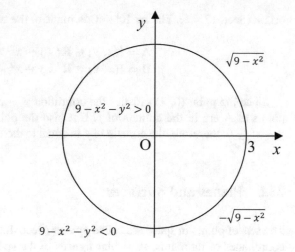

Fig. 23.3 Half-planes
identified on the plane xy by
the line $y = x + 1$

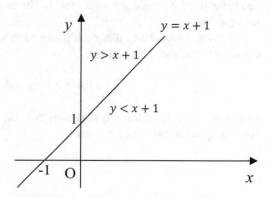

$$x - y + 1 > 0 \qquad\qquad (23.5)$$

i.e., $y < x + 1$. The point $(0, 0)$, that lies on the half-plane below the line represented by $y = x + 1$, satisfies (23.5); and, therefore, the region that contains $(0, 0)$ and lies below the line is the domain of the function $\ln(x - y + 1)$.

Exercise 23.1 Find the domain of the function

$$f(x, y) = \sqrt{y-x^2}$$

The square root is defined if the radicand is nonnegative:

$$y - x^2 \geq 0$$

The solutions of the equation $g(x, y) = y - x^2 = 0$ provide the set C of points, represented by the equation of parabola $y = x^2$ with vertical axis and vertex at the origin (Sect. 17.4.1). The set $\mathbf{R}^2 - C$ is made of the two disjoint regions:

$$A = \{(x, y) \in \mathbf{R}^2 : y - x^2 > 0\}$$
$$B = \{(x, y) \in \mathbf{R}^2 : y - x^2 < 0\}$$

Since the point $(0, 1)$ satisfies the condition $y - x^2 > 0$, then it is in A, so the points of A are in the domain of f. But also the points of C verify the condition $y - x^2 \geq 0$: therefore, the domain of f is equal to the union of the two sets A and C.

23.6 Planes and Surfaces

The sets of points in space are called *figures* (Sect. 4.2). A figure is identified by the coordinates of its points. Particular figures of the space are the coordinate planes, i.e., is the three planes that contain two of the coordinate axes x, y, z; i.e., the planes xy, xz and yz.

We know (Sect. 10.4) that each plane of space is represented by a first degree equation in the variables x, y, z:

$$ax + by + cz + d = 0 \tag{23.6}$$

with a, b, c, not all null. If c is non-null, then (23.6) assumes the form $z = f(x, y)$. For example, the plane of equation

$$x + y + z - 1 = 0$$

is represented also by the equivalent equation $z = 1 - x - y$, which has the form $z = f(x, y)$

23.7 Level Curves

We have observed (Sect. 23.4) that the geometrical representation of the function $z = f(x, y)$ can take the form of a surface. In the study of surfaces that arise in practice, such as those of a mountainous relief or a seabed, or, in general, cartographic surfaces, the *level curves*, also called *contour lines*, drawn on the map highlight and interrelate elements of information.

The contour lines, which connect the points of the surface of the earth that have the same altitude, are called *contour maps*; the contour lines that represent the points on the surface of the earth that have equal depths below the sea level are called *depth contours*.

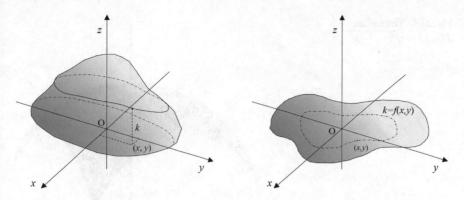

Fig. 23.4 a Section γ of the surface $z = f(x, y)$ with the plane $z = k$; **b** level curve $k = f(x, y)$, projection of γ on the xy plane

The *isobars* are the curves joining places with the same atmospheric pressure at a particular time.

In geographic or topographic maps the contour lines are drawn at intervals, for example on a map of a mountainous region we find the level curves: 100 m, 200 m, 300 m, ... above sea level; the appropriate numerical indication is shown at each level line (Fig. 23.4).

Let us go back to our surface, the graph of $z = f(x, y)$.

Definition 23.2 Let D be the domain of the function $z = f(x, y)$. The *curve of level* k of the function f is, by definition, the orthogonal projection on the xy plane of the intersection of the graph of f with the plane $z = k$, with k real number. Each level curve k of f lies on D: indeed, it is the set C_k of the points $(x, y, 0)$ in D, such that $f(x, y) = k$.

Example 23.5 Let us consider the function $f(x, y) = \sqrt{9 - x^2 - y^2}$ (Example 23.4). The *curve of level k* of function f is the curve of the plane xy having equation

$$\sqrt{9 - x^2 - y^2} = k \tag{23.7}$$

The square root has value $k \geq 0$. If $k = 0$, the level curve 0 has equation $x^2 + y^2 = 9$, i.e., the circumference C_0 which lies in the xy plane, has center O and radius 3. For $k > 0$ the Eq. (23.7) is equivalent to

$$x^2 + y^2 = 9 - k^2$$

For every k in the interval $[0, 3)$ the level curve C_k is a circumference, whose radius $\sqrt{9 - k^2}$ is a decreasing function of k. If $k = 3$, then $\sqrt{9 - k^2} = 0$. The plane $z = 3$ touches the surface $z = f(x, y)$ only at the point $(0, 0, 3)$, in fact, for $k = 3$, the equation $x^2 + y^2 = 9 - k^2$ becomes $x^2 + y^2 = 0$, whose unique real solution is the point $(x, y) = (0, 0)$. These considerations make us perceive the shape of the surface

Fig. 23.5 The surface
$f(x, y) = \sqrt{9-x^2-y^2}$.

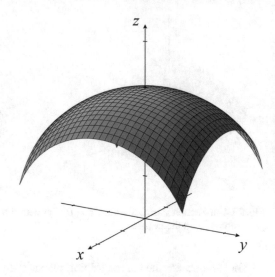

$z = \sqrt{9-x^2-y^2}$,which is the surface of the hemisphere centered at the origin and having radius 3, let say a dome resting on the xy plane (Fig. 23.5).

Example 23.6 Let us find the contour lines of the surface

$$z = e^{-(x^2+y^2)}$$

The curves $k = e^{-(x^2+y^2)}$, i.e., $\ln k = -(x^2 + y^2)$ or $x^2 + y^2 = r^2$, where $r^2 = -\ln k$, are the concentric circumferences in the xy plane, centered at the origin having radius r. Admissible values for k are $0 < k \le 1$.

Let us get an idea of the shape of $z = e^{-(x^2+y^2)}$. The graph of $z = e^{-(x^2+y^2)}$ is a surface of rotation around the z axis. If the point $(x, y, 0)$ moves away indefinitely from the origin $(0, 0, 0)$, the distance z of the point $P(x, y, z)$ of the surface from the xy plane becomes smaller and smaller, remaining positive.

The surface is all in the layer between the $z = 0$ and $z = 1$ planes, the $z = 1$ plane touches the surface at the point $(0, 0, 1)$; in fact, for $k = 1$, we have $0 = \ln 1 = -(x^2 + y^2)$. Therefore, $x = y = 0$ (Fig. 23.6).

23.8 Upper Bounded and Lower Bounded Functions

The concepts of boundedness of a real-valued function of one or two variables are similar (Sect. 7.1.2). Let f be a real-valued function of the real variables x and y and A the domain of f.

Definition 23.3 The function f is said to be *upper bounded* in A if there exists a real number k such that $f(x, y) \le k$, for every point (x, y) in A.

Fig. 23.6 The surface
$z = e^{-(x^2+y^2)}$

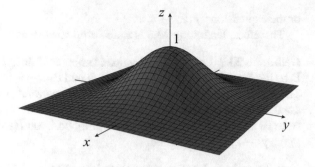

Definition 23.4 The function f is said to be *lower bounded* in A if there exists a real number h such that $f(x, y) \geq h$ for every point (x, y) in A.

Definition 23.5 The function f is said to be *bounded* in A if it is upper bounded and lower bounded in A.

Let $f(A)$ denote the range of f. If the function f is upper bounded in A, the supremum of the range $f(A)$ is defined as the *supremum* of f in A and is denoted by the symbol $\sup_A f$.

If the function f is lower bounded in A, the infimum of the set $f(A)$ is defined as the *infimum* in A of the function f, denoted $\inf_A f$.

The function $f(x, y) = \sqrt{9-x^2-y^2}$ is bounded in its domain and $\sup_A f = 3$, $\inf_A f = 0$.

The function $f(x, y) = e^{-(x^2+y^2)}$ is bounded in its domain. The range of the function $f(x, y) = e^{-(x^2+y^2)}$ is the interval $(0, 1]$, $f(x, y)$ attains its maximum 1 at $(0, 0)$, $f(0, 0) = 1$, while the range has infimum zero.

Remark 23.2 The concepts of boundedness and infimum and supremum clearly extend to the real-valued functions of n real variables.

23.9 Limits

Definition 23.6 Let f be a real-valued function of the two real variables x and y, defined in the subset D of $\mathbf{R}^2$. Let $P(x, y)$ be a point of D and $P_0(x_0, y_0)$ an accumulation point for the set D. The real number ℓ is called *limit* of f (P) as P *approaches* P_0, if for every neighborhood $J(\ell)$ of ℓ there exists a neighborhood of the point P_0, $I(P_0)$, such that for every point P, distinct from P_0, belonging to D and $I(P_0)$, it follows that f (P) belongs to $J(\ell)$.

To denote that ℓ is *limit* of f(P) *as* P *approaches* P_0, the symbols are used

$$\lim_{P \to P_0} f(P) = \ell \quad \text{or} \quad \lim_{(x,y) \to (x_0, y_0)} f(P) = \ell$$

or the expression "$f(P) \to \ell$ as $P \to P_0$".

Therefore, Definition 23.6 is stated equivalently as follows:

Definition 23.7 Let f be a real-valued function of the two real variables x and y. Let D be the domain of f, $P(x, y)$ a point of D and $P_0(x_0, y_0)$ an accumulation point of D. The real number ℓ is called *limit* of $f(P)$ *as* P *approaches* P_0 if, for every real number $\varepsilon > 0$ there exists a real number $\delta > 0$ such that for every point $P(x, y) \in D$, distinct from P_0 and belonging to the circular neighborhood $I(P_0)$ of radius δ, the inequality $|f(x, y) - \ell| < \varepsilon$ is fulfilled.

If the real number ℓ matches Definition 23.6 and Definition 23.7, the function f is said to be *convergent* to ℓ at the point P_0.

Remark 23.3 The limit of the function f as P approaches P_0, exists if and only if both limits

$$\lim_{x \to x_0} f(x, y) \text{ and } \lim_{y \to y_0} f(x, y)$$

exist separately and independent of each other.

We considered the case of finite limit $\ell \in \mathbf{R}$, i.e., the finite case. Now we deal with the case of infinite limit.

Definition 23.8 Let f be a real-valued function of two real variables x and y, having domain D. Let $P(x, y)$ a point of D and $P_0(x_0, y_0)$ an accumulation point of D. We say that the function $f(P)$ has *limit plus infinity*, $+\infty$, or $f(P)$ *diverges positively*, *as* P *approaches* P_0, or *as* $P \to P_0$, and we write

$$\lim_{P \to P_0} f(P) = +\infty \quad \text{or} \quad \lim_{(x,y) \to (x_0, y_0)} f(x, y) = +\infty$$

if, however fixed the real number $k > 0$, there exists a neighborhood I of P_0, such that, for every point $P \in I \cap D$, $P \neq P_0$, the inequality holds:

$$f(P) > k \quad \text{or} \quad f(x, y) > k$$

We say that the function $f(P)$ has *limit minus infinity*, $-\infty$, or $f(P)$ *diverges negatively*, *as* P *approaches* P_0, or *as* $P \to P_0$, and we write

$$\lim_{P \to P_0} f(P) = -\infty \quad \text{or} \quad \lim_{(x,y) \to (x_0, y_0)} f(x, y) = -\infty$$

if, however fixed the real number $h < 0$, there exists a neighborhood I of P_0, such that, for every point $P \in I \cap D$, $P \neq P_0$, the inequality holds:

$$f(P) < h \quad \text{or} \quad f(x, y) < h$$

We have defined the limit of a real-valued function of a real variable as $x \to +\infty$ and $x \to -\infty$ (Sect. 18.2). The definition is due to the total ordering defined by the relation $\leq$ on the real line.

A total ordering cannot be defined in the plane $\mathbf{R}^2$ and the concepts of $+\infty$ and $-\infty$ are not defined there. We admit the existence in the plane of a single *point at infinity*, denoted ∞. A *neighborhood of infinity in* $\mathbf{R}^2$ (or a *neighborhood of the point at infinity*) is, by definition, the region of the points which are exterior to a circle centered at the origin.

Definition 23.9 A subset of $\mathbf{R}^n$ is said to be a *bounded* set if there exists a circular neighborhood which contains it. A subset of $\mathbf{R}^n$ is said to be an *unbounded* set if it is not a bounded set.

Let D be a *bounded subset* of $\mathbf{R}^2$ and I a circular neighborhood that contains D. If h is the radius of I, let P and Q be any two points in D having distance d(P, Q). Therefore, d(P, Q) $\leq 2h$. Hence, the numerical set described by d(P, Q) as the points P and Q vary in D, is upper bounded and its supremum is called the *diameter* of the set D.

If $D \subseteq \mathbf{R}^2$ is unbounded also the above-mentioned numerical set is unbounded and the diameter of D is assumed to be $+\infty$.

Definition 23.10 Let f be a real-valued function of two real variables, defined in the unbounded subset $D \subseteq \mathbf{R}^2$ and let P(x, y) be a point of D. The function f (P) is said to have limit ℓ as P $\to \infty$ (or $(x, y) \to \infty$), and we write.

$$\lim_{P \to \infty} f(P) = \ell \quad \text{or} \quad \lim_{(x,y) \to \infty} f(x, y) = \ell$$

if, for every real number $\varepsilon > 0$, there is a neighborhood I of the infinity, such that, for all points $P \in I \cap D$ the following inequality is verified

$$|f(P) - \ell| < \varepsilon \quad \text{or} \quad |f(x, y) - \ell| < \varepsilon$$

We have, $\lim_{(x,y) \to \infty} e^{-(x^2 + y^2)} = 0$ (see Example 23.6).

Definition 23.11 Let f (P) $= f(x, y)$ be defined in an unbounded subset $D \subseteq \mathbf{R}^2$. The function f(P) is said to have limit plus infinity, or to diverge positively, as P $\to \infty$ and we write

$$\lim_{P \to \infty} f(P) = +\infty \quad \text{or} \quad \lim_{(x,y) \mapsto \infty} f(x, y) = +\infty$$

if, for every real $k > 0$, there is a neighborhood I of the infinity such that, for all points $P \in I \cap D$ the following inequality is verified

$$f(P) > k \quad \text{or} \quad f(x, y) > k$$

The function $f(P)$ is said to have limit minus infinity, or to diverge negatively, as $P \to \infty$ and we write

$$\lim_{p \to \infty} f(P) = -\infty \quad \text{or} \quad \lim_{(x,y) \to \infty} f(x, y) = -\infty$$

if, for every real $h < 0$, there is a neighborhood I of the infinity such that, for all points in $I \cap D$ the following inequality is verified

$$f(P) < h \text{ or } f(x, y) < h$$

The theorem of uniqueness of the limit, the theorems on the operations with limits, the permanence of the sign theorem (see Chap. 18) hold. Furthermore, let us mention the following:

Theorem 23.1 (Theorem of the comparison) *Let be given the real-valued functions f and g of two real variables. If $\lim_{P \to P_0} f(P) = \ell$ and $\lim_{P \to P_0} f(P) = m$ and if $f(P) \leq g(P)$, in a neighborhood of P_0, then $\ell \leq m$.*

The concept of limit extends to the real-valued functions of several variables.

23.10 Continuity

As it was already noted in the case of functions of one variable, the limit of a function $f(P)$ as P tends to P_0 and the value $f(P_0)$ are completely independent values. Indeed, while for the existence of the limit it is necessary that P_0 is an accumulation point of the domain D of f, in order to calculate $f(P_0)$ it is necessary instead that P_0 belongs to D, and each of the two circumstances can occur by itself without the other occurring. However, if P_0 is an accumulation point which belongs to D, then the limit of the function as $P \to P_0$, and the value of the function at P_0 can be considered simultaneously and if they are equal, i.e.,

$$\lim_{P \to P_0} f(P) = f(P_0)$$

then the function $f(P)$ is said to be *continuous* at P_0 and the point P_0 is called a *point of continuity* for the function f.

The function f continuous at any point of its domain D is said to be *continuous* in D.

Example 23.7 The function $f(P) = f(x, y) = x^2 + y^2$ is continuous at the point $P_0(0, 0)$. Indeed,

(0,0) belongs to the domain of f, $\text{Dom}(f) = \mathbf{R}^2$;

(0,0) is an accumulation point of $\text{Dom}(f)$;

there exists the limit of the function as P approaches $P_0(0,0)$ and this limit equals 0;

0 is the value of the function at $(0, 0)$.

Also the real-valued functions of two real variables, composite of elementary functions, i.e., powers, exponentials, logarithms, circular functions and their inverse, are continuous in the respective domains.

Example 23.8 Let us evaluate the limits:

$$\lim_{(x,y)\to(\pi,0)} x \cos\left(\frac{x-y}{2}\right) = \pi \cos\frac{\pi}{2} = 0$$

$$\lim_{(x,y)\to(-3,2)} (xy^3 - y^2) = -24 - 4 = -28$$

Definition 23.12 Let f be a real-valued function of two real variables, which has domain D and let P_0 be an accumulation point of D. If the function f is continuous at $D - \{P_0\}$ and there exists the finite limit

$$\lim_{P\to P_0} f(P)$$

then the new function which has domain $D \cup \{P_0\}$, is equal to f at every point of D and is continuous in P_0, i.e., it assumes the value $\lim_{P\to P_0} f(P)$ at P_0, is called the *extension by continuity* of the function f to the set $D \cup \{P_0\}$.

Example 23.9 The function $f(P) = f(x, y) = \frac{\sin(x+y)}{x+y}$ is continuous at any point of $\mathbf{R}^2$ with the exception of the point $(0, 0)$, where f is not defined. The function f can be extended by continuity at the point $(0, 0)$; in fact, as (x, y) approaches $(0, 0)$ on the line $y = mx$, for every $m \in \mathbf{R}$, then by (21.5)

$$\lim_{(x,y)\to(0,0)} \frac{\sin(x + y)}{x + y} = \lim_{x\to 0} \frac{\sin((1 + m)x)}{(1 + m)x} = 1$$

Therefore, the function defined on the whole plane, that assumes the value 1 at the origin and coincides with $\frac{\sin(x+y)}{x+y}$ elsewhere, is the extension by continuity of f to $\mathbf{R}^2$.

Example 23.10 The function $f(P) = f(x, y) = \frac{xy}{x^2+y^2}$ is continuous at any point of $\mathbf{R}^2$ with the exception of the point $(0, 0)$, where it is not defined. The function f does not allow extension by continuity at the point $(0, 0)$; in fact, along each line $y = mx$ the value of the function is constant

$$\frac{xy}{x^2 + y^2} = \frac{m}{1 + m^2}$$

and depends only on m. At the coordinate axes the function is null. The function, having no limit as (x, y) approaches $(0, 0)$, cannot be extended by continuity at the point $(0, 0)$.

In analogy to the case of functions of one variable, the following theorems hold.

Theorem 23.2 (Theorem of the permanence of the sign) *If the real function f is defined on the subset A of $\mathbf{R}^2$ and continuous at the accumulation point P_0 of A and if $f(P_0) \neq 0$, then there exists a neighborhood of P_0 such that, for every point P in the neighborhood, the values $f(P)$ and $f(P_0)$ have the same sign, i.e. $f(P)f(P_0) > 0$.*

Theorem 23.3 (Weierstrass' theorem) *If the real-valued function $f(P)$ is continuous in a closed and bounded set $A \subseteq \mathbf{R}^2$, then f assumes the maximum value M and the minimum value m in A.*

Theorem 23.4 *Let f be a continuous function in a closed and connected set $A \subseteq \mathbf{R}^2$. If f assumes two distinct values α and β, then it assumes all the values of the interval $[\alpha, \beta]$.*

The definition of continuity for functions of two variables extends to functions of $n \ (\geq 3)$ variables; let us stress that the neighborhoods of a point P of $\mathbf{R}^n$ are open balls centered at P.

Definition 23.13 Let f be a real-valued function of n real variables defined on D. If we fix the values of the variables $x_1, x_2, \ldots, x_{k-1}, x_{k+1}, \ldots, x_n$ of the function $f(x_1, x_2, \ldots, x_n)$, we obtain a function of the single variable x_k,

$$g(x_k) = f(a_1, a_2, \ldots, a_{k-1}, x_k, a_{k+1}, \ldots, a_n)$$

where the a_i's, with $i \neq k$, are constants that replace the x_i's. The one-variable function $g(x_k)$ is called a *partial function* of f (Fig. 23.7).

Theorem 23.5 *Let f be a real-valued function of n real variables, defined on D $\subseteq \mathbf{R}^n$, continuous at $(a_1, a_2, \ldots, a_{k-1}, a_k, a_{k+1}, \ldots, a_n)$. Then the partial function $g(x_k) = f(a_1, a_2, \ldots, a_{k-1}, x_k, a_{k+1}, \ldots, a_n)$ is continuous at a_k.*

Remark 23.4 With reference to the real-valued functions of two variables, the level curves point out the behavior of horizontal sections and the partial functions show the features of vertical sections of the graphs.

Remark 23.5 The Theorem 23.5, for $n = 2$, states that if $f(x, y)$ is continuous at $P_0(x_0, y_0)$, then $f(x, y)$ is continuous at P_0 with respect to each of the variables. However, vice versa is not true: it is possible for $f(x, y)$ to be continuous with respect to x and to be continuous with respect to y, but $f(x, y)$ is not continuous at P_0. It is the case of the function

$$f(P) = f(x, y) = \frac{x^2}{x^2 + y^2}$$

Fig. 23.7 $g(x)$ partial function of $f(x, y)$

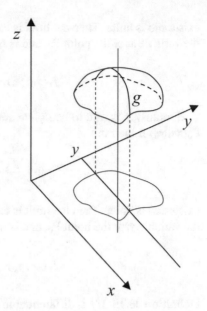

defined on the xy plane, except the point O(0,0), which is an accumulation point for the domain of f. We wonder if there exists the limit of f as $(x, y) \to (0, 0)$. Let us observe first that the function does not assume negative values. Moreover, x^2 is not greater than $x^2 + y^2$, so the function does not assume values greater than 1. Therefore, the limit, if it exists, cannot be greater than 1, nor less than 0. But the limit cannot be even between 0 and 1. Indeed, if a limit ℓ, would exist and it were between 0 and 1, for every $\varepsilon > 0$ there should exist a number $\delta > 0$ such that, for every point P distinct from the origin O(0, 0), such that the distance $d(P, O) < \delta$, $|f(P) - \ell| < \varepsilon$ should occur. Now, if the point P of the neighborhood $I(O, \delta)$, centered in O with radius δ, is on the x axis, then $y = 0$ and $f(x, 0) = 1$; if P is on the y axis, then $x = 0$ and $f(0, y) = 0$; therefore, in any neighborhood of O there is a segment of the x axis, where $f(P) = 1$, and a segment of the y axis, where $f(P) = 0$. Hence, the inequality $|f(P) - \ell| < \varepsilon$ would not be verified for every $\varepsilon > 0$. In conclusion, f has no limit as P approaches $(0, 0)$, and, therefore, f cannot be continuous at $(0, 0)$.

23.11   Partial Derivatives

Definition 23.14 Let f be a real-valued function of two real variables x and y defined on an open interval $I \subseteq \mathbf{R}^2$. If $P_0(x_0, y_0)$ is a point in I, the function f is said to be *differentiable* with respect to the variable x at the point P_0 if the limit

$$\lim_{x \to x_0} \frac{f(x, y_0) - f(x_0, y_0)}{x - x_0}$$

exists and is finite. Then the limit is called the *partial derivative* of f with respect to the variable x, at the point P_0, and is indicated by one of the symbols

$$f_x(x_0, y_0), \ f_x(P_0), \ \frac{\partial f}{\partial x}(x_0, y_0)$$

Similarly, f is said to be *differentiable* with respect to the variable y at the point $P_0(x_0, y_0)$ if the limit

$$\lim_{y \to y_0} \frac{f(x_0, y) - f(x_0, y_0)}{y - y_0}$$

exists and is finite. Then the limit is called the *partial derivative* of f with respect to the variable y, at the point P_0, and is indicated by one of the symbols

$$f_y(x_0, y_0), \ f_y(P_0), \ \frac{\partial f}{\partial y}(x_0, y_0)$$

Definition 23.15 If f is differentiable with respect to x (or y) at any point of the open interval I, then f is said to be *differentiable* with respect to x (or y) in the interval.

Remark 23.6 The partial derivative of $f(x, y)$ with respect to x is calculated by considering y as a constant and the partial derivative of f w. r. to y considering x as a constant.

Exercise 23.2 Compute the partial derivatives:

$f(x, y) = x^2 - 2xy + 5y^2 - 1; \ f_x(x, y) = 2x - 2y; \ f_y(x, y) = -2x + 10y$

$g(x, y) = \ln(x^2 + y^2); \ g_x(x, y) = \frac{2x}{x^2 + y^2}; \ g_y(x, y) = \frac{2y}{x^2 + y^2}$

$h(x, y) = x \sin y - y \sin x; \ h_x(x, y) = \sin y - y \cos x; \ h_y(x, y) = x \cos y - \sin x$

Given a function f of two variables, the geometric interpretation of the partial derivatives of f w. r. to x is illustrated in (Fig. 23.8), where the curve G is the section of the surface, representing the graph of f, with the plane passing through $P(x_1, y_1)$ and parallel to the plane xz. If a tangent line to G exists at each point, the partial derivative $f_x(x_1, y_1)$ is the value $\tan \gamma$, the tangent of the angle γ, whose sides are the line t, tangent to the curve G at $Q(x_1, y_1, f(x_1, y_1))$, and its orthogonal projection t' on the plane xy. Similar considerations apply to the curve H and the partial derivative of f w. r. to y.

Fig. 23.8 Geometric meaning of the partial derivative of f with respect to x; $\tan\gamma = f_x(x_1, y_1)$

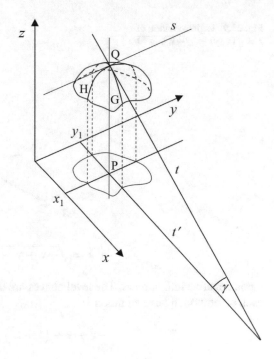

23.12 Domains and Level Curves

Exercise 23.3 Find the domain of the function

$$z = f(x, y) = \sqrt{-x + y - 1}$$

and determine the level curves.

Solution The function f is defined in the set of points of the plane xy whose coordinates (x, y) satisfy the condition: $-x + y - 1 \geq 0$. Let us determine the domain of f in the plane xy.

The equation $-x + y - 1 = 0$ represents a line of the plane xy (Fig. 23.3), which divides the plane into two half-planes, upper and lower, which have the line as a common boundary. The points of the line lie on the domain of f. Only one of the two half-planes, $-x + y - 1 > 0$, $-x + y - 1 < 0$, is in the domain of f. To locate it, we choose any point on the plane that is not on the line, for example the point $(0, 0)$. The coordinates $(0, 0)$ do not satisfy the inequality $-x + y - 1 > 0$ because $-0 + 0 - 1 = -1 < 0$. Then the lower half plane, in which the point $(0, 0)$ lies, is not included in the domain of f; therefore, the domain of f is formed by the union of the line $-x + y - 1 = 0$ with the half plane above it. The generic curve of level k of the surface $z = f(x, y) = c$ is

Fig. 23.9 Representation of $z = f(x, y) = \sqrt{-x + y - 1}$

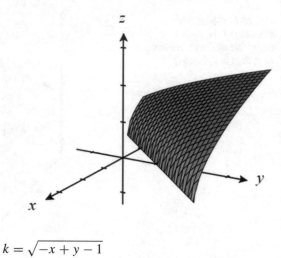

$$k = \sqrt{-x + y - 1}$$

k non negative real number. The level curves are the lines in the plane xy, parallel to each other, which have equation

$$-x + y - \left(1 + k^2\right) = 0$$

The surface $z = \sqrt{-x + y - 1}$, although made of (parallel) lines, is not a plane, but a surface of $\mathbf{R}^3$ that belongs to the class of *ruled surfaces*.

The equation $z = \sqrt{-x + y - 1}$ assumes the form $z^2 = -x + y - 1$ with the condition $-x + y - 1 \geq 0$. Precisely, the surface is a portion of a cylinder (Fig. 23.9).

Exercise 23.4 Find the domain of the function

$$z = f(x, y) = \sqrt{y - x^2}$$

and determine the level curves.

Solution The function f is defined in the set of points of the plane whose coordinates (x, y) satisfy the condition: $y - x^2 \geq 0$. This inequality is sufficient to identify the domain of $f(x, y) = \sqrt{y - x^2}$. The curve $y = x^2$ is a well known parabola (Fig. 23.10). This curve is the common boundary of two regions of the plane xy: the region A where $y > x^2$ and region B, where $y < x^2$. In the region A we find, for example, the point $(0, 1)$, in fact: if $(x, y) = (0, 1)$, we get $y - x^2 = 1 - 0 = 1 > 0$. Every point of the set B, for example, $(x, y) = (0, -1)$,, satisfies the inequality $y - x^2 < 0$.

Hence, the domain of the function is the union of the region A and the parabola $y = x^2$. The curve of level k of the surface $z = f(x, y) = \sqrt{y - x^2}$ has equation $k = \sqrt{y - x^2}$, k non negative real.

The surface $z = \sqrt{y - x^2}$ is a part of the surface $z^2 = y - x^2$, called *elliptic paraboloid* (Fig. 23.10).

Fig. 23.10 The surface
$z = \sqrt{y-x^2}$

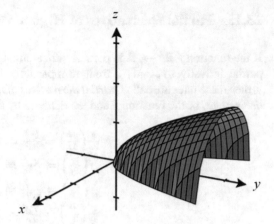

23.13 Solved Problems

Calculate the partial derivatives of the functions:

a. $f(x, y) = x^2 - 3x^2y - 4y^2$
b. $f(x, y) = \frac{x^2}{y} - \frac{y^2}{x}$
c. $f(x, y) = \sin(3x + 5y)$
d. $f(x, y) = e^{x^2+xy}$
e. $f(x, y) = \mathrm{atan}xy^2 - \mathrm{atan}yx^2$

Answers:

a. $f_x(x, y) = 2x - 6xy;\ f_y(x, y) = -3x^2 - 8y$
b. $f_x(x, y) = \frac{2x}{y} + \frac{y^2}{x^2};\ f_y(x, y) = -\frac{x^2}{y^2} - \frac{2y}{x}$
c. $f_x(x, y) = 3\cos(3x + 5y);\ f_y(x, y) = 5\cos(3x + 5y)$
d. $f_x(x, y) = e^{x^2+xy}(2x + y);\ f_y(x, y) = xe^{x^2+xy}$
e. $f_x(x, y) = \frac{1}{1+x^2y^4}y^2 - \frac{1}{1+x^4y^2}2xy;\ f_y(x, y) = \frac{1}{1+x^2y^4}2xy - \frac{1}{1+x^4y^2}x^2$

23.14 Partial Derivatives of the Functions of Several Variables

The concept of partial derivative of a function of two variables extends to functions of n variables, $n \geq 3$. Let f be a real-valued function defined in an open set A of $\mathbf{R}^n$ and $\overline{P}(\overline{x}_1, \overline{x}_2, \ldots, \overline{x}_n)$ a point of A.

The function f is said to be *partially differentiable* with respect to a fixed variable x_k at the point $\overline{P}$ if the function $f(\overline{x}_1, \overline{x}_2, \ldots, \overline{x}_{k-1}, x_k, \overline{x}_{k+1}, \ldots, \overline{x}_n)$ of the single variable x_k is differentiable at the point $\overline{x}_k$.

23.15 Partial Derivatives of Higher Order

If the function $f:\mathbf{R}^2 \to \mathbf{R}$ is partially differentiable in an open set I of $\mathbf{R}^2$, and the partial derivatives f_x and f_y, are in turn partially differentiable with respect to x and y, then these ones are called *partial derivatives of the second order*, or *second partial derivatives*, of the function f and are denoted by the symbols:

$$\frac{\partial}{\partial x}\left(\frac{\partial f}{\partial x}\right) = \frac{\partial^2 f}{\partial x^2} = f_{xx}$$

$$\frac{\partial}{\partial x}\left(\frac{\partial f}{\partial y}\right) = \frac{\partial^2 f}{\partial x \partial y} = f_{xy}$$

$$\frac{\partial}{\partial y}\left(\frac{\partial f}{\partial x}\right) = \frac{\partial^2 f}{\partial y \partial x} = f_{yx}$$

$$\frac{\partial}{\partial y}\left(\frac{\partial f}{\partial y}\right) = \frac{\partial^2 f}{\partial y^2} = f_{yy}$$

The partial derivatives of higher order than the second are similarly defined.

Exercise 23.5

1. Calculate the first and second partial derivatives of the functions:

    a. $f(x, y) = x^2 - 3xy - 4y^2$;
    b. $g(x, y) = x^2 - 2y^2$;
    c. $h(x, y) = x^3 - y$;

    Answers

    a. $f_x(x, y) = 2x - 3y; f_y(x, y) = -3x - 8y;$
        $f_{xx}(x, y) = 2; \quad f_{xy}(x, y) = -3; f_{yx}(x, y) = -3; f_{yy}(x, y) = -8$

    b. $gx(x, y) = 2x; gy(x, y) = -4y;$
        $g_{xx}(x, y) = 2; g_{xy}(x, y) = 0; g_{yx}(x, y) = 0; g_{yy}(x, y) = -4$

    c. $h_x(x, y) = 3x^2; h_y(x, y) = -1;$
        $h_{xx}(x, y) = 6x; h_{xy}(x, y) = 0; h_{yx}(x, y) = 0; h_{yy}(x, y) = 0$

2. Calculate the first and second partial derivatives of the function $f(x, y) = e^{xy}$.

$$f_x(x, y) = ye^{xy}; f_y(x, y) = xe^{xy}$$

$$f_{xx}(x, y) = yye^{xy} = y^2 e^{xy}$$

$$f_{xy}(x, y) = e^{xy} + yxe^{xy} = e^{xy}(1 + xy)$$

$$f_{yx}(x, y) = e^{xy} + xye^{y} = e^{xy}(1 + yx)$$

$$f_{yy}(x, y) = x^2 e^{xy}$$

3. Calculate the first and second partial derivatives of the function $f(x, y) = \sin xy$

$$f_x(x, y) = y \cos xy; \ f_y(x, y) = x \cos xy$$

$$f_{xx}(x, y) = yy(-\sin xy) = -y^2 \sin xy$$

$$f_{xy}(x, y) = 1 \cos xy + y(-\sin xy)x = \cos xy - xy \sin xy$$

$$f_{yx}(x, y) = 1 \cos xy + x(-\sin xy)y = \cos xy - xy \sin xy$$

$$f_{yy}(x, y) = -x^2 \sin xy$$

4. Calculate the first and second partial derivatives of the function $f(x, y) = x^2 + xy - 5xy^7$

$$f_x(x, y) = 2x + y - 5y^7$$

$$f_y(x, y) = x - 35xy^6$$

$$f_{xx}(x, y) = 2$$

$$f_{xy}(x, y) = 1 - 35y^6$$

$$f_{yx}(x, y) = 1 - 35y^6$$

$$f_{yy}(x, y) = -210xy^5$$

In the Exercises the equality $f_{xy}(x, y) = f_{yx}(x, y)$ occurs. This is not a coincidence. In fact, the following theorem provides a sufficient condition for the invertibility of the order of differentiation.

Theorem 23.6 (Schwarz's theorem). *If the function f is endowed with the derivatives f_x, f_y, f_{xy} in any point of the open set A where f_{xy} is continuous, then also the second derivative f_{yx} exists and $f_{xy} = f_{yx}$ in A.*

The functions of Exercises satisfy the hypotheses of Schwarz's Theorem.

Bibliography

Adams, R.A., Essex, C.: Calculus. Several Variables. Pearson Education Canada, Toronto (2003)

Anton, H.: Calculus. Wiley, New York (1980)

Chinn, W.G., Steenrod, N.E.: First Concepts of Topology. Random House, New York (1966)

Lax, P., Burnstein, S., Lax, A.: Calculus with Applications and Computing. Springer-Verlag, New York (1976)

Minnaja, C.: Matematica due. Decibel Editrice, Padova (1994)

Patterson, E.M.: Topology. Oliver and Boyd LTD, Edinburgh (1967)

Royden, H.L., Fitzpatrick, P.M.: Real Analysis. Pearson, Toronto (2010)

Spiegel, M.R.: Advanced Calculus. McGraw-Hill, New York (1963)

Thomas, G.B., Jr., Finney, R.L.: Calculus and Analytic Geometry. Addison-Wesley Publishing Company Inc., Reading, Massachusetts (1979)

Chapter 24
Curves and Implicit Functions

24.1 Curves and Graphs

The term *curve* has been used to indicate the graph of real-valued continuous function of a single variable (see Chaps. 18 and 19). However, the concepts of curve and graph do not coincide. For instance, the circumference is not a graph of a function: the unit circumference $x^2 + y^2 = 1$ is the union of the graphs of the functions $y = \sqrt{1 - x^2}$ and $y = -\sqrt{1 - x^2}$.

We introduced (Sect. 7.2.4) the parametric equations of the line

$$x = x_0 + mt, \; y = y_0 + ny \tag{24.1}$$

The slope of the line (24.1) is $\frac{n}{m}$ and (m, n) is a couple of direction numbers; the line passes through the point (x_0, y_0) and $t \in \mathbf{R}$ is a parameter. A parametric representation of a segment of the line (24.1) is obtained joining the constraint $a \leq t \leq b$ to the Eq. (24.1).

The elimination of the parameter t allows to obtain the *ordinary* or *cartesian* equation of the line $ax + by + c = 0$.

Furthermore, we considered in (Sect. 8.1.3) a parametric representation of the circumference with center at the origin and radius 1:

$$x = \cos t, \; y = \sin t \tag{24.2}$$

$t \in [0, 2\pi]$. By squaring and adding Eq. (24.2), since $\sin^2 t + \cos^2 t = 1$, the *ordinary* or *cartesian* equation of the circumference $x^2 + y^2 = 1$ is obtained.

The following equations

$$x = t, \; y = t^2 \tag{24.3}$$

$t \in \mathbf{R}$, are an example of a parametric representation of a set of points. We can eliminate the parameter t from (24.3) by replacing x with t in the second equation:

© The Author(s), under exclusive license to Springer Nature Switzerland AG 2023 451
A. G. S. Ventre, *Calculus and Linear Algebra*,
https://doi.org/10.1007/978-3-031-20549-1_24

$$y = x^2$$

which is the ordinary equation of a parabola (Sects. 17.4.1 and 21.8).

In general, the graph of any real-valued function of a real variable $y = f(x)$ coincides with the set of points represented by the parametric equations

$$x = t$$
$$y = f(t)$$

where t belongs to the domain of f.

Let us consider some curves represented by ordinary equations $f(x, y) = 0$.

Examples

- The equation $f(x, y) = x^2 + y^2 - 1 = 0$ represents the circumference with center $(0, 0)$ and radius 1;
- the equations $y - \sqrt{1-x^2} = 0$ and $y + \sqrt{1-x^2} = 0$ represent two semi circumferences included in the circumference $x^2 + y^2 = 1$;
- the equation $f(x, y) = x^2 - y = 0$ represents a parabola with vertical axis;
- the equations $y - \sqrt{x} = 0 = 0$ and $y + \sqrt{x} = 0$ represent two arcs of parabola (Fig. 24.1);
- the equation $f(x, y) = \frac{1}{x} - y = 0$ represents the equilateral hyperbola (Sect. 17.4.2).

Remark 24.1 Changes among different forms of representation of a given curve are possible. A parametric representation $x = f(t)$, $y = g(t)$ defines an orientation in the curve, whereas the equation $f(x, y) = 0$ does not. This circumstance implies that, when passing from the parametric representation to the form $f(x, y) = 0$, some elements of information can be lost. If the representation of the curve cannot reveal the ordering, the curve does not describe a trajectory of a point; in fact, the motion of the point is described if the position of the point is known at any time. For example, the variable t in the parametric representation

Fig. 24.1 The graph of the function $y = \sqrt{x}$ lies in the first quadrant, the graph of $y = -\sqrt{x}$ lies in the fourth quadrant

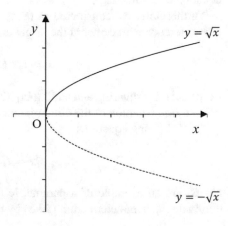

$$x = t^2, y = t$$

indicates the instant in which the current point of the curve is at the position (x, y), while the equation $y^2 = x$ of the curve does not show where the point is at time t.

24.2 Regular Curves

Definition 24.1 Given the closed interval $[a, b]$ in **R** and a couple of real-valued functions (f, g) both continuous in $[a, b]$, the set γ of the points (x, y) such that

$$\gamma) \quad x = f(t), y = g(t) \tag{24.4}$$

$t \in [a, b]$, is called a *continuous plane curve*, the Eq. (24.4) are called *parametric equations* of γ, the functions f and g the *components* of the curve γ.

Usually, $x(t)$ and $y(t)$ denote the components of γ and the curve γ is represented by the parametric equations:

$$\gamma) \quad x = x(t), y = y(t) \tag{24.5}$$

$t \in [a, b]$. The equalities (24.5) are the *parametric equations* of the curve γ and t is the *parameter*

The points $A = (x(a), y(a))$ and $B = (x(b), y(b))$ are named the *first endpoint* and the *second endpoint* of γ, respectively. The set of points on the curve is *ordered*, or *oriented*, by the relation:

$$P_1(x(t_1), y(t_1)) \text{ precedes } P_2(x(t_2), y(t_2)) \text{ if and only if } t_1 < t_2.$$

The set of points of the curve γ having the reverse orientation is denoted by $-\gamma$.

Definition 24.2 The curve γ is said to be a *regular curve* if

1. the component functions $x = x(t)$ and $y = y(t)$ are continuous, differentiable with continuous derivatives $x'(t)$ and $y'(t)$,
2. there is not any t in $[a, b]$ such that $x'(t) = y'(t) = 0$. The proposition is equivalent to inequality $\left(x'(t)\right)^2 + \left(y'(t)\right)^2 > 0$, for every $t \in [a, b]$.

A curve γ of the space $\mathbf{R}^3$ is defined as the set of points (x, y, z) such that the components $x = x(t), y = y(t), z = z(t)$ are continuous functions of the variable $t \in [a, b]$. The definition of regular curve in $\mathbf{R}^3$ extends the requirements (1) and (2) to the third components.

A curve which does not lie in a plane is called a *skew curve*.

Remark 24.2 The curves can be generated by the intersection of two surfaces. Contour lines are orthogonal projections onto $\mathbf{R}^2$ of plane curves, namely, the intersections of the graph of $f(x, y)$ and a plane $z = k$.

24.2.1   Tangent Line to a Regular Curve

Definition 24.3 The *tangent line to a regular curve* at the point $(x(t), y(t))$ is defined as

i.   the line that passes through the point $(x(t), y(t))$ and
ii. has direction numbers proportional to $(x'(t), y'(t))$

An alternative definition of tangent line to a regular curve is obtained assuming the condition

ii'. has slope $\frac{y'(t)}{x'(t)}$

instead of (ii).

If $x'(t) = 0$ the tangent line is vertical.
The tangent line is an oriented line with the same orientation of the curve.
Each regular curve has a tangent line at each point.
Consider, for example, the point $(x(t), y(t))$ of the circumference

$$x = \cos t, \, y = \sin t$$

calculated at $t = \frac{\pi}{4}$; the tangent line to the circumference at the point $\left(\cos\frac{\pi}{4}, \sin\frac{\pi}{4}\right) = \left(\frac{\sqrt{2}}{2}, \frac{\sqrt{2}}{2}\right)$ has direction numbers proportional to $(x'(t), y'(t)) = (-\sin t, \cos t)$. If $t = \frac{\pi}{4}$, then a couple of direction numbers is $(x'(t), y'(t)) = \left(-\frac{\sqrt{2}}{2}, \frac{\sqrt{2}}{2}\right)$, proportional to $(-1, 1)$. Therefore, the parametric equations of the tangent line to the circumference are

$$x = \frac{\sqrt{2}}{2} - t, \quad y = \frac{\sqrt{2}}{2} + t$$

Remark 24.3 The Definition 24.2 requires that $x'(t)$ and $y'(t)$ are not both null, which is a necessary condition for $x'(t)$ and $y'(t)$ to be direction numbers (Sect. 7.3).

Remark 24.4 The Definition 24.3 includes the notion of tangent line to the graph of the differentiable function $y = f(x)$. Indeed, the slope of the tangent line to the graph of f at x is $f'(x)$, and direction numbers are proportional to $(x'(t), y'(t))$ and $(1, f'(x))$.

Fig. 24.2 Angle of two
regular oriented plane curves

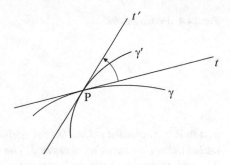

Fig. 24.3 Tangent and
normal at P to the curve

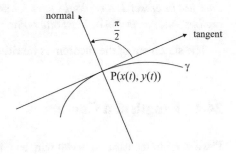

Definition 24.4 The *angle* of two regular oriented plane curves, which meet at a point P, is defined as the angle formed between their two tangents at point P (Fig. 24.2).

Definition 24.5 The *normal* line to the regular plane curve γ at P is, by definition, the line through P orthogonal to the tangent line at P. The normal line is oriented so that the angle between the oriented normal and the oriented tangent is $\frac{\pi}{2}$. It follows that any couple of direction numbers of the normal line is proportional to $(y'(t), -x'(t))$ (Fig. 24.3).

24.3 Closed Curves

An oriented plane curve represented by the parametric equations

$$x = x(t), \, y = y(t)$$

$t \in [a, b]$, whose endpoints coincide, $x(a) = x(b)$ and $y(a) = y(b)$, is called a *closed curve*. If, for every $t' \neq t''$, except possibly the endpoints a, b, it is verified the inequality $(x(t'), y(t')) \neq (x(t''), y(t''))$ the curve is called a *simple curve*. A closed plane simple curve is called a *Jordan curve*.

Theorem 24.1 (Jordan's theorem) *If J is a Jordan curve, then the set $\mathbf{R}^2 - J$, obtained by removing the curve J from the plane, consists of two disjoint and open*

Fig. 24.4 Jordan curve J

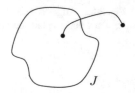

sets in $\mathbf{R}^2$: a bounded set, called the region inside the curve J and an unbounded set, called the region outside the curve J. The two regions have the common boundary J. Two points, however chosen in the same region, can be joined by a plane curve that does not intersect J, any plane curve joining two points, however chosen in different regions, intersects J in at least one point.

The statement of the theorem is intuitive (Fig. 24.4), but the proof is not easy.

24.4   Length of a Curve

If we want to measure a curved wire as a first step we can stretch and straighten the wire. Furthermore, a useful information is that a curve is longer than its chord, which is the segment joining the endpoints of the curve.

Definition 24.6 Let γ be an oriented regular plane curve of equations

$$x = x(t), \, y = y(t)$$

$t \in [a, b]$. Consider a decomposition D of the interval $[a, b]$ by means of n points:

$$t_0 = a < t_1 < t_2 < \ldots < t_n = b$$

Let $p(D)$ be the length of the polygonal that consecutively joins the points $(x(t_i), y(t_i))$, $i = 0, 1, 2, \ldots, n$. As the decomposition D of the interval $[a, b]$ varies, the number $p(D)$ describes a numerical set A. If A is bounded, then $supA$ is called the *length of curve γ* and the curve is said *rectifiable*.

Theorem 24.2 *A regular curve is rectifiable and its length L is equal to the definite integral:*

$$L = \int_a^b \sqrt{H(t)} \, dt \tag{24.6}$$

where

$$H(t) = \left(x'(t)\right)^2 + \left(y'(t)\right)^2 \tag{24.7}$$

If the curve is the graph of the function $y = f(x)$ defined on $[a, b]$, the couple $(1, f'(x))$ replaces $(x'(t), y'(t))$ (see Remark 24.4). Then (24.7) assumes the form:

$$H(t) = 1 + \left(f'(x)\right)^2 \qquad (24.8)$$

Given a curve γ of parametric equations

$$x = x(t), \ y = y(t)$$

$t \in [a, b]$, the concept of restriction of γ to the interval $[c, d]$ properly included in $[a, b]$, is similar to that of restriction of a function. The new curve $x = x(t), \ y = y(t)$, $t \in [a, b]$, is called an arc of the curve γ.

24.4.1 Problems

a. Using formula (24.6) find the length of the circumference γ with center $(0, 0)$ and radius r.

A parametric representation of the circumference γ is

$$x = r \cos t, \ y = r \sin t,$$

$t \in [0, 2\pi]$. Therefore, the length of the circumference is

$$L = \int_0^{2\pi} \sqrt{(r\sin t)^2 + (-r\cos t)^2}\, dt = 2\pi r$$

b. Find the arc length of the *catenary* (Fig. 24.5)

whose equation is $f(x) = \frac{1}{2}\left(e^x + e^{-x}\right)$ in the interval $[-2, 2]$.
Since $f'(x) = \frac{1}{2}\left(e^x - e^{-x}\right)$ by (24.8)

$$H(t) = 1 + \left(f'(x)\right)^2 = 1 + \frac{1}{4}\left(e^{2x} - 2 + e^{-2x}\right) = \frac{1}{4}\left(e^x + e^{-x}\right)^2$$

which implies

$$\sqrt{H(t)} = \frac{1}{2}\left(e^x + e^{-x}\right)$$

Therefore, from (24.6) the arc length is obtained:

Fig. 24.5 Arc length of the
catenary, $x \in [-2, 2]$

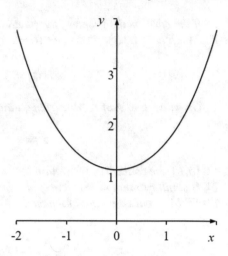

$$L = \frac{1}{2} \int\limits_{-2}^{2} (e^x + e^{-x})dx = \frac{1}{2}[e^x - e^{-x}]_{x=-2}^{x=2} = \frac{1}{2}\left(e^2 - \frac{1}{e^2}\right)$$

c.   Calculate the arc length L of the parabola $y = f(x) = x^2$, $x \in [0, 1]$.

   Hint: Use the formula (24.7): $H(t) = 1 + (f'(x))^2 = 1 + 4x^2$. Then calculate L
   $= \int_0^1 \sqrt{1 + 4x^2}dx$. To this aim use the substitution: $x = \frac{1}{2}\tan t$ (Sect. 22.10.1).

   Ans. $\sqrt{\frac{5}{2}} + \frac{1}{4}\ln\left(2 + \sqrt{5}\right)$.

24.5   Curvilinear Abscissa

Let the oriented regular curve γ have parametric equations

$$x = x(t), \ y = y(t),$$

$t \in [a, b]$, and let $P(x(t), y(t))$ be a point in the curve γ. Let us consider the function
$s(t)$, the length of the arc with first endpoint $(x(a), y(a))$ and second endpoint $(x(t),$
$y(t))$. The domain of $s(t)$ is the interval $[a, b]$. By (24.6) the value of the arc length is

$$s(t) = \int\limits_{a}^{t} \sqrt{H(v)}dv$$

where $H(v) = (x'(v))^2 + (y'(v))^2$ and v denotes the variable between a and t.
The function $s(t)$ is called *curvilinear abscissa* of $P \in \gamma$. For $v > 0$, the integrand is

positive, strictly increasing in $[a, b]$, $s(a) = 0$ and $s(b) = L = $ length of the curve γ. Since the function $s(t)$ is strictly increasing in its domain $s(t)$ is invertible in $[a, b]$. Let us denote the inverse of s with $t = q(s)$ whose domain is $[0, L]$, the range of $s(t)$. Then the regular curve γ has a new parametric representation in terms of the variable s:

$$x = x(q(s)), \, y = y(q(s)),$$

$s \in [0, L]$, that is called the *parametric representation in function of the curvilinear abscissa*. As a consequence, each point P of the curve γ is identified by the length of the arc that reaches P. This representation allows to express in a simple way the magnitudes *intrinsically* related with the curve γ. Evidently, the length of curve γ is

$$L = \int_0^L ds$$

24.6 Derivative of the Composite Functions

We dealt with the differentiation of composite functions in the case of functions of a single variable (Sect. 20.6). Even for the real-valued functions of two real variables $f(x, y)$, it may happen that each of the two variables is a function of other variables. Let x and y be functions of a real variable.

Definition 24.7 Let $f(x, y)$ be a real-valued function of two real variables having domain $D \subseteq \mathbf{R}^2$. Let $x = x(t)$ and $y = y(t)$ be functions of the variable t defined in the interval $[a, b]$, such that, if $t \in [a, b]$, then the point $(x(t), y(t))$ belongs to D. The function $f(x(t), y(t))$ defined in $[a, b]$ is called a *composite function* through the functions $x(t), y(t)$.

The composite function $f(x(t), y(t))$ is a function of a variable which shares the same properties of functions of a single variable. For instance, we can calculate its derivative.

Remark 24.5 The composite function $f(x(t), y(t))$ is the restriction of $f(x, y)$ to the curve $x = x(t), y = y(t), t \in [a, b]$, contained in the region D.

The functions $x = x(t), y = y(t)$ are called the *components* of the composite function. The geometric meaning is illustrated in Fig. 24.6.
The curve γ

$$x = x(t), \, y = y(t),$$

Fig. 24.6 The surface represents the graph of $f(x, y)$ and the curve γ', with endpoints A′ and B′, represents the graph of the composite function $f(x(t), y(t))$

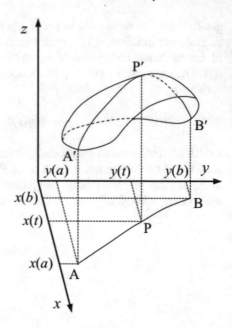

$t \in [a, b]$, lies on the domain D of f, $D \subseteq \mathbf{R}^2$ and has endpoints A and B. The graph of f, i.e., the set of the points in the space $\mathbf{R}^3$ having coordinates $(x, y, f(x, y))$, (x, y) belonging to D, is represented by the graph of the function $z = f(x, y)$ which represents a surface.

The surface $z = f(x, y)$ contains the curve γ', possibly skew, with endpoints A′, B′, consisting of all the points P′ with coordinates $(x(t), y(t), f(x(t), y(t)))$. The orthogonal projection of γ' onto the plane xy is the curve γ with endpoints A, B.

The following theorem states a sufficient condition for the differentiability of the composite function $f(x(t), y(t))$.

Theorem 24.3 (Chain rule) *Let $f(x, y)$ be endowed with continuous first partial derivatives at every point of an open set A of $\mathbf{R}^2$. If the functions $x(t)$, $y(t)$ are defined in $[a, b]$ and differentiable in the open interval (a, b) and, for every $t \in [a, b]$, the point $(x(t), y(t))$ belongs to A, then the composite function $f(x(t), y(t))$ is differentiable in (a, b) and the following equality holds:*

$$f'(x(t), y(t)) = f_x(x(t), y(t))x'(t) + f_y(x(t), y(t))y'(t) \qquad (24.9)$$

where f' denotes the derivative of the composite function of the only variable t. Equality (24.9) takes the alternative form:

$$\frac{df}{dt} = \frac{\partial f}{\partial x}\frac{dx}{dt} + \frac{\partial f}{\partial y}\frac{dy}{dt}$$

Example 24.1 Calculate the derivative of the function $f(x, y) = x^2 + 3xy$ composite of

$$x = x(t) = 1 - t$$
$$y = y(t) = t^2$$

$$\frac{df}{dt} = [2x + 3y]_{x=1-t, y=t^2}(-1) + [3x]_{x=1-t, y=t^2}(2t) =$$
$$-\left(2(1-t) + 3t^2\right) + (3-3t)2t = -2 + 8t - 9t^2$$

We obtain the same result by constructing the function

$$f(x(t), y(t)) = x(t)^2 + 3x(t)y(t) = (1-t)^2 + 3(1-t)t^2 =$$
$$1 - 2t + 4t^2 - 3t^3$$

whose derivative is $f'(x(t), y(t)) = -2 + 8t - 9t^2$.

24.7 Implicit Functions

We know that a real valued function f of a real variable x

$$y = f(x)$$

includes the procedure to calculate *explicitly* the value y, given x. Along with $y = f(x)$ also the notation

$$y = y(x)$$

is used.

However, in some cases an equation

$$f(x, y) = 0$$

that relates the variables x and y is given.

Definition 24.8 Let $f(x, y)$ be a real-valued function defined in an interval $A = [a, b] \times [c, d] \subseteq \mathbf{R}^2$. If for every $x \in [a, b]$ there exists exactly one $y \in [c, d]$ such that

$$f(x, y) = 0 \tag{24.10}$$

then y is the value of a function of the variable x; such a function assigns to any $x \in [a, b]$ the element $y \in [c, d]$ such that $f(x, y) = 0$. The Eq. (24.10) is said to *implicitly define y as a function of x.* The domain of the implicitly defined function is the set of those x such that there is a unique y for which (24.10) is satisfied. The function y is said to be an *implicit function* of the variable x.

Example 24.2 The equation

$$f(x, y) = x^2 + y^2 - 1 = 0$$

defines implicitly, for every $x \in [-1, 1]$ the two functions $y = +\sqrt{1 - x^2}$ and $y = -\sqrt{1 - x^2}$.

Example 24.3 The equation

$$f(x, y) = x^2 + y^2 + 1 = 0$$

does not define implicitly any real-valued function since the equation is not satisfied by any couple (x, y) of real numbers.

Referring to Fig. 24.7 the curve C is represented by the implicit Eq. (24.10) which implicitly defines the function $y = y(x)$ in the rectangular neighborhood $I \times J$.

Example 24.4 The equation

$$xy - 2x - y - 3 = 0 \qquad\qquad (24.11)$$

may be solved for y:

$$y = \frac{3 + 2x}{x - 1}$$

The domain of $y = \frac{3+2x}{x-1}$ is $\mathbf{R} - \{1\}$. Equation (24.11) implicitly defines y as a function of x in each rectangular neighborhood of a point (x_0, y_0) solution of (24.11). Observe that the points of abscissa 1 in the plane xy are not solutions of the Eq. (24.11).

Fig. 24.7 The part of the curve C inside $I \times J$ is intersected in at most one point by any vertical line

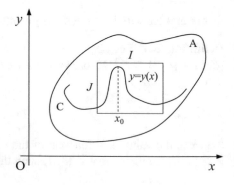

Example 24.5 The equation of the unit circumference $x^2 + y^2 = 1$ does not implicitly define y as a function of x in a neighborhood of the point $P(1,0)$ because the part of the circumference contained in each rectangular neighborhood of P is intersected at two points by some vertical line (Fig. 24.8).

The equation $x^2 + y^2 = 1$ implicitly defines y as a function of x in a neighborhood of the point $(1/\sqrt{2}, 1/\sqrt{2})$ (Fig. 24.9).

Remark 24.6 The expression "f is continuous along with its first partial derivatives" means "f is continuous and its first partial derivatives are continuous too".

A sufficient condition to ensure the existence of a function $y(x)$ defined in a neighborhood of x_0 and implicitly defined by Eq. (24.10) is yielded by the following:

Theorem 24.4 (Dini's theorem) *Let $f(x, y)$ be a real-valued function of two real variables, continuous along with its first partial derivatives in an open set* $A \subseteq \mathbf{R}^2$. *If (x_0, y_0) is a point in A such that*

$$f(x_0, y_0) = 0, \ f_y(x_0, y_0) \neq 0$$

Fig. 24.8 The part of the circumference contained in each rectangular neighborhood of $(1, 0)$ is intersected in two points by some vertical line

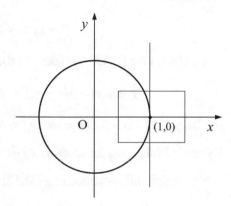

Fig. 24.9 $y(x)$ implicitly defined in a neighborhood of $(1/\sqrt{2}, 1/\sqrt{2})$

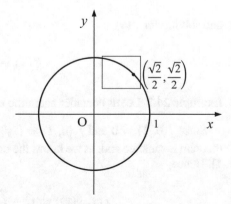

then there exist a neighborhood I of x_0 and a neighborhood J of y_0, such that $I \times J \subseteq A$ and a unique continuous function $y(x)$ with domain I and range J such that $y = y(x_0)$ and $f(x, y(x)) = 0$, for every $x \in I$; furthermore, the function $y(x)$ is differentiable with continuous derivative and the equality holds

$$f_x(x, y(x)) + f_y(x, y(x))y' = 0 \qquad (24.12)$$

which can be written:

$$y'(x) = \frac{dy}{dx} = -\frac{f_x(x, y(x))}{f_y(x, y(x))} \qquad (24.13)$$

The equalities (24.12) and (24.13) yield the implicit differentiation rule, i.e., the way to determine the derivative $y'(x)$ (in a neighborhood).

Property 24.1 *Under the hypotheses of Theorem 24.4, if $y(x)$ is the function implicitly defined by (24.10) in a neighborhood of x_0, let γ be the set of points P such that $f(x, y(x)) = 0$. The set γ is a regular curve and the tangent line to γ at $P_0(x_0, y_0)$ has the equation*

$$y - y_0 = y'(x_0)(x - x_0) \qquad (24.14)$$

By (24.12) the Eq. (24.14) takes the form

$$f_x(x_0, y(x_0))(x - x_0) + f_y(x_0, y(x_0))(y - y_0) = 0 \qquad (24.15)$$

The Eq. (24.15) is the equation of the *tangent line* at $P_0(x_0, y_0)$ to the curve γ.

Example 24.6 Given the equation $xy - 3x - y - 2 = 0$, find $y'(x)$.

By implicit differentiation rule (24.12) we obtain:

$$xy'(x) + 1y - 3 - y'(x) = y'(x)(x - 1) + y - 3 = 0$$

and solving for $y'(x)$,

$$y'(x) = \frac{3 - y}{x - 1}$$

Example 24.7 Let us consider again the equation $f(x, y) = x^2 + y^2 - 1 = 0$.

Since $f(0, 1) = 0$ and $f_y(0, 1) = [2y]_{x=0,y=1} = 2 \neq 0$ the hypothesis of Dini's theorem is satisfied and, as we know, the domain of function $y(x) = \sqrt{1 - x^2}$ is $[-1, 1]$. Hence,

$$f(x, y(x)) = x^2 + \left(\sqrt{1 - x^2}\right)^2 - 1 = 0$$

for every x. The right-hand side of (24.13) yields

$$-\frac{f_x(x, y(x))}{f_y(x, y(x))} = -\frac{2x}{\sqrt{1-x^2}}$$

which is, as it is easily seen, the derivative of $\sqrt{1-x^2}$.

Example 24.8 Let us find the equation of the tangent line to the $f(x, y) = e^x - e^y - xy = 0$ at the point $(0, 0)$.

The given equation implicitly defines y in function of x. The hypotheses of Theorem 24.4 are fulfilled; indeed $f_x(x, y) = e^x - y$, $f_y(x, y) = -e^y - x$ and $f_y(0, 0) = -e^0 = -1 \neq 0$. As $e^y + x \neq 0$, by implicit differentiation (24.12),

$$y'(x) = -\frac{f_x(x, y(x))}{f_y(x, y(x))} = -\frac{e^x - y}{-e^y - x} = \frac{e^x - y}{e^y + x}$$

The slope of the tangent at $(0, 0)$ is $y'(0) = 1$. Hence the equation of the tangent to the curve is $y = x$.

24.7.1 Higher Order Derivatives

Consider (24.12) in the form

$$f_x + f_y y' = 0$$

and differentiate it

$$f_{xx} + f_{xy} y' + f_{xy} y' + f_{yy}(y')^2 + f_y y'' = 0 \qquad (24.16)$$

that from (24.12), yields

$$y''(x) = -\frac{f_{xx} + 2f_{xy} y' + f_{yy}(y')^2}{f_y} = \frac{f_{xx}(f_y)^2 - 2f_{xy} f_x + f_{yy}(f_x)^2}{(f_x)^3} \qquad (24.17)$$

Equations (24.16) and (24.17) show that the first and second derivatives (and, reiterating, all the higher ones) of the implicitly defined function $y(x)$ can be calculated directly starting from the partial derivatives of the function f, without the need to obtain the expression of $y(x)$: the function $y(x)$ can be studied with regard to the properties of monotonicity, extremes, concavity, convexity, etc., even if $y(x)$ is not known.

Example 24.9 From Examples 24.4 and 24.6 the equation $xy - 3x - y - 2 = 0$ defines implicitly y as a function of x

$$y = \frac{3 + 2x}{x - 1}$$

whose first derivative is

$$y'(x) = \frac{3 - y}{x - 1} \tag{24.18}$$

The second order derivative is obtained from (24.18), taking into account that y is not an independent variable but a function of x, i.e., $y = y(x)$:

$$y'' = D_x \left(\frac{3 - y}{x - 1} \right) = \frac{(x - 1)(-y') - 1(3 - y)}{(x - 1)^2} = \frac{(x - 1)\frac{y-3}{x-1} - 3 + y}{(x - 1)^2}$$

$$= \frac{y - 3 - 3 + y}{(x - 1)^2} = \frac{2y - 6}{(x - 1)^2}$$

where the symbol D_x denotes differentiation w. r. to x.

Example 24.10 Find the equations of the tangent lines with slope $k = -1$ to the circumference $x^2 + y^2 - 1 = 0$.

Apply implicit differentiation: $2x + 2yy' = 0$. Hence

$$y' = -\frac{x}{y}$$

So, at the point of tangency (x_0, y_0), we obtain $-1 = k = y'(x_0) = -\frac{x_0}{y_0}$. Then

$$x_0 = y_0 \tag{24.19}$$

The point (x_0, y_0) is on the circumference, from where $x_0^2 + y_0^2 - 1 = 0$. Then, by (24.19), $2x_0^2 = 1$. Therefore, the abscissas of the two points of tangency are $\frac{1}{\sqrt{2}}$ and $-\frac{1}{\sqrt{2}}$ and, by (24.19), the points of tangency are $\left(\frac{1}{\sqrt{2}}, \frac{1}{\sqrt{2}} \right)$ and $\left(-\frac{1}{\sqrt{2}}, -\frac{1}{\sqrt{2}} \right)$.

Bibliography

Adams, R.A.: Calculus of Several Variables. Pearson Education Canada Inc., Toronto (2003)
Ayres, F., Jr., Mendelson, E.: Calculus. McGraw-Hill, New York (1999)
Cacciafesta, F.: Matematica generale. Giappichelli Editore, Torino (2004)

Chapter 25
Surfaces

25.1 Introduction

Let us define some classes of surfaces represented by particular implicit equations:

$$F(x, y, z) = 0 \tag{25.1}$$

If $F(x, y, z)$ is a polynomial of degree n and three variables, then the Eq. (25.1) is said to represent an *algebraic surface of order n*. An algebraic surface of order n and a generic line that does not lie on it intersect in at most n points of the space $\mathbf{R}^3$. The plane is an algebraic surface of order 1, the sphere is an algebraic surface of order 2.

The equation $F(x, y, z) = x^2 + y^2 + z^2 - 9 = 0$ is the *implicit*, or *ordinary*, representation of the spherical surface with center at the origin and radius 3 (Sect. 9.10).

A surface can be represented by equations with two real parameters, u, v:

$$x = g_1(u, v)$$
$$y = g_2(u, v)$$
$$z = g_3(u, v)$$

If the parameters u and v may be eliminated an ordinary representation is obtained.

The given definition is quite broad. Sometimes regularity features are required on a surface: for example, the functions F, g_1, g_2, g_3 could be continuous or differentiable.

Example 25.1 Let a surface S be represented by the parametric equations:

$$x = 2u - v$$
$$y = -u + 2v$$
$$z = -u$$

© The Author(s), under exclusive license to Springer Nature Switzerland AG 2023
A. G. S. Ventre, *Calculus and Linear Algebra*,
https://doi.org/10.1007/978-3-031-20549-1_25

Replacing $-z$ for u in the first two equations, we get: $x = -2z - v$, $y = z + 2v$; hence, setting $v = -x - 2z$ in the previous equation we obtain: $y = z + 2(-x - 2z)$, i.e., $2x + y + 3z = 0$. Therefore, the three parametric equations are equivalent to the ordinary equation of the plane $2x + y + 3z = 0$.

25.2 Cylinder

From a descriptive point of view, the simplest surfaces, after the planes, are the cylinders.

Definition 25.1 Given a plane curve γ, let r be a line non parallel to the plane of γ. The lines s of the space, parallel to r and passing through the point P, as P varies on γ, form a surface called a *cylinder* (or *cylindrical surface*). The curve γ is called the *directrix* of the cylinder and each line s is called a *generatrix* of the cylinder (Fig. 25.1).

We can think of the cylinder as a surface generated by a line (generatrix) that moves leaning on the points of the directrix always remaining parallel to itself.

The *right circular cylinder* has a circumference as directrix and generatrices perpendicular to the plane of the circumference.

Consider the implicit equation

$$x^2 + y^2 = a^2 \tag{25.2}$$

$a > 0$, that represents the circumference in the plane xy with center $(0, 0)$ and radius a. Equation (25.2) makes sense also in a coordinate system of the space xyz. Indeed, it is easy to recognize that the Eq. (25.2) represents a right circular cylinder having the circumference $x^2 + y^2 = a^2$ of the plane xy as directrix and the generatrices parallel

Fig. 25.1 Cylinder with the curve γ as directrix and some generatrices, i.e., the lines s passing through the points P and parallel to r

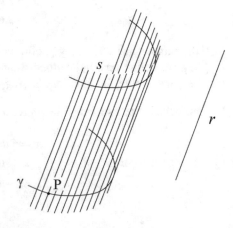

Fig. 25.2 Cylinder $x^2 + y^2$ $= a^2$

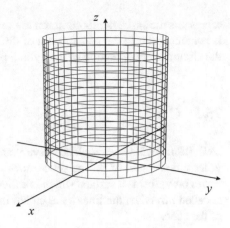

to the z axis. In fact, any couple (x_0, y_0) is a solution of the Eq. (25.2), if and only if, whatever z is, the triple (x_0, y_0, z) is a solution of the equation $x^2 + y^2 + 0z = a^2$ (Fig. 25.2).

Remark 25.1 It is known that any curve $f(x, y) = 0$ in a coordinate system of the xy plane defines a cylinder of space which also has the equation $f(x, y) = 0$ in a coordinate system of the space xyz and consists of the lines which lean on the curve and are parallel to the z axis.

Remark 25.2 If a cylinder has the generatrices parallel to the coordinate z (or y, or x) axis (Figs. 25.2 and 25.3), then it is represented by an equation in which the variable z (or y, or x) does not appear, i.e., has coefficient zero, and vice versa.

The surface $y - \sin x = 0$ (Fig. 25.3) is non-algebraic.
The system of equations

$$x^2 + y^2 = a^2$$
$$z = k$$

Fig. 25.3 Cylinder having directrix on the xy plane the graph of $y = \sin x$, restricted to the interval $[-\pi, \pi]$ and vertical generatrices. The cylinder has equation $y - \sin x = 0$, i.e., $y - \sin x + 0z = 0$

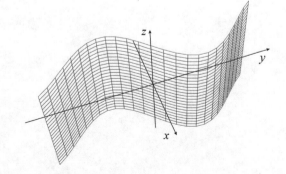

represents the set of points common to the cylinder (25.2) and the plane $z = k$, which is perpendicular to the generatrices of the cylinder. The system therefore represents the circumference, section of the cylinder and the plane $z = k$.

25.3 Cone

Definition 25.2 Given a plane curve γ and a point V not belonging to the plane of γ, let P be a point of γ. The set of lines s passing through V and the point P, as P varies on γ, forms a surface which is called a *cone* (or *conical surface*). The curve γ is called *directrix*, the lines s are called the generatrices and the point V the *vertex* of the cone.

We can think of the cone as a surface generated by a line (generatrix) that moves leaning on the points of the directrix and passing through a given point V (Fig. 25.4).

The *circular cone* has a circumference as directrix. A circular cone is *right* if the orthogonal projection of the vertex on the plane of the directrix coincides with the center of the circumference (Fig. 25.5).

We wonder how to recognize in what conditions the equation

$$F(x, y, z) = 0 \qquad (25.3)$$

represents a cone. To this aim we must find a point $V(a, b, c)$ such that, for every point $P_0(x_0, y_0, z_0)$ solution of (25.3) distinct from V, every point P of the line VP_0 is a solution of (25.3). Before carrying out some examples, let us remember (Sect. 10.5) that the equations

Fig. 25.4 Cone with vertex V, directrix γ and generatrices

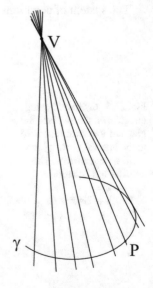

Fig. 25.5 Right circular cone

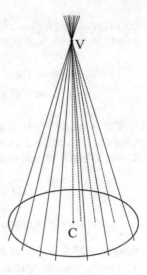

$$x = a + t(x_0 - a)$$
$$y = b + t(y_0 - b)$$
$$z = c + t(z_0 - c)$$

with t real parameter, are parametric equations of the line VP_0.

Example 25.2 Let us verify that the surface of equation

$$(x - a)^2 + (y - b)^2 = z^2 \qquad (25.4)$$

is a cone with vertex $V(a, b, 0)$. Indeed, if $P_0(x_0, y_0, z_0)$ is a solution of (25.4), i.e., $(x_0 - a)^2 + (y_0 - b)^2 = z_0^2$, then every point

$$x - a = t(x_0 - a)$$
$$y - b = t(y_0 - b)$$
$$z = t z_0$$

of the line VP_0, is a solution of (25.4), for every real t, as we can verify plugging the right-hand sides of the previous system in Eq. (25.4). The cone (25.4) is circular because any section with a plane $z = h$, for every non-null h, is a circumference.

25.3.1 *Homogeneous Polynomial*

A polynomial of degree m and n variables, in which only addends (monomials) of degree m appear, is called a *homogeneous polynomial* of degree m. For example, the polynomial $F(x, y, z) = x^2 - y^2 + 2xz + z^2$ is a homogeneous polynomial of degree 2 and variables x, y, z. For every $t \in \mathbf{R}$, the equalities are verified:

$$F(tx, ty, tz) = (tx)^2 - (ty)^2 + 2t^2xz + (tz)^2 = t^2(x^2 - y^2 + 2xy + z^2)$$
$$= t^2 F(x, y, z)$$

The property holds in general. Indeed, for every t and for every n-tuple of real numbers $(tx_1, tx_2, \ldots, tx_n)$, if $f(tx_1, tx_2, \ldots, tx_n)$ is a homogeneous polynomial of degree m and n variables, then the equality holds:

$$f(tx_1, tx_2, \ldots, tx_n) = t^m f(x_1, x_2, \ldots, x_n)$$

The following properties apply (see Example 25.2).

Property 25.1 *If a cone has the vertex at the point* $(0, 0, 0)$*, then it is represented by a homogeneous equation with coordinates* x, y, z*; vice versa, every homogeneous equation with coordinates* x, y, z *represents a cone with vertex at the point* $(0, 0, 0)$*.*

Property 25.2 *If a cone has its vertex at point* $V(a, b, c)$*, then it is represented by a homogeneous equation with the differences* $(x - a)$*,* $(y - b)$*,* $(z - c)$ *as variables; vice versa, every homogeneous equation with variables* $(x - a)$*,* $(y - b)$*,* $(z - c)$ *represents a cone with vertex* (a, b, c)*.*

25.4 Exercises

1. Examine the surface of the space xyz:

$$y = x^2 \tag{25.5}$$

As a first step let us observe that in a coordinate system of the plane xy the equation $y = x^2$ represents a parabola (Sect. 17.4.1). The Eq. (25.5) in a coordinate system xyz is independent of the variable z and by Remark 25.1 represents a cylinder with the generatrices parallel to the z axis. The intersection of the cylinder (25.5) and the xy plane coincides with the solution set of the system of equations

$$y = x^2$$
$$z = 0$$

which is, of course, a parabola of the plane xy. The surface (25.5) is called a *parabolic cylinder* (Fig. 25.6)

Fig. 25.6 Parabolic cylinder
$y = x^2$

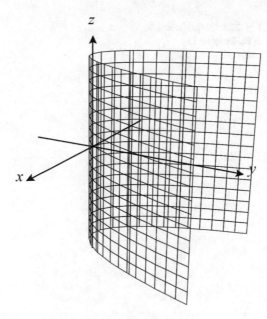

2. Analogously, the equation

$$z = x^2$$

represents a parabolic cylinder with generatrices parallel to the y axis (Fig. 25.7). The equations $x = z^2$, $x = y^2$, $z = y^2$ and $x = z^2$ represent parabolic cylinders.

Fig. 25.7 Parabolic cylinder
$z = x^2$

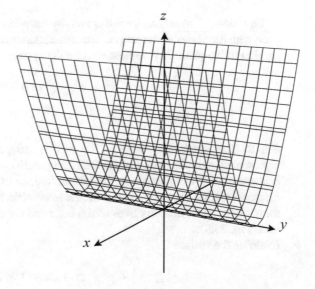

Fig. 25.8 Hyperbolic
cylinder $xy = 1$

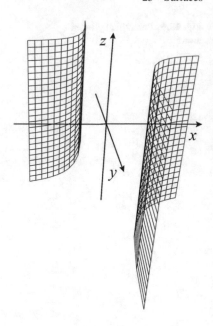

3. The surface of equation $xy = 1$ in the space xyz is a cylinder with the generatrices
    parallel to the z axis and intersects the plane xy in the equilateral hyperbola $y = \frac{1}{x}$
    (Sect. 17.4.2). The cylinder $xy = 1$ in the space xyz is called *hyperbolic cylinder*
    (Fig. 25.8).
4. Examine the surface

$$x^2 + y^2 - z^2 = 0 \tag{25.6}$$

Equation (25.6) is homogeneous and, by Property 25.1, it is the equation of a
cone with the vertex at the origin. The intersections of the surface and the planes
parallel to the plane xy is represented by the system of equations

$$x^2 + y^2 - z^2 = 0$$
$$z = h$$

$h \in \mathbf{R}$. The equation $x^2 + y^2 - h^2 = 0$ is independent of the variable z, it
represents a cylinder with the generatrices parallel to the z axis. The plane $z = h$ intersects the cylinder along the circumference of radius h and center $(0, 0, h)$; this circumference coincides with the intersection of the cone (25.6) with the
same plane. The symmetry axes of the cone and the cylinder coincide with the z
axis (Fig. 25.9).
5. Examine the surface

$$x^2 + y^2 - z^2 + 2x + 1 = 0 \tag{25.7}$$

Fig. 25.9 Cone $x^2 + y^2 - z^2 = 0$

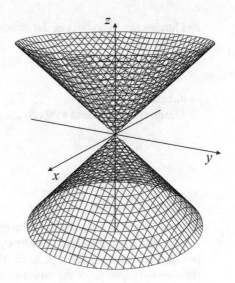

The equation can be rewritten as:

$$(x + 1)^2 + y^2 - z^2 = 0$$

and represents, by Property 25.2, a cone with the vertex at $(-1, 0, 0)$. The sections of the cone and the planes $z = h$ not passing through the vertex, i.e., $h \neq 0$, are circumferences

$$(x + 1)^2 + y^2 = h^2$$
$$z = h$$

intersections of the right circular cylinder $(x + 1)^2 + y^2 = h^2$ with the plane $z = h$. The cone (25.7) is a surface of rotation around the line a parallel to the z axis and passing through the point $(-1, 0, 0)$ in the plane xy. The line a, symmetry axis of the cone, is represented as the intersection of the two planes:

$$x = -1$$
$$y = 0$$

6. Find the equation of the cylinder with the generatrices parallel to the line r of equations

$$x - 2z = 0$$
$$y - z - 3 = 0$$

and having the hyperbola of the plane xy

$$xy - 1 = 0$$

as directrix.

Observe that the lines s represented by the equations

$$x = 2z - p \qquad (25.8)$$

$$y = z - q \qquad (25.9)$$

for every p and q, are parallel to the line r.

The line s is a generatrix of the cylinder if and only if the point of s which lies on the plane xy, i.e. the point $(x, y, z) = (-p, -q, 0)$ with third coordinate 0, satisfies the equation of the directrix $xy - 1 = 0$, hence $pq = 1$.

Eliminate the parameters p and q by replacing $p = \frac{1}{q}$ in the Eq. (25.8) to get $\frac{1}{q} = 2z - x$ and $q = \frac{1}{2z-x}$; then plugging this value of q into the Eq. (25.9), obtain $y = z - \frac{1}{2z-x}$.

Therefore, the equation $(y - z)(2z - x) = -1$ represents the required cylinder which is an algebraic surface of order 2 (Sect. 25.1). (The equation can have the form $2yz - xy - 2z^2 + xz + 1 = 0$.)

Bibliography

Hartshorne, R.: Geometry: Euclid and Beyond. Springer, New York (2000)

Hilbert, D.: Grundlagen der Geometrie. G.B. Teubner, Stuttgart (1968)

Lax, P., Burnstein, S., Lax, A.: Calculus with Applications and Computing. Springer-Verlag, New York (1976)

Lobačevskij, N.I.: Nuovi principi della geometria. Boringhieri, Torino (1965)

Chapter 26
Total Differential and Tangent Plane

26.1 Introduction

The linear function (Sect. 7.2) in the variable x has the form $y = ax + b$; the numbers a and b are the *coefficients* of the linear function. We defined (Sect. 20.13) the differential of a real-valued function f of a real variable and we proved that the differential exists if and only if the derivative of the function f exists. We also know that a differentiable function at a point is continuous at that point. Furthermore, the graph of a differentiable function f at x_0 admits the tangent line at x_0 (Sect. 20.3).

We have also shown that the increment of the function f, relative to the increment Δx of the independent variable x, and the differential df at x differ by a function ω of x and Δx, $\omega(x, \Delta x)$, which is an infinitesimal of higher order than Δx, as $\Delta x \to 0$:

$$\Delta f = df + \omega(x, \Delta x)$$

The differential df at the point x is defined by the equality $df = f'(x)\Delta x$. So the differential df is a linear function of Δx. Therefore, Δf in turn is also a linear function of Δx, whose coefficients are $f'(x)$ and $\omega(x, \Delta x)$.

26.2 Total Differential

The concepts of limit and continuity of the real-functions of several variables develop in analogy with those related to the functions of a single variable. However, we have observed that substantial differences exist, for example, in the Definition 23.10 of limit as $P \to \infty$.

Let $f: D \subseteq \mathbf{R}^2 \to \mathbf{R}$ be a real-valued function of the two real variables x and y. The fact that the partial derivatives $f_x(x_0, y_0), f_y(x_0, y_0)$ at the point P_0 (x_0, y_0) exist,

© The Author(s), under exclusive license to Springer Nature Switzerland AG 2023   477
A. G. S. Ventre, *Calculus and Linear Algebra*,
https://doi.org/10.1007/978-3-031-20549-1_26

does not imply that the function f is continuous at P_0. In fact, there are functions with partial derivatives at a point, which are not continuous at this point.

For example, consider the function $g(x, y)$, which assumes value 0 on both coordinate axes and 1 on the remaining part of the plane $\mathbf{R}^2$. The function g has no limit when P approaches the origin $(0, 0)$, because its restriction to the axes has limit 0 at each point P_0 of the axes, when P $(\neq P_0)$ approaches P_0 moving on the axis that contains P_0; whereas, if P approaches P_0 in any way, i. e., in an open disk centered at P_0, then $g(x, y)$ has limit 1. However, the limit of the difference quotient calculated on the x axis exists and is zero, because the difference quotient is constantly zero and therefore its limit is null. The same goes for the y axis. Therefore, the first partial derivatives of $g(x, y)$ at $(0, 0)$ exist. The function $g(x, y)$ is an example of a function not endowed with a limit at a point, and therefore not continuous, but partially differentiable with respect to the variables x and y.

We conclude that if f is a real-valued function defined in an open set D of $\mathbf{R}^2$, the existence of the partial derivatives at a point $P(x, y)$ of D does not imply the continuity of f at P. However, the implication holds by strengthening some hypothesis on partial derivatives. In fact, the following theorem applies.

Theorem 26.1 Let f be a real-valued function of the two real variables x, y, endowed with partial derivatives f_x and f_y in the open set D $\subseteq \mathbf{R}^2$. If the partial derivatives are bounded in an open neighborhood of the point P $\in$ D, then f is continuous at P.

Proof Let $P(x, y) \in$ D be a point satisfying the hypothesis and $I(P)$ the neighborhood of P, included in D and including an open rectangular interval H $= (x - k, x + k)$ $\times (y - k, y + k)$ with half-widths equal to $k > 0$, so that there exists a positive real number h, such that, for every point Q in H

$$|f_x(Q)| \le h, \quad |f_y(Q)| \le h, \tag{26.1}$$

Then, for every point $Q(x + \Delta x, y + \Delta y)$ belonging to H, i. e., $|\Delta x| < k$ and $|\Delta y| < k$, as a result $(x, y + \Delta y) \in$ H. Hence:

$$\Delta f = f(Q) - f(P) = f(x + \Delta x, y + \Delta y) - f(x, y) = f(x + \Delta x, y + \Delta y)$$
$$- f(x, y + \Delta y) + f(x, y + \Delta y) - f(x, y) \tag{26.2}$$

The function $f(t, y + \Delta y)$ of the single variable $t, t \in (x - k, x + k)$, is differentiable in $(x - k, x + k)$, and hence continuous; by Lagrange's theorem (Sect. 21.4), there exists $r \in (0,1)$ such that:

$$f(x + \Delta x, y + \Delta y) - f(x, y + \Delta y) = f_x(x + r\Delta x, y + \Delta y)\Delta x \tag{26.3}$$

Similarly, fixed x the function $f(x, t)$, $t \in (y - k, y + k)$ is a differentiable and continuous function of the second variable; by Lagrange's theorem there exists $s \in (0,1)$ such that:

$$f(x, y + \Delta y) - f(x, y) = f_y(x, y + s\Delta y)\Delta y \tag{26.4}$$

Therefore, plugging the right-hand sides of (26.3) and (26.4) into (26.2), we get

$$\Delta f = f_x(x + r\Delta x, y + \Delta y)\Delta x + f_y(x, y + s\Delta y)\Delta y$$

where the points $(x + r\Delta x, y + \Delta y)$ and $(x, y + s\Delta y)$ belong to H. Therefore, in virtue of (26.1) the inequality:

$$|\Delta f| \le h(|\Delta x| + |\Delta y|)$$

is obtained. Hence, if Q approaches P, then $(\Delta x, \Delta y)$ approaches $(0, 0)$ and then $\Delta f \to 0$. Thus, the continuity of f at P follows, as we wanted to show.   □

Let f be a real-valued function of the two real variables x and y defined on an open set $D \subseteq \mathbf{R}^2$, let Δx and Δy be increments given to x_0 and y_0, respectively, such that $P(x_0, y_0)$ and $Q(x_0 + \Delta x, y_0 + \Delta y)$ are points in the domain D of f. The quantity

$$\Delta f = f(Q) - f(P) = f(x_0 + \Delta x, y_0 + \Delta y) - f(x_0, y_0)$$

is said to be the *increment* of f at (x_0, y_0) relative to the increments Δx and Δy of the independent variables.

Definition 26.1 The function f is said to be *differentiable* at $P(x_0, y_0) \in D$ if it is endowed with first partial derivatives f_x and f_y at (x_0, y_0) and the function.

$$\omega = \Delta f - (f_x(x_0, y_0)\Delta x + f_y(x_0, y_0)\Delta y) \tag{26.5}$$

is an infinitesimal of higher order than $\rho = |PQ| = \sqrt{\Delta x^2 + \Delta y^2}$, namely

$$\lim_{\Delta x \to 0, \Delta y \to 0} \frac{\omega}{\rho} = 0$$

Let $f(x, y)$ be differentiable at $P(x_0, y_0)$. The linear combination

$$df = f_x(x_0, y_0)\Delta x + f_y(x_0, y_0)\Delta y \tag{26.6}$$

is called the *total differential* or simply *differential of f* at (x_0, y_0).

Let us apply the definition of total differential to the functions $f(x, y) = x$ and $f(x, y) = y$; we obtain $dx = \Delta x$ and $dy = \Delta y$. Then (26.6) can be rewritten

$$df = f_x(x_0, y_0)dx + f_y(x_0, y_0)dy$$

By definition of total differential, Eq. (26.5) yields

$$\Delta f = df + \omega \tag{26.7}$$

where ω is an infinitesimal of higher order than $\sqrt{\Delta x^2 + \Delta y^2}$. Equation (26.7) is formally equal to Eq. (20.18) found in the case of a function of a single variable. However, we observe an essential difference between the two cases: while in the case of a single variable the Eq. (26.7) (along with ω infinitesimal of higher order than Δx) is obtained by assuming only the existence of the derivative of f, in the case of two variables, the hypothesis of partial differentiability, w. r. to x and y, of f is not sufficient to guarantee the validity of (26.7) (along with ω infinitesimal of higher order than $\sqrt{\Delta x^2 + \Delta y^2}$). Indeed, there are examples of real-valued functions of two variables such that the difference $\Delta f - df$ it is not an infinitesimal of higher order than $\sqrt{\Delta x^2 + \Delta y^2}$.

Definition 26.2 A function f (x, y) defined in the open set D $\subseteq \mathbf{R}^2$, is said to be of *class C^m(D)* or simply *to be C^m*, m non negative integer, if it is continuous in D and has continuous all the partial derivatives up to the order m. A function of *class C^0* is, by definition, a continuous function.

So the class C^1(D) consists of the continuous functions whose partial derivatives f_x and f_y are continuous in D; such functions are also called *continuously differentiable* in D.

Theorem 26.2 Let f be a real-valued function of the two real variables x and y. If f (x, y) is defined in the open set D $\subseteq \mathbf{R}^2$ and is of class C^1(D), then the function f (x, y) is differentiable at any point of D.

Remark 26.1 The inverse of Theorem 26.2 does not hold; indeed, a function can be differentiable in D without being C^1(D).

26.3   Vertical Sections of a Surface

Let us consider some geometric aspects of the continuous function f (x, y) with continuous first partial derivatives at the point P(x_0, y_0). Let us intersect the surface, graph of $f(x, y)$, and the plane α of equation x $= x_0$, which is the plane parallel to the plane yz containing the points of first coordinate x_0 (Fig. 26.1).

The plane α intersects the surface in a curve G, which is the set of the points $(x_0, y, f$ $(x_0, y_0))$, with variable y. The curve G has the tangent line t at y_0; in fact, $f_y(x_0, y_0)$ exists and this means that the partial function $g(y) = f$ (x_0, y) is differentiable at y_0.

The line t tangent to the curve G at the point Q$(x_0, y_0, f$ $(x_0, y_0))$ also lies on the plane α; the angle δ that t forms with the plane xy, namely with the orthogonal projection t' of t on the plane xy, has the tangent, $\tan\delta$, which equals $f_y(x_0, y_0)$.

The existence of the partial derivative f_y at the point (x_0, y_0) indicates that the curve G has the tangent line at the point Q$(x_0, y, f$ $(x_0, y_0))$, laying on the plane $x = x_0$.

Fig. 26.1 The point P(x_0, y_0) belongs to the domain of f; the coordinates of the point Q are (x_0, y_0, f (x_0, y_0)); α is the plane tP, whose equation is $x = x_0$; t' is the orthogonal projection of the line t on the plane xy, δ is the angle of the lines t and t'; the plane β contains P and is orthogonal to the y axis, β has equation $y = y_0$; the curve G = [graph of f]$\bigcap$[plane α]; curve H = [graph of f]$\bigcap$[plane β]; the line s is the tangent to the curve H at the point Q

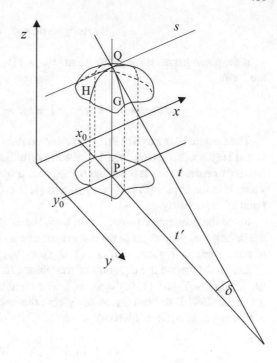

26.4 The Tangent Plane to a Surface

Let us consider a real-valued function f having domain $D \subseteq \mathbf{R}^2$ and suppose f differentiable at the point (x_0, y_0) of D. We want to stress the relation between the geometric properties of the surface S, graph of the equation $z = f$ (x, y), and the differential of f at (x_0, y_0).

To this aim let us define the *tangent plane* to the surface S at the point P(x_0, y_0) [we can also say: tangent plane to the surface at the point Q(x_0, y_0, f(x_0, y_0)], starting from its geometric construction: intersect the surface S with any plane passing through the point P and parallel to the z axis; there are infinite planes that form a bundle of planes, that have in common the line PQ. Each of these planes and the surface intersect in a curve γ passing through the point Q. Observe that if f is differentiable at P(x_0, y_0), it can be shown that all the tangent lines at Q to the curves γ are on a plane called the *tangent plane* at P to the surface.

Theorem 26.3 If $f(x, y)$ is differentiable at the point P(x_0, y_0), then the equation of the tangent plane at P to the surface S is.

$$z - f(x_0, y_0) = f_x(x_0, y_0)(x - x_0) + f_y(x_0, y_0)(y - y_0) \qquad (26.8)$$

Proof Let S be the surface graph of the equation $z = f$ (x, y). The equation of the generic plane in a coordinate system of the space xyz is (see Sect. 10.4)

$$ax + by + cz + d = 0 \tag{26.9}$$

If the plane passes through the point $(x_0, y_0, f(x_0, y_0))$, then the Eq. (26.9) assumes the form

$$a(x - x_0) + b(y - y_0) + c(z - f(x_0, y_0)) = 0 \tag{26.10}$$

The tangent plane to S at $P(x_0, y_0)$ must contain the tangent lines s, t to the curves G and H (Sect. 26.3) (Fig. 26.2). Since the orthogonal projection of each segment of the line t on the x axis is 0 and the orthogonal projection of each segment of s on the y axis is 0, the triples $(0, 1, f_y(x_0, y_0))$ and $(1, 0, f_x(x_0, y_0))$, are direction numbers of t and s, respectively.

Since the orthogonal projection of each segment of the line t on the x axis is 0 and the orthogonal projection of each segment of s on the y axis is 0, direction numbers of t and s are $(0, 1, f_y(x_0, y_0))$ and $(1, 0, f_x(x_0, y_0))$ (Sect. 20.3), respectively.

Let us determine the equation of the plane (26.10) parallel to the two directions $(0, 1, f_y(x_0, y_0))$ and $(1, 0, f_x(x_0, y_0))$. The condition of parallelism (Sect. 15.4) of the plane (26.10) to both directions yields the homogeneous linear system in three unknowns a, b, c (Sect. 14.10.6):

$$0a + 1b + f_y(x_0, y_0)c = 0$$
$$1a + 0b + f_x(x_0, y_0)c = 0$$

Fig. 26.2 Tangent plane to the surface at Q

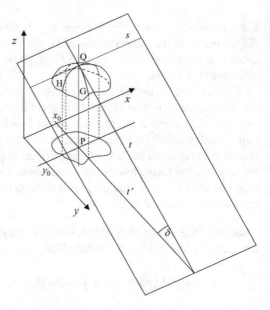

which is easily solved by setting $c = 1$, to get $b = -f_y(x_0, y_0)$, and $a = -f_x(x_0, y_0)$ (see Sect. 14.9). Therefore, the Eq. (26.8) represents the required tangent plane. □

Exercise 26.1 Given the surface

$$f(x, y) = x^2 - 2xy - y^2 - 2x \qquad (26.11)$$

find the equation of the tangent plane to the surface at the point P $(1, -1)$. First observe that the domain of f is $\mathbf{R}^2$. Furthermore, f is of class C^1 in $\mathbf{R}^2$: indeed,

$$f_x(x, y) = 2x - 2y - 2$$
$$f_y(x, y) = -2x - 2y$$

are continuous in $\mathbf{R}^2$. Therefore, f is differentiable in $\mathbf{R}^2$ and the tangent plane at P to the surface (26.11 has equation

$$z - f(1, -1) = f_x(1, -1)(x - 1) + f_y(1, -1)(y + 1) \qquad (26.12)$$

where $f(1, -1) = 0, f_x(1, -1) = 2, f_y(1, -1) = 0$ and the Eq. (26.12) simplifies: $2x - z + 2 = 0$.

Remark 26.2 We know that the tangent line to a curve of the plane can leave the curve, in a neighborhood of the contact point, in a single half-plane or, as in the case of the tangent line in an inflection point (Sect. 21.11.2), it can cross the curve itself. Also, the plane tangent to a surface at one point can leave parts of the surface in one and the other half-space identified by the tangent plane in a neighborhood of the contact point. An example will come in handy.

Example 26.1 Consider the surface of equation $z = xy$. The surface is called a *hyperbolic paraboloid* or *saddle paraboloid* (Fig. 26.3).

The origin $(0, 0, 0)$ belongs to the surface $z = xy$. The plane tangent to the surface at the origin is obtained from (26.8 where $(x_0, y_0) = (0, 0), f(x_0, y_0) = 0, f_x(x_0, y_0) = 0, f_y(x_0, y_0) = 0$. Therefore, the tangent plane α at the origin to the surface $z = xy$ has equation $z = 0$. The system of equations

$$z = xy$$
$$z = 0$$

represents the intersection of the surface and the tangent plane at the origin. This intersection consists of two lines: the y axis represented by the system

$$x = 0$$

Fig. 26.3 Saddle paraboloid
of equation $z = xy$

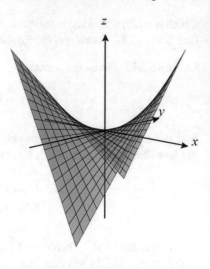

$$z = 0$$

and the x axis

$$y = 0$$
$$z = 0$$

If we fix any neighborhood of the origin, which is the contact point of the surface and the tangent plane α, we find points of the surface that belong to one of the two half-spaces (see Sect. 4.3.1) of origin α and points that belong to the other half-space.

Bibliography

Anton, H.: Calculus. Wiley, New York (1980)
Ghizzetti, A., Rosati, F.: Analisi matematica I. Zanichelli Editore, Bologna (2005)
Spiegel, M.R.: Advanced Calculus. McGraw-Hill Book Company, New York (1963)

Chapter 27
Maxima and Minima. Method of Lagrange Multipliers

27.1 Relative and Absolute Extrema of Functions of Two Variables

In Sect. 21.2 we defined absolute and relative extrema of a real-valued function of a real variable. Upper bounded and lower bounded real-valued functions of two variables were defined in Sect. 23.8. Let us now deal with the extrema of the real-valued functions of two real variables. The topic develops analogously to what already done above.

Definition 27.1 Let f be a real-valued function defined and continuous in the open set $A \subseteq \mathbf{R}^2$. The point $(x_0, y_0) \in A$ is called a *relative maximum (minimum) point* of f in A if there exists a circular neighborhood I of the point (x_0, y_0) such that

$$f(x, y) \leq f(x_0, y_0) \quad (f(x, y) \geq f(x_0, y_0)) \tag{27.1}$$

whatever the point $(x, y) \in I \cap A$ is.

If (27.1) holds for every point (x, y) of the domain A, then (x_0, y_0) is called a *maximum (minimum) point* or an *absolute maximum (minimum) point* of f. The value attained by the function at a relative maximum (minimum) point is called a *relative maximum (minimum)* of the function; the value attained by the function at an absolute maximum point is said to be the *absolute maximum*, or the *maximum*, of f in A; the value attained by f at an absolute minimum point of f in A is said to be the *absolute minimum*, or the *minimum*, of f in A.

Maxima and minima (relative maxima and minima) of a function are named *extrema (relative extrema)* and the maximum and minimum (relative maximum and minimum) points are named *extremum (relative extremum) points* of the function.

Example 27.1 The function $f(x, y) = e^{-(x^2+y^2)}$ (see Example 23.6; Sect. 23.7) is bounded, its range is the left-open interval $(0, 1]$, the point $(0, 0)$ is the absolute maximum point of f. The function f has infimum 0.

© The Author(s), under exclusive license to Springer Nature Switzerland AG 2023
A. G. S. Ventre, *Calculus and Linear Algebra*,
https://doi.org/10.1007/978-3-031-20549-1_27

Fermat's theorem (Sect. 21.3) for real-valued functions of two variables holds.

Theorem 27.1 [Fermat's theorem]. Let f be a real-valued function defined on an open set $A \subseteq \mathbf{R}^2$. If the point (x_0, y_0) of A is a relative extremum point of f and f is endowed with first partial derivatives at (x_0, y_0), then.

$$f_x(x_0, y_0) = f_y(x_0, y_0) = 0 \tag{27.2}$$

Let us observe that under the hypotheses of Fermat's theorem, if f is differentiable (Sect. 26.2) in the open set A the total differential of f is zero at the relative extremum points and the tangent plane (26.8) to the surface $z = f(x, y)$ at a relative extremum point (x_0, y_0) has equation $z = f(x_0, y_0)$; therefore, the tangent plane is parallel to the xy plane.

Definition 27.2 A point (x_0, y_0) in the domain A of f that satisfies the conditions (27.2) is called a *critical point* of f.

Example 27.2 Consider the function $f(x, y) = x^2 + 2y^2$, whose domain is $\mathbf{R}^2$. We wonder if the function has relative extremum points. We first check if the function has critical points. For this purpose, let's see if there are points at which the first derivatives of f are zero:

$$f_x(x, y) = 2x = 0$$
$$f_y(x, y) = 4y = 0$$

Therefore, $(0, 0)$ is the only critical point of f. Is $(0, 0)$ a relative maximum or minimum point of f? We observe that $f(x, y) = x^2 + 2y^2 \geq f(0, 0) = 0$, for every (x, y). Therefore $(0, 0)$ is a relative minimum point of f. The point $(0, 0)$ is also an absolute minimum point, since the function does not assume negative values.

Each relative extremum point (x_0, y_0) of the function f is a critical point of f. Vice versa is not always true: there are critical points that are neither relative maximum points nor relative minimum points of f.

The following example shows that, although $f_x(x, y) = f_y(x, y) = 0$, the point (x, y) is not a relative extremum point.

Example 27.3 Consider the function $f(x, y) = xy$, defined on $\mathbf{R}^2$ (see Example 26.1), whose graph is a hyperbolic paraboloid. Since $f_x(x, y) = y$ and $f_y(x, y) = x$ the function has a unique critical point $(0, 0)$. However, it happens that in every neighborhood of $(0, 0)$, there are points (x, y) such that.

$$f(x, y) = xy > f(0, 0) = 0$$

and points (x, y) such that

$$f(x, y) = xy < f(0, 0) = 0$$

Hence the point $(0, 0)$ is neither a relative maximum point nor a relative minimum point of f. In other words: if x and y have the same sign, i. e., the point (x, y) is in the I or III quadrant, then $xy > 0$; if x and y have opposite signs, i. e., the point (x, y) is in the II or IV quadrant, then $xy < 0$.

Definition 27.3 Let the function f be defined in the open set A. The critical point (x_0, y_0) of f is called a *saddle point* for f if in every neighborhood of (x_0, y_0) there are points (x, y) such that $f(x, y) < f(x_0, y_0)$ and points (x, y) such that $f(x, y) > f(x_0, y_0)$.

For example, the point $(0, 0)$ is a saddle point for the function $f(x, y) = xy$.

27.2 Exercises

Find the critical points of the functions:

a. $f(x, y) = x^2 - 3xy - 4y^2$
b. $g(x, y) = x^2 - 2y^2 - 2x$
c. $h(x, y) = xy^2$

Solutions

a. The critical points of the function $f(x, y) = x^2 - 3xy - 4y^2$ are the solutions (x, y) of the system of equations

$$f_x(x, y) = 2x - 3y = 0$$
$$f_y(x, y) = -3x - 8y = 0$$

This system has the unique solution $(x, y) = (0, 0)$, which is the only critical point of the function f.

b. The critical points of the function $g(x, y) = x^2 - 2y^2 - 2x$ are the solutions (x, y) of the system of equations

$$g_x(x, y) = 2x - 2 = 0$$
$$g_y(x, y) = -4y = 0$$

This system has the unique solution $(x, y) = (1, 0)$, which is the only critical point of g.

c. The critical points of the function $h(x, y) = xy^2$ are the solutions (x, y) of the system of equations

$$h_x(x, y) = y^2 = 0$$
$$h_y(x, y) = 2xy = 0$$

The partial derivatives vanish simultaneously at the points $(x, 0)$, for every real number x. Therefore, all the points of the x axis are critical points of h.

27.3   Search for Relative Maxima and Minima

The occurrence of the equalities (27.2) is a necessary condition in order that (x_0, y_0) be a relative extremum point of the function f.

Let us now state sufficient conditions for the existence of relative extremum points.

Definition 27.4 Let $f(x, y)$ be a real-valued function of two real variables endowed with second partial derivatives. The function H, defined by:

$$H(f(x, y)) = f_{xx}(x, y) f_{yy}(x, y) - \left(f_{xy}(x, y) \right)^2$$

that can be put in the form of a determinant

$$H(f(x, y)) = \begin{vmatrix} f_{xx}(x, y) & f_{xy}(x, y) \\ f_{xy}(x, y) & f_{yy}(x, y) \end{vmatrix}$$

is called the *Hessian determinant*, or simply the *Hessian* of f at (x, y).

Theorem 27.2 Let f be a real-valued function of two real variables defined on an open set $D \subseteq \mathbf{R}^2$, with continuous first and second partial derivatives at the point (x_0, y_0) which belongs to the domain of f. If $f_x(x_0, y_0) = 0$ and $f_y(x_0, y_0) = 0$, one of the following cases occurs:

case 1. if $H(f(x_0, y_0)) > 0$ and $f_{xx}(x_0, y_0) > 0$, then (x_0, y_0) is a relative minimum point of f;

case 2. if $H(f(x_0, y_0)) > 0$ and $f_{xx}(x_0, y_0) < 0$, then (x_0, y_0) is a relative maximum point of f;

case 3. if $H(f(x_0, y_0)) < 0$, then (x_0, y_0) is neither a relative maximum nor a relative minimum point. If $H(f(x_0, y_0)) < 0$, then (x_0, y_0) is a *saddle point* of f;

case 4. if $H(f(x_0, y_0)) = 0$, then the problem remains open. In this case one can try to operate directly on the function; for instance, the study of the sign of the difference $f(x_0, y_0) - f(x, y)$ near (x_0, y_0) can give indications.

Exercise 27.1 Find the relative extremum points of the function

$$f(x, y) = x^2 - 3xy + 4y^2 + 2x + 4y$$

Solution . The domain of f is $\mathbf{R}^2$. The critical points are the solutions of the system:

$$f_x(x, y) = 2x - 3y + 2 = 0$$
$$f_y(x, y) = -3x + 8y + 4 = 0$$

that has the unique solution $(-4, -2)$. Compute the value $H(f(-4, -2))$:

$$f_{xx}(-4, -2) = 2, \ f_{xy}(-4, -2) = -3, \ f_{yy}(-4, -2) = 8,$$

then $H(f(-4, -2)) = 16 - 9 = 7 > 0$. By Theorem 27.2, $(-4, -2)$ is a relative minimum point and $f(-4, -2) = -8$ is the value of f at the relative minimum point.

Exercise 27.2 Find the relative extremum point of the function

$$f(x, y) = 2xy - x^2 \tag{27.3}$$

Solution The function has domain $\mathbf{R}^2$. The first partial derivatives are

$$f_x(x, y) = -2x + 2y$$
$$f_y(x, y) = 2x$$

There exists a unique critical point $(0, 0)$. The second partial derivatives are:

$$f_{xx}(x, y) = -2, \ f_{xy}(x, y) = 2, \ f_{yy}(x, y) = 0$$

The value of the Hessian of function (27.3) at the point $(0, 0)$ is

$$H(f(0, 0)) = f_{xx}(0, 0) f_{yy}(0, 0) - f_{xy}^2(0, 0) = -2(0) - 4 = -4 < 0$$

The point $(0, 0)$ is a saddle point for f. The function $f(x, y)$ has no relative extremum points.

Exercise 27.3 Find the critical points of the function $f(x, y) = x^3 + y^3 - 3xy$ and examine the boundedness of f.

Solution The domain of f is $\mathbf{R}^2$. Compute the first derivatives of f and equal them to zero:

$$f_x(x, y) = 3x^2 - 3y = 0$$
$$f_y(x, y) = 3y^2 - 3x = 0$$

The system of the two equations takes the equivalent form:

$$x^2 = y$$
$$y^2 - x = 0 \tag{27.4}$$

Replace x^2 with y in the second equation:

$$x^2 = y$$

$$x(x^3 - 1) = 0$$

The system decomposes into two systems:

$$x^2 = y$$
$$x = 0$$

that has the solution $(0, 0)$, and

$$x^2 = y$$
$$x^3 - 1 = 0$$

whose solution is $(1, 1)$. Therefore, the first partial derivatives of f vanish at the points $(0, 0)$ and $(1, 1)$. The second partial derivatives are:

$$f_{xx}(x, y) = 6x$$
$$f_{xy}(x, y) = -3$$
$$f_{yy}(x, y) = 6y$$

The value of the Hessian at the point $(0, 0)$ is

$$H(f(0, 0)) = f_{xx}(0, 0) f_{yy}(0, 0) - f_{xy}^2(0, 0) = -9 < 0$$

By Theorem 27.2, $(0, 0)$ is a saddle point for f. The value of the Hessian at the point $(1, 1)$ is:

$$H(f(1, 1)) = f_{xx}(1, 1) f_{yy}(1, 1) - f_{xy}^2(1, 1) = 27 > 0$$

By Theorem 27.2, $(1, 1)$ is a relative minimum point of f and $f(1, 1) = -1$ is the value of the relative minimum of f.

In order to examine the boundedness of f in $\mathbf{R}^2$, calculate the limits

$$\lim_{\substack{x \to +\infty \\ y \to +\infty}} (x^3 + y^3 - 3xy) = +\infty$$

$$\lim_{\substack{x \to -\infty \\ y \to -\infty}} (x^3 + y^3 - 3xy) = -\infty$$

The function $f(x, y) = x^3 + y^3 - 3xy$ is neither upper bounded, nor lower bounded. Therefore, absolute extrema of f in $\mathbf{R}^2$ do not exist.

Remark 27.1 The system (27.4) represents the intersection of the parabolas $x^2 - y = 0$ and $y^2 - x = 0$, both passing through the points (0, 0) and (1, 1): what suggests an easier way to solve the system of Eqs. (27.4).

27.4 Absolute Maxima and Minima in R^2

We studied a procedure for determining the possible relative maxima and relative minima of a real-valued function of two real variables.

We now describe a procedure that allows to determine the possible absolute maxima and minima of a real-valued function of two real variables $f(x, y)$ defined in $D \subseteq \mathbf{R}^2$. We observe that the function $f(x, y) = x^3 + y^3 - 3xy$ examined in the Exercise 27.3, has a relative minimum and it is not endowed with a maximum or a minimum.

Weierstrass' theorem (Sect. 19.2.1) assures that a continuous function defined in a closed and bounded set D has a maximum value and a minimum value in D.

Exercise 27.4 Let us consider the restriction of the function

$$f(x, y) = xy$$

to the closed circle D having center O(0, 0) and radius 1, defined by the inequality $x^2 + y^2 \leq 1$, restriction which we call $f(x, y)$ again. By Weierstrass' theorem, f has minimum and maximum in D. We found (Example 27.3) that the only critical point of f is the saddle point (0, 0). So the minimum and maximum points of f in D must be found in the boundary of D, namely the circumference $x^2 + y^2 = 1$.

Let us consider the function $f(x, y) = xy$ with (x, y) belonging to the circumference $x^2 + y^2 = 1$. The parametric equations of the circumference (Sect. 8.1.3) are:

$$x = \cos t$$
$$y = \sin t$$

$t \in [0, 2\pi]$. Let us consider the extrema of the composite function

$$f(\cos t, \sin t) = \cos t \sin t$$

which is a function $g(t)$ of a single real variable t, continuous in the closed and bounded interval $[0, 2\pi]$. The extrema of g are, by Weierstrass' theorem, its maximum and its minimum. Therefore, we set

$$g(t) = \cos t \sin t$$

and find the maximum and minimum of the function $g(t)$ in $[0, 2\pi]$. The relative extrema of g lie among the points at which the derivative vanishes:

$$g'(t) = -\sin t \sin t + \cos t \cos t = \cos^2 t - \sin^2 t$$

Hence, $g'(t) = 0$ if and only if $1 - 2\sin^2 t = 0$, i. e.,

$$\sin t = \pm \frac{1}{\sqrt{2}}$$

whose four solutions are

$$t = \frac{\pi}{4}, \quad t = \frac{3\pi}{4}, \quad t = \frac{5\pi}{4}, \quad t = \frac{7\pi}{4}$$

and among which the extremum points of $g(t)$ must be found:

$$g\left(\tfrac{\pi}{4}\right) = \cos\left(\tfrac{\pi}{4}\right)\sin\left(\tfrac{\pi}{4}\right) = \tfrac{1}{\sqrt{2}}\tfrac{1}{\sqrt{2}} = \tfrac{1}{2}$$
$$g\left(\tfrac{3\pi}{4}\right) = \cos\left(\tfrac{3\pi}{4}\right)\sin\left(\tfrac{3\pi}{4}\right) = -\tfrac{1}{2}$$
$$g\left(\tfrac{5\pi}{4}\right) = \cos\left(\tfrac{5\pi}{4}\right)\sin\left(\tfrac{5\pi}{4}\right) = \tfrac{1}{2}$$
$$g\left(\tfrac{7\pi}{4}\right) = \cos\left(\tfrac{7\pi}{4}\right)\sin\left(\tfrac{7\pi}{4}\right) = -\tfrac{1}{2}$$

The maximum of g is attained at the points $\frac{\pi}{4}$ and $\frac{5\pi}{4}$ and the minimum at $\frac{7\pi}{4}$ and $\frac{3\pi}{4}$ .The values $\frac{1}{2}$ and $-\frac{1}{2}$ are the maximum and the minimum of the restriction of f $(x, y) = xy$ to the circle $x^2 + y^2 \leq 1$.

27.5 Search for Extrema of a Continuous Function

If the real-valued function of two real variables $f(x, y)$ is defined on the closed and bounded set D where it is continuous, then there exist the maximum and the minimum of f in D. If partial derivatives of f exist in the interior of D and if a maximum (minimum) point of f is in the interior of D, then the partial derivatives at this point are equal to zero. If no maximum (minimum) point of f is in the interior of D, then this extremum point has to be found on the boundary of D. In order to identify the maximum and minimum points of the continuous function f in the closed and bounded set D we adopt the following procedure:

- determine the critical points and relative extrema of f in the interior of D;
- determine the maximum and minimum of f on the boundary of D;
- compare the values: the maximum of the function is the greatest of the extrema obtained in the two previous phases, the minimum of the function is the least of the extrema obtained in the two previous phases.

Exercise 27.5 Find the extrema of the function $z = f(x, y) = e^{-(x^2+y^2)}$.

Solution The function $e^{-(x^2+y^2)}$ was studied in the Example 23.6. Now we want to determine the extrema of the restriction of the function to the unit circle $x^2 + y^2 \leq 1$. The function is positive in the domain $\mathbf{R}^2$. The partial derivative of f are:

$$f_x(x, y) = -2xe^{-(x^2+y^2)}$$
$$f_y(x, y) = -2ye^{-(x^2+y^2)}$$

Moreover, $f_x(x, y) = 0$ if and only if $x = 0$ and $f_y(x, y) = 0$ if and only if $y = 0$. The point $O(0, 0)$ is the unique critical point of f. The function $e^{-(x^2+y^2)}$ is differentiable up to the order 2 in $\mathbf{R}^2$.

$$f_{xx}(x, y) = -2e^{-(x^2+y^2)} + 4x^2e^{-(x^2+y^2)}$$
$$f_{xy}(x, y) = 4xye^{-(x^2+y^2)}$$
$$f_{yy}(x, y) = 4y^2e^{-(x^2+y^2)} - 2e^{-(x^2+y^2)}$$

Since $f_{xx}(0, 0) = -2, f_{xy}(0, 0) = 0, f_{yy}(0, 0) = -2$, and

$$H(f(0, 0)) = f_{xx}(0, 0)f_{yy}(0, 0) - f_{xy}^2(0, 0) = 4 > 0$$

The case 2 of Theorem 27.2 occurs and the point $(0, 0)$ is a relative maximum point at which $e^{-(x^2+y^2)}$ takes value $e^0 = 1$. At every point (x, y) in the boundary $x^2 + y^2 = 1$ the value of function f is

$$f(x, y) = e^{-(x^2+y^2)} = e^{-1}$$

Thus, the function f assumes absolute maximum value $\frac{1}{e} < 1$ at every point of the circumference $x^2 + y^2 = 1$.

27.6 Constrained Extrema. Method of Lagrange Multiplier

We will deal with the problem of knowing the extrema of a restriction of a given function. It happens along a mountain road, that we want to know the maximum altitude we have reached: we do not want to know how high the mountain is, but the maximum altitude we have reached in the walk. Still, the mayor has a plan to improve waste collection efficiency, but budget constraints exist. Almost all acts, individual or collective, are immersed in a context and subject to some conditions.

We will study some problems of *constrained*, or *conditional extrema*. Let a real-valued function $z = f(x, y)$ of two real variables be defined on an open set $A \subseteq \mathbf{R}^2$. We want to calculate the maxima and minima of $f(x, y)$, when x and y are not independent but linked by a relation, a *constraint*, of the type $\varphi(x, y) = 0$, which implicitly defines y as a function of x (Sect. 24.7). From a geometric point of view, we consider values of z corresponding to the points of a curve C contained in the set A. In this case a point (x_0, y_0) of the curve will be a maximum (minimum) point if there is a neighborhood of it such that the values of z in all the points belonging to the intersection of this neighborhood with the curve are less (or greater) than the value $z = f(x_0, y_0)$. These maxima or minima, which are called *constrained* or *conditional extrema*, can be found by a procedure known as *Lagrange's multiplier method*.

Let specify the data of the problem. We suppose f and φ are functions of class C^1 in A (Sect. 26.2), i. e., the functions f and φ and their partial derivatives are continuous in A. Suppose that the set C of points of A, represented by the equation $\varphi(x, y) = 0$, is a plane curve. The extrema of the restriction of f to the curve C are called *constrained* or *conditional extrema* and the equation $\varphi(x, y) = 0$ is called the *constraint*.

In order to find constrained relative maximum points and relative minimum points of the restriction of f to C the following procedure is implemented:

1. find the maxima and minima of the following function, called *the Lagrangian*,

$$L(x, y, \lambda) = f(x, y) + \lambda\varphi(x, y) \tag{27.5}$$

where λ is a parameter, that fulfils the system of necessary conditions on partial derivatives:

$$L_x = 0, \ L_y = 0, \ L_\lambda = 0 \tag{27.6}$$

2. examine the values of f on the points of the boundary of A that belong to the curve C.

Example 27.4 Find the extrema of the function $f(x, y) = \sqrt{4-x^2-y^2}$ subject to the constraint $\varphi(x, y) = y - x^2 - 1 = 0$.

The domain of f is the set of points (x, y) such that

$$x^2 + y^2 - 4 \le 0$$

i. e., the circle with center $(0, 0)$ and radius 2. The curve $y - x^2 - 1 = 0$ is the convex parabola with vertical axis and vertex $(0, 1)$. We form the auxiliary function (27.5)

$$L(x, y, \lambda) = \sqrt{4-x^2-y^2} + \lambda\left(y-x^2-1\right)$$

subject to the conditions (27.6):

$$L_x = \frac{-2x}{2\sqrt{4-x^2-y^2}} - 2\lambda x = 0 \tag{27.7}$$

$$L_y = \frac{-2y}{2\sqrt{4-x^2-y^2}} + \lambda = 0 \tag{27.8}$$

$$L_\lambda = y - x^2 - 1 = 0 \tag{27.9}$$

Let us solve the system of the equations in the open circle $x^2 + y^2 - 4 < 0$; this implies $\sqrt{4-x^2-y^2} > 0$.

From (27.8) we obtain

$$\lambda = \frac{y}{\sqrt{4-x^2-y^2}}$$

which plugged in (27.7),

$$\frac{-x}{\sqrt{4-x^2-y^2}} - \frac{2xy}{\sqrt{4-x^2-y^2}} = 0$$

yields $x + 2xy = 0$, i. e.,

$$x(1 + 2y) = 0$$

Set $x = 0$ in (27.9) to obtain $y = 1$; so the point C(0, 1) is a relative or absolute extremum (maximum or minimum) point in the interior of the circle; if we put $y = -\frac{1}{2}$ in the same equation, then we obtain the equation system:

$$y = -\frac{1}{2}$$
$$x^2 = -\frac{3}{2}$$

that is inconsistent because $x^2 \geq 0$. Let us find the common points to the constraint and the boundary of the circle:

$$y = x^2 + 1$$
$$x^2 + y^2 - 4 = 0$$

Replace x^2 with $y - 1$ in the second equation:

$$x^2 = y - 1$$
$$y - 1 + y^2 - 4 = 0$$

Solve the equation $y^2 + y - 5 = 0$:

$$y = \frac{-1 \pm \sqrt{21}}{2}$$

As $x^2 = y - 1$, the value $x^2 = \frac{-1-\sqrt{21}}{2} - 1 < 0$ must be excluded. So

$$x = \pm \sqrt{\frac{-3 + \sqrt{21}}{2}}$$

Thus C(0, 1) is the extremum point in the interior of the circle, and

$$A = \left(-\sqrt{\frac{-3 + \sqrt{21}}{2}}, \frac{-1 + \sqrt{21}}{2} \right) \quad \text{and} \quad B = \left(\sqrt{\frac{-3 + \sqrt{21}}{2}}, \frac{-1 + \sqrt{21}}{2} \right)$$

are the extrema points in common to the constraint and the boundary of the circle. In conclusion, from the comparison of the values:

$$f(C) = f(0, 1) = \sqrt{3}$$
$$f(A) = f(B) = 0$$

we obtain the maximum $\sqrt{3}$ and the minimum 0 of the given problem of constrained extrema.

Example 27.5 Find the extrema of the function $f(x, y) = 10x^2 - 16xy + 10y^2$ subject to the constraint $\varphi(x, y) = x^2 + y^2 - 1 = 0$.

The domain of f is $\mathbf{R}^2$. Let us search for the possible maxima and minima of f on the unit circumference $x^2 + y^2 - 1 = 0$. The Lagrangian (27.5) is

$$L(x, y, \lambda) = 10x^2 - 16xy + 10y^2 + \lambda(x^2 + y^2 - 1)$$

subject to the conditions

$$L_x = 20x - 16y + 2\lambda x = 0 \tag{27.10}$$

$$L_y = 20y - 16x + 2\lambda y = 0 \tag{27.11}$$

$$L_\lambda = x^2 + y^2 - 1 = 0 \tag{27.12}$$

We must solve the system (27.10) to (27.12) in the unknowns x, y, λ. From (27.10) and (27.11) we obtain the equations

$$x(10 + \lambda) = 8y$$
$$y(10 + \lambda) = 8x$$

that yield:

$$\frac{10 + \lambda}{8} = \frac{y}{x} \quad \text{and} \quad \frac{8}{10 + \lambda} = \frac{y}{x}$$

Hence the equality of the left-hand sides:

$$\frac{10 + \lambda}{8} = \frac{8}{10 + \lambda}$$

that for $\lambda \neq -10$, implies

$$\lambda^2 + 20\lambda + 36 = 0$$

whose roots are $\lambda = -10 \pm \sqrt{100 - 36} = -10 \pm 8$, i. e., $\lambda_1 = -2$ and $\lambda_2 = -18$.
If we set $\lambda_1 = -2$ in the system of equations then:

$$20x - 16y - 4x = 0$$
$$x^2 + y^2 - 1 = 0$$

Hence

$$16x - 16y = 0$$
$$x^2 + y^2 - 1 = 0$$

and

$$x = y$$
$$2x^2 = 1$$

The last two equations lead to the values $x = \pm \frac{1}{\sqrt{2}}$ and $y = \pm \frac{1}{\sqrt{2}}$ and determine
the points $\left(\frac{1}{\sqrt{2}}, \frac{1}{\sqrt{2}}\right)$, $\left(-\frac{1}{\sqrt{2}}, -\frac{1}{\sqrt{2}}\right)$ in the interior of the domain of f.
If we set $\lambda_1 = -18$, then

$$20y - 16x - 4y = 0$$
$$x^2 + y^2 - 1 = 0$$

Hence

$$-16y - 16x = 0$$
$$x^2 + y^2 - 1 = 0$$

and

$$y = -x$$
$$2x^2 = 1$$

The last two equations lead to the values $x = \pm \frac{1}{\sqrt{2}}$ and $y = \pm \frac{1}{\sqrt{2}}$ and determine

the points $\left(\frac{1}{\sqrt{2}}, -\frac{1}{\sqrt{2}}\right)$, $\left(-\frac{1}{\sqrt{2}}, \frac{1}{\sqrt{2}}\right)$ in the interior of the domain of f.

So we have found four points that are candidates to solve our problem of constrained extrema. Let us make the comparisons of the values of f at these points.

$$f\left(\tfrac{1}{\sqrt{2}}, \tfrac{1}{\sqrt{2}}\right) = 10\left(\tfrac{1}{\sqrt{2}}\right)^2 - 16\left(\tfrac{1}{\sqrt{2}}\right)^2 + 10\left(\tfrac{1}{\sqrt{2}}\right)^2 = 5 - 8 + 5 = 2$$

$$f\left(-\tfrac{1}{\sqrt{2}}, -\tfrac{1}{\sqrt{2}}\right) = 10\left(\tfrac{-1}{\sqrt{2}}\right)^2 - 16\left(\tfrac{-1}{\sqrt{2}}\right)^2 + 10\left(\tfrac{-1}{\sqrt{2}}\right)^2 = 2$$

$$f\left(\tfrac{1}{\sqrt{2}}, -\tfrac{1}{\sqrt{2}}\right) = 10\left(\tfrac{1}{\sqrt{2}}\right)^2 - 16\left(\tfrac{1}{\sqrt{2}}\tfrac{-1}{\sqrt{2}}\right) + 10\left(\tfrac{-1}{\sqrt{2}}\right)^2 = 5 + 8 + 5 = 18$$

$$f\left(\tfrac{-1}{\sqrt{2}}, \tfrac{1}{\sqrt{2}}\right) = 10\left(\tfrac{-1}{\sqrt{2}}\right)^2 - 16\left(\tfrac{-1}{\sqrt{2}}\tfrac{1}{\sqrt{2}}\right) + 10\left(\tfrac{1}{\sqrt{2}}\right)^2 = 18$$

Thus, the constrained maximum of f is 18, the constrained minimum is 2.

Remark 27.2 Example 27.5 deals with a system made of three equations of degree two in the unknowns x, y, λ. This means that the system of the Eqs. (27.10) to (27.12) is not linear: precisely, it has degree eight; indeed, the degree of an equation system is defined as the product of the degrees of the equations. A nonlinear system of equations can be inconsistent or consistent. In the latter case the system can have a finite number of solutions or infinite solutions.

27.7 Method of Lagrange Multipliers

The procedure for finding constrained relative maxima and minima used in (Sect. 27.6) can be generalized. Let us consider, for instance, the function f (x, y, z) subject to the constraint conditions $g(x, y, z) = 0$, $h(x, y, z) = 0$ and form the Lagrangian

$$L(x, y, z, \lambda, \mu) = f(x, y, z) + \lambda g(x, y, z) + \mu h(x, y, z)$$

Let us search for the triples that make extreme the function f (x, y, z) subject to the constraints $g(x, y, z) = 0$, $h(x, y, z) = 0$, among the points that are the critical points (x, y, z, λ, μ) of the Lagrangian. We must solve the system of equations

$$L_x = f_x(x, y, z) + \lambda g_x(x, y, z) + \mu h_x(x, y, z) = 0$$
$$L_y = f_y(x, y, z) + \lambda g_y(x, y, z) + \mu h_y(x, y, z) = 0$$
$$L_z = f_z(x, y, z) + \lambda g_z(x, y, z) + \mu h_z(x, y, z) = 0$$
$$L_\lambda = g(x, y, z) = 0$$
$$L_\mu = h(x, y, z) = 0$$

where λ and μ, which are independent of x, y, z, are the *Lagrange multipliers.*

Bibliography

Anton, H.: Calculus. Wiley, New York (1980)

Lax, P., Burnstein, S., Lax, A.: Calculus with Applications and Computing. Springer, New York (1976)

Royden, H.L., Fitzpatrick, P.M.: Real Analysis. Pearson, Toronto (2010)

Spivak, M.: Calculus. Cambridge University Press (2006)

Chapter 28
Directional Derivatives and Gradient

28.1 Directional Derivatives

We saw the geometric meaning of the partial derivatives (Sect. 23.11). Referring to Fig. 28.1 let us intersect the surface S graph of the function $f(x, y)$ with the plane passing through the point $Q(x_0, y_0, f(x_0, y_0))$ and parallel to z and x axes, and with the plane passing through Q and parallel to z and y axes. There exist other planes which are parallel to z axis and passing through Q each cutting the plane xy in a line.

Definition 28.1 Let A be an open set of $\mathbf{R}^2$ and f a function of class C^1 in A. Let r be an oriented line with unit vector $\mathbf{r}$ and $P_0(x_0, y_0)$ and $P(x, y) = (x_0 + \Delta x, y_0 + \Delta x)$ distinct points of r. If (PP_0) denote the length of the oriented segment PP_0 of r, the ratio.

$$\frac{f(P) - f(P_0)}{(PP_0)}$$

is called the *incremental ratio* of f from the starting point P_0. Put $\Delta f = f(P) - f(P_0)$ and $\Delta \rho = (PP_0)$, if the limit

$$\lim_{\Delta \rho \to 0} \frac{\Delta f}{\Delta \rho}$$

exists and is finite, then it is called the *directional derivative of f* at $P_0(x_0, y_0)$ *in the direction and sense of r*, or *with respect to the unit vector $\mathbf{r}$*, denoted by

$$\left(\frac{df}{dr}\right)_{P_0}$$

Let θ be the oriented angle $\widehat{xr}$ (Fig. 28.1) and, by (Sect. 8.4.1),

© The Author(s), under exclusive license to Springer Nature Switzerland AG 2023
A. G. S. Ventre, *Calculus and Linear Algebra*,
https://doi.org/10.1007/978-3-031-20549-1_28

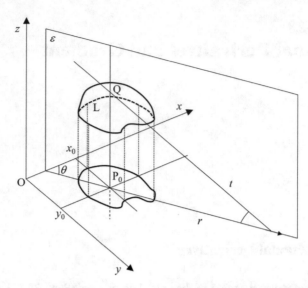

Fig. 28.1 Directional derivative of f in P w. r. to the unit vector **r**. The point $P_0(x_0, y_0)$ belongs to the domain of f; the point Q has coordinates $(x_0, y_0, f(x_0, y_0))$; the plane ε passes through $P_0(x_0, y_0)$, is perpendicular to the xy plane and forms the angle $\theta = \widehat{xr}$ with the x axis; the dashed curve $L = [\text{graph of } f] \cap [\text{plane } \varepsilon]$; the line r contains the orthogonal projection of the curve L on the xy plane; t tangent line to the curve L at Q; the value $\tan \widehat{xr}$ is the value of the directional derivative of f at P_0 w. r. to the unit vector **r**

$$\mathbf{r} = (\cos \theta, \sin \theta) = (\cos \widehat{xr}, \cos \widehat{yr})$$

the unit vector of the oriented line r (see Sect. 10.3.1).

In relation to the functions of class C^1 in an open of $\mathbf{R}^2$ the following property holds.

If f is C^1, then each plane of the bundle of planes parallel to the z axis and passing through Q cuts the surface S in a curve, contains the tangent line to the surface at the point Q and all these tangent lines lie on the tangent plane to the surface at Q. The slope of the tangent line t is equal to $\left(\frac{df}{dr}\right)_{P_0}$.

The following theorem holds:

Theorem 28.1 If f is C^1 in an open subset A of $\mathbf{R}^2$ and the point P_0 belongs to A then there exists the directional derivative of f at $P_0(x_0, y_0)$ with respect to any oriented line passing through P_0 and it results.

$$\frac{df}{dr} = \frac{\partial f}{\partial x} \cos \widehat{xr} + \frac{\partial f}{\partial y} \cos \widehat{yr}.$$

28.2 Gradient

Definition 28.2 If f has the first partial derivatives at the point (x, y), the vector of $\mathbf{R}^2$

$$\left(f_x(x, y),\, f_y(x, y)\right)$$

is called the *gradient* of f (x, y), denoted with the symbol grad$f(x, y)$ or ∇f (x, y). (The symbol ∇ is a *vector operator* that reads "nabla".)

Other expressions of the gradient in terms of unit vectors of the coordinate axes are:

$$\operatorname{grad} f(x, y) = \left(f_x(x, y),\, f_y(x, y)\right) = f_x(x, y)\boldsymbol{x} + f_y(x, y)\boldsymbol{y}$$

and, in form of scalar product:

$$\operatorname{grad} f(x, y) = \left(f_x(x, y),\, f_y(x, y)\right) \cdot (\boldsymbol{x}, \boldsymbol{y}).$$

Let γ be a regular plane curve of parametric equations

$$x = x(t)$$
$$y = y(t)$$

with t in $[a, b]$.

Recall (Sect. 24.2) that the line r tangent to the curve γ at the point P $P_0(x_0, y_0) = (x(t_0), y(t_0))$ has direction numbers $x'(t_0), y'(t_0)$. Therefore, the parametric equations of the line r are:

$$x = x_0 + x'(t_0)t$$
$$y = y_0 + y'(t_0)t$$

with t in $[a, b]$.

If the regular curve γ is represented by the implicit form $f(x, y) = 0$, then the equation of the line r tangent to γ at P_0 is

$$\left(\frac{\partial f}{\partial x}\right)_{P_0} (x - x_0) + \left(\frac{\partial f}{\partial y}\right)_{P_0} (y - y_0) = 0.$$

Therefore, the direction numbers of the tangent line r are:

$$\left(\frac{\partial f}{\partial y}\right)_{P_0},\, -\left(\frac{\partial f}{\partial x}\right)_{P_0}.$$

It follows that the normal line n to γ at $P_0(x_0, y_0)$ (Sect. 24.2.1) has direction numbers:

$$\left(\frac{\partial f}{\partial x}\right)_{P_0}, \left(\frac{\partial f}{\partial y}\right)_{P_0}.$$

Hence, the normal line n to γ at $P_0(x(t_0), y(t_0))$ has direction numbers $y\prime(t_0), -x\prime(t_0)$

Therefore, we state:

Theorem 28.2 If the partial derivatives of f (x, y) at $P_0(x_0, y_0)$ exists and grad$f(x_0, y_0) \neq (0, 0)$, then grad$f(x_0, y_0)$ is a vector orthogonal to the tangent line to the level curve $k = f(x_0, y_0)$ at the point (x_0, y_0).

Exercise 28.1 Let r be the line $y = x$. If x and y are the unit vectors of the coordinate axes x and y, calculate grad$f(1, 1)$ and the derivative of $f(x, y) = x^2 + y^2$ at the point $(1, 1)$ w.r. to the unit vector r of r.

Solution The components of the unit vector of r are:

$$r = \frac{x + y}{|x + y|} = \frac{(1, 0) + (0, 1)}{\sqrt{2}} = \left(\frac{1}{\sqrt{2}}, \frac{1}{\sqrt{2}}\right)$$

and grad$f(f(1, 1) = f_x(x, y)x + f_y(x, y)y = (2x, 2y)_{(x,y)=(1,1)} = (2, 2)$.
Then

$$\left(\frac{df(x, y)}{dr}\right)_{(1,1)} = \operatorname{grad} f(1, 1) \cdot r = (2, 2) \cdot \left(\frac{1}{\sqrt{2}}, \frac{1}{\sqrt{2}}\right) = \frac{4}{\sqrt{2}} = 2\sqrt{2}.$$

Example 28.1 Given the surface $z = f(x, y) = x^2 + y^2$, let us check that the vector grad$f(2, 1)$ is perpendicular to the level curve $k = f(2, 1)$. In fact, grad$f$ $(2, 1) = (2x, 2y)_{(x,y)=(2,1)} = (4, 2)$. Furthermore, the level curve $f(2,1) = 4 + 1 = 5$ is the circumference $\gamma : x^2 + y^2 = f(2, 1) = 5$ the line r tangent to γ at $(2, 1)$ has direction numbers (Fig. 28.2):

$$\left(\frac{\partial f}{\partial y}\right)_{(2,1)} = (2y)_{(2,1)} = 2, \quad -\left(\frac{\partial f}{\partial x}\right)_{(2,1)} = (-2x)_{(2,1)} = -4.$$

The vectors grad $f(2,1) = (4,2)$ and $(2, -4)$ are perpendicular since the scalar product $(4, 2) \times (2, -4)$ is null.

The following theorems correlate the gradient with the directional derivative.

Theorem 28.3 Given the function $f(x, y)$ and the vector grad$f(x, y)$, the directional derivative $\frac{df}{dr}$ with respect to the oriented direction of the unit vector r equals the

Fig. 28.2 Grad$f$ (2,1) = (4, 2) is orthogonal to the line $2x + y = 5$, tangent at P(2, 1) to the circumference $x^2 + y^2 = 5$

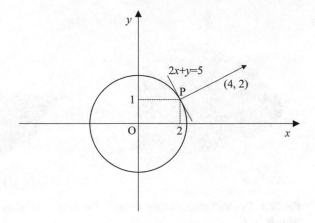

projection of the vector grad$f$ on the line r. The relation is expressed by the scalar product:

$$\frac{df}{dr} = \text{grad} f \cdot (\cos \widehat{xr}, \cos \widehat{yr}) = f_x(x, y) \cos \widehat{xr} + f_y(x, y) \cos \widehat{yr}.$$

Exercise 28.2 Let r be the bisecting line of the first and third quadrant and $\boldsymbol{x}$ and $\boldsymbol{y}$ the unit vectors of the coordinate axes x and y. Calculate grad$f(1, 1)$ and the derivative of $f(x, y) = x^2 + y^2$ at the point $(1, 1)$ w.r. to the unit vector $\boldsymbol{r}$ of r.

Solution The unit vector of r has component:

$$r = \frac{x + y}{|x + y|} = \frac{(1, 0) + (0, 1)}{\sqrt{2}} = \left(\frac{1}{\sqrt{2}}, \frac{1}{\sqrt{2}} \right)$$

and grad$f$ $f(1, 1) = (2x, 2y)_{(x, y)=(1,1)} = (2, 2)$.
    Then

$$\left(\frac{df(x, y)}{dr} \right)_{(1.1)} = \text{grad} f(1, 1) \cdot r = (2, 2) \cdot \left(\frac{1}{\sqrt{2}}, \frac{1}{\sqrt{2}} \right) = \frac{4}{\sqrt{2}} = 2\sqrt{2}.$$

Theorem 28.4 Let f be of class $C^1(A)$, $A \subseteq \mathbf{R}^2$. The derivative at a point w.r. to the unit vector $\boldsymbol{r}$ has a maximum if the direction of $\boldsymbol{r}$ coincides with that of the gradient: the value of the maximum is the modulus $|\text{grad}f|$.

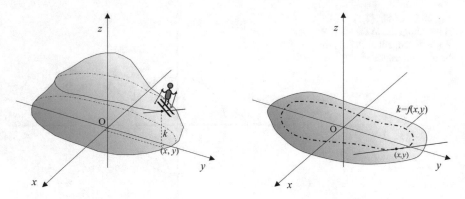

Fig. 28.3 Downhill racer steepest descent. The racer at (x, y) moves perpendicular to the level curve at (x, y)

28.2.1 Steepest Descent

The above considerations lead to a procedure of intuitive appeal, called the *gradient method*, for finding a relative extrema or critical points.

Let us start from a point with the objective of reaching a minimum point: we give an increase in the direction of the gradient; the point moves in a direction perpendicular to the level line in which it is located (Theorem 28.4). From the reached point the procedure iterates: each point reached in consequence of the increase in the direction of the gradient is the origin of the path oriented in the direction and sense of the fastest decrease: the procedure is known as the *steepest descent* algorithm (Fig. 28.3).

A downhill racer (Fig. 28.3) chooses the direction and the sense of—grad$f$ (x, y) at each point (x, y), where f (x, y) measures the altitude.

The streams represented on a topographic map flow in the direction of the steepest slope, which, at each point (x, y), is that of—grad$f$ (x, y).

Bibliography

Anton, H.: Calculus. Wiley, New York (1980)

Ayres, F., Jr., Mendelson, E.: Calculus. McGraw-Hill, New York (1999)

Lax, P., Burnstein, S., Lax, A.: Calculus with Applications and Computing. Springer, New York (1976)

Jr, Thomas, G.B., Finney, R.L.: Calculus and Analytic Geometry. Addison-Wesley Publishing Company Inc., Reading, Massachusetts USA (1979)

Chapter 29
Double Integral

29.1 Area of a Plane Set

We defined the concepts of measurability and area of a plane set starting from the area of the polygon (see Sect. 22.3). We now will modify the definition of measurability based on the simpler notion of area of the rectangle. The modification will ease the extension of the concept of *measure* to three-dimensional space.

The intervals of the plane $\mathbf{R}^2$ are rectangles with sides parallel to the coordinate axes (Sect. 23.1). A set of the points of the plane that is the union of a finite number of intervals of $\mathbf{R}^2$, two by two without interior common points, is called a *pluri-interval* of the plane. The area of a pluri-interval is defined as the sum of the areas of the rectangles, two by two without interior common points, which constitute it.

A bounded plane set A endowed with interior points is said to be *measurable* if the numerical set of the areas of the pluri-intervals contained in A and the numerical set of the areas of the pluri-intervals containing A are contiguous (Sect. 6.9). If A is measurable, the separation element a of the two numerical sets is called the *area* of the set A, $a = areaA$, or the *measure* of A. A bounded plane set A without interior points is said to be *measurable* with *measure zero* if the set of the areas of the pluri-intervals containing A has infimum zero. The definition of measurability given in terms of pluri-intervals is not substantially different from that of measurability in terms of polygons. In fact, every pluri-interval is also a polygon and, therefore, every measurable set according to the new definition is also so according to the previous one; and vice versa. Hence, both definitions lead to the same value of the area of a measurable set.

Definition 29.1 A subset A of $\mathbf{R}^2$, or $\mathbf{R}^3$, is called a *domain* if the following conditions are met:

- A is closed;
- each point of A is an accumulation point for the interior of A.

© The Author(s), under exclusive license to Springer Nature Switzerland AG 2023   507
A. G. S. Ventre, *Calculus and Linear Algebra*,
https://doi.org/10.1007/978-3-031-20549-1_29

Example 29.1 A closed 3-dimensional sphere (Sect. 23.1) is a domain of $\mathbf{R}^3$, while the union of a closed 3-dimensional sphere and an isolate point exterior to the sphere is not a domain.

Remark 29.1 It can be shown that if A is a domain, then

- A is the closure of an open set and
- A is a closed set and is the union of an open set and its boundary ∂A.

Definition 29.2 The *diameter* of a bounded subset A of $\mathbf{R}^2$, or $\mathbf{R}^3$, is, by definition, the supremum of the numerical set of the distances of any two points of A.

Theorem 29.1 Any bounded and measurable domain A of $\mathbf{R}^2$ can be decomposed into a finite number of measurable domains, having diameter less than any fixed positive real ε, two by two without interior points in common and whose union equals A.

29.2   Volume of a Solid

We dealt with the intervals of $\mathbf{R}^3$ (see Sect. 23.1). A closed interval of $\mathbf{R}^3$ is a rectangle parallelepiped with the sides parallel to the coordinates axes and whose points P have coordinates x, y, z that belong to non-degenerate closed intervals of $\mathbf{R}$. If $a \leq x \leq b$, $c \leq y \leq d$ and $e \leq z \leq f$, the *volume*, or *measure*, of the interval $I = [a, b] \times [c, d] \times [e, f]$ of $\mathbf{R}^3$ is defined by the product

$$\mathrm{vol} I = (b - a)(d - c)(f - e)$$

We learned from elementary geometry how to calculate the volumes of some solids such as parallelepipeds, cylinders and circular cones, pyramids, some polyhedra, the sphere. Let us now define a generalization of the concept of volume.

To this aim we extend the concept of pluri-interval to the space $\mathbf{R}^3$ and define a particular class of polyhedra, called *pluri-rectangles* or *pluri-intervals*. A *pluri-interval*, or *pluri-rectangle*, of $\mathbf{R}^3$ is, by definition, a polyhedron which is the union of a finite number of rectangles of $\mathbf{R}^3$ (i. e., parallelepipeds with the edges parallel to the coordinate axes) two by two without interior points in common. The volume of a pluri-interval of $\mathbf{R}^3$ is the sum of the volumes of the parallelepipeds which constitute it. (Fig. 29.1).

If a parallelepiped of $\mathbf{R}^3$ is degenerate, i. e., it reduces to a point, or a segment, or a plane rectangle, its volume is zero.

Every pluri-interval Y of $\mathbf{R}^3$ can be decomposed into a finite number of closed parallelepipeds $I_1, I_2, \ldots, I_n$, included in Y, two by two without interior points in common and such that $Y = \cup_{i=1,\ldots,n} I_i$.

Such a decomposition is not unique. Indeed, each closed parallelepiped A is the union of two closed parallelepipeds without interior points in common.

Fig. 29.1 A pluri-interval of
R³

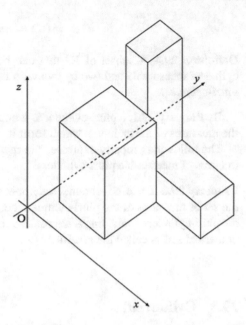

However, from elementary geometry we know that the sum of the volumes of two parallelepipeds without interior points in common equals the volume of their union.

Definition 29.3 Let A be a bounded domain of **R**³ endowed with interior points. The set A is said to be *measurable* if the set of the volumes of the pluri-rectangles containing A and the set of the volumes of the pluri-rectangles contained in A are contiguous. Then the element of separation of the two numerical sets is called the *volume* of A, or the *measure* of A, and is denoted with the symbols vol(A), or $m(A)$. The bounded subset A of **R**³ without interior points is said to be *measurable* with *measure zero* if the set of the volumes of the pluri-rectangles containing A has infimum zero.

Property 29.1 If the subset A of **R**², or **R**³, is the union of two bounded and measurable domains A_1 and A_2 without interior points in common, then A is measurable and

$$m(A) = m(A_1) + m(A_2)$$

Let A be a bounded domain in the plane xy. If $[a, b]$ is a closed interval, the set C of the points in the space **R**³ having coordinates (x, y, z) such that $(x, y) \in A$ and $z \in [a, b]$, i. e., the set $C = A \times [a, b]$, is called a *cylinder* of **R**³. The domains of the planes $z = a$ and $z = b$ which have A as orthogonal projection on the plane xy are called the *bases* and the number $h = b - a$ the *height* of the cylinder.

Property 29.2 If the plane domain A is measurable, then the cylinder C of base A and height h is measurable and

$$m(C) = hm(A)$$

Definition 29.4 A subset of $\mathbf{R}^3$ that can be decomposed into a finite number of cylinders measurable and two by two with no interior points in common is called a *pluri-cylinder*.

By Property 29.1, a pluri-cylinder is a domain whose measure equals the sum of the measures of the cylinders which form it.

The following property expresses the measurability criterion of a domain of $\mathbf{R}^3$ in terms of measurable pluri-cylinders.

Property 29.3 Let A be a bounded domain of $\mathbf{R}^3$. The domain A is measurable if the set of measures of the pluri-cylinders contained in A and the set of measures of the pluri-cylinders containing A are contiguous. The element of separation of the two numerical sets is called the *measure* of A.

29.3 Cylindroid

Definition 29.5 Let $f(P) = f((x, y)$ be a real-valued function of the two real variables x and y, continuous and non-negative in the bounded and measurable domain $A \subseteq \mathbf{R}^2$. The *cylindroid* of base A relative to the function f is, by definition, the set of points $(x, y, z) \in \mathbf{R}^3$, such that

$$(x, y) \in A \text{ and } 0 \leq z \leq f(x, y)$$

whose boundary is made of the union of the graph of $f(P)$, the domain A and a portion of the cylindrical surface formed by the lines parallel to z axis and passing through the points of the boundary of A. Of course, if $f((x, y)$ is constant the cylindroid is a cylinder of base A.

Property 29.4 If A is a bounded and measurable domain of $\mathbf{R}^2$, the cylindroid of base A relative to the continuous and positive function f is a measurable domain.

29.4 Double Integral

If A is a domain of $\mathbf{R}^2$ and $A_1, A_2, \ldots, A_n$ are domains of $\mathbf{R}^2$ two by two without interior points in common such that $A = A_1 \cup A_2 \cup \ldots \cup A_n,$, the set of domains $\{A_1, A_2, \ldots, A_n\}$ is called a *decomposition* of A into partial domains. Such a decomposition is denoted $D(A_1, A_2, \ldots, A_n)$ or simply D.

If A is a bounded domain the greatest diameter of the domains A_i is named the *size* or the *diameter* of the decomposition D.

Let f be a real-valued function of two real variables continuous in a bounded and measurable domain A of $\mathbf{R}^2$. Given a decomposition $D(A_1, A_2, \ldots, A_n)$ of A into measurable domains, in each domain A_i, $i = 1,\ldots, n$, let us choose a point $P_i(x_i, y_i)$ and calculate the sum

$$\sigma_D = \sum_{i=1}^{n} f(P_i)m(A_i).$$

It can be shown that there exists a real number J such that, for every $\varepsilon > 0$ there exists a number $\delta(\varepsilon) > 0$, such that, for every decomposition of the domain A into a finite number of measurable domains of size less than $\delta(\varepsilon)$, it follows:

$$|\sigma - J| < \varepsilon.$$

The number J is called the *double integral*, or *integral*, of the function $f(x, y)$ *extended* to the domain A and is denoted by $\int_A f(P)dA$ or $\iint_A f(x, y)dxdy$.

Let us state the following theorem.

Theorem 29.2 If $f(x, y)$ is a function continuous and positive in the bounded and measurable domain A of the plane, then the volume of the cylindroid of base A relative to the function f is equal to the integral of the function f extended to the domain A.

Remark 29.2 Consider the restriction of the function $f(x, y) = 1$ to the bounded and measurable domain A. We have:

$$\iint_A f(x, y)dxdy = \iint_A 1dxdy$$

$$= \text{ volume of the cylinder of base } A \text{ and height } 1$$

$$= m(A) = \text{ area of the domain } A.$$

Let the functions $f(x, y)$ and $g(x, y)$ be continuous in the bounded and measurable domain A of the plane, such that $f(x,y) < g(x, y)$ in $A - \partial A$. The set of points (x, y, z) of $\mathbf{R}^3$ such that (x, y) belongs to A and $f(x, y) \leq z \leq g(x, y)$ is a domain called a *normal domain* with respect to the xy plane, relative to the functions f and g.

It can be shown that the normal domain with respect to the xy plane, relative to the continuous functions f and g is a measurable domain whose volume is given by

$$\iint_A (g(x, y) - f(x, y))dxdy \tag{29.1}$$

29.5 Properties of the Double Integral

The double integral enjoys properties similar to those related to the integral of the real-valued functions of a single variable.

If f and g are real-valued functions of two real variables continuous in the bounded and measurable domain A of $\mathbf{R}^2$, then the following properties hold:

Property 29.5 If f and g are real-valued functions of two real variables continuous in the bounded and measurable domain A of $\mathbf{R}^2$, then whatever the real numbers h and k are, the following equality holds.

$$\int_A (hf(P) + kg(P))dA = h \int_A f(P)dA + k \int_A g(P)dA.$$

Property 29.6 If f is a continuous function in the bounded and measurable domain A and if A is the union of two measurable domains A_1, A_2 without interior points in common, then

$$\int_A f(P)dA = \int_{A_1} f(P)dA + \int_{A_2} f(P)dA$$

Property 29.7 (Mean value theorem) If f is a continuous function in the bounded and measurable domain A, if e' and e'' are the minimum and maximum of f in A, respectively, then

$$e'm(A) \le \int_A f(P)dA \le e''m(A).$$

If, moreover, A is a connected set (see Definition 23.1), then there exists in A a point Q such that:

$$\int_A f(P)dA = f(Q)m(A).$$

Let A be a bounded and measurable domain of $\mathbf{R}^2$.

Property 29.8 If the function f is continuous and non-negative in the bounded and measurable domain A, then

$$\int_A f(P)dA \ge 0.$$

Property 29.9 If f and g are continuous in A and $f(P) \geq g(P)$ in A, then

$$\int_A f(P)dA \geq \int_A g(P)dA.$$

Property 29.10 If f is continuous and non-negative in A and if B is a measurable domain contained in A, then

$$\int_B f(P)dA \leq \int_A f(P)dA.$$

Property 29.11 If f is continuous in A, then

$$\left| \int_A f(P)dA \right| \leq \int_A |f(P)|dA.$$

29.6   Double Integral Reduction Formulas

We describe a procedure for calculating a class of double integrals. For this purpose, some *reduction formulas* will be used, which allow the calculation of the double integral to be traced back to that of several integrals of functions of a single variable.

We state a theorem that provides a reduction formula that can be applied if the bounded and measurable integration domain A is normal with respect to one of the coordinate axes (Sect. 22.9).

Theorem 29.3 Let $p(x)$ and $q(x)$ be continuous functions in the closed interval $[a, b]$ such that, for every $x \in [a, b]$, $p(x) \leq q(x)$ and, for every $x \in (a, b)$, $p(x) < q(x)$. If A is the normal domain with respect to the x axis, defined by

$$A = \left\{ (x, y) \in \mathbf{R}^2 : a \leq x \leq b; \, p(x) \leq y \leq q(x) \right\}$$

and if the function $f(x, y)$ is continuous in A, then

$$\iint_A f(x, y)dxdy = \int_a^b \left[\int_{p(x)}^{q(x)} f(x, y)dy \right] dx. \tag{29.2}$$

A formula similar to (29.2) applies if A is a normal domain with respect to the y axis:

$$A = \{(x, y) \in \mathbf{R}^2 : c \le y \le d; u(y) \le x \le v(y)\}$$

where $u(y)$, $v(y)$ are continuous functions in $[c, d]$ such that $u(y) < v(y)$ in (c, d). Indeed,

$$\iint\limits_{A} f(x, y)dxdy = \int\limits_{c}^{d} \left[\int\limits_{u(y)}^{v(y)} f(x, y)dx \right] dy. \qquad (29.3)$$

Exercise 29.1 Calculate the volume of the cylindroid of base

$$A = \{(x, y) : 0 \le x \le 1; x^2 \le y \le x\}$$

relative to the constant function $f (x, y) = 1$.

Solution From (29.2) and Remark 29.2, we obtain:

$$\iint\limits_{A} 1dxdy = \int\limits_{0}^{1} \left[\int\limits_{x^2}^{x} dy \right] dx = \int\limits_{0}^{1} [y]_{x^2}^{x} dx = \int\limits_{0}^{1} (x - x^2)dx$$

$$= \left[\frac{x^2}{2} - \frac{x^3}{3} \right]_{0}^{1} = \frac{1}{6}$$

Exercise 29.2 Calculate the area of the domain A whose boundary is made of the parabola $y = x^2$ and the lines $y = 2x$ and $x = 1$ (Fig. 29.2).

Solution 1. The domain is normal w. r. to the x axis. So, by (29.2) we have:

Fig. 29.2 The domain bounded by the parabola, the line $y = 2x$ and the line $x = 1$

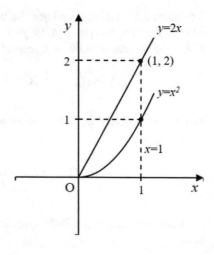

$$\iint\limits_{A} 1\,dxdy = \int\limits_{0}^{1}\left[\int\limits_{x^2}^{2x} dy\right]dx = \left[x^2 - \frac{x^3}{3}\right]_{0}^{1} = 1 - \frac{1}{3} = \frac{2}{3}.$$

Solution 2. The domain is normal w. r. to the y axis. In order to apply (29.3), subdivide the domain A into the domains A_1 above the line $y = 1$ and A_2 below the line $y = 1$. By Property 29.6 and Remark 29.2, we obtain:

$$\iint\limits_{A} dxdy = \int\limits_{0}^{1}\left[\int\limits_{\frac{y}{2}}^{\sqrt{y}} dx\right]dy + \int\limits_{1}^{2}\left[\int\limits_{\frac{y}{2}}^{1} dx\right]dy$$

$$= \frac{2}{3} - \frac{1}{4} + \int\limits_{1}^{2}[x]_{x=\frac{y}{2}}^{x=1}\,dy = \frac{5}{12} + \frac{1}{4} = \frac{2}{3}$$

Exercise 29.3 Find the volume of the subset D of $\mathbf{R}^3$ bounded by the right circular cylinder $x^2 + y^2 = 4$, by the plane $y + z = 4$ and the plane $z = 2$.

Solution 1. The restrictions of the functions $z = f(x, y) = 2$, $z = g(x, y) = 4 - y$ to the circle A of the xy plane of equation $x^2 + y^2 \leq 4$, define a domain normal with respect to the xy plane. Indeed, for each (x, y) of the circle A, $f(x, y) = 2 \leq z \leq g(x, y) = 4 - y$, and $f(x, y) < g(x, y)$ in the open circle $x^2 + y^2 < 4$. To calculate the volume of domain D apply (29.1):

$$\mathrm{vol}D = \iint\limits_{A} (g(x, y) - f(x, y))dxdy = \int\limits_{-2}^{2}\left[\int\limits_{-\sqrt{4-y^2}}^{\sqrt{4-y^2}} (2 - y)dx\right]dy$$

$$= \int\limits_{-2}^{2}\left[(2 - y)\int\limits_{-\sqrt{4-y^2}}^{\sqrt{4-y^2}} dx\right]dy = \int\limits_{-2}^{2} (2 - y)[x]_{-\sqrt{4-y^2}}^{\sqrt{4-y^2}}dy$$

$$= \int\limits_{-2}^{2} (2 - y)2\sqrt{4 - x^2}dy = 2\int\limits_{-2}^{2} (2 - y)\sqrt{4 - x^2}dy.$$

Recall $\int_{-2}^{2}(2-x)\sqrt{4-x^2}dx = 4\pi$ (Sect. 22.10.1). Hence $\mathrm{vol}D = 8\pi$.

Solution 2. The exercise can be solved in an elementary way (Fig. 29.3). Let us consider a cube with the side of length 4 and two parallel sides a, b not lying on the same face. Let the plane α pass through a and b. Inscribe a right circular cylinder of height 4 in the cube: the bases of the cylinder lie in parallel faces of the cube. The plane α cuts the cube into two congruent figures and the cylinder into two congruent

Fig. 29.3 The line a, parallel to the y axis is the line common to the plane $z = 2$ and the α plane of equation $z = 4 - y$; the upper face of the cube is on the $z = 2$ plane, the lower face on the xy plane

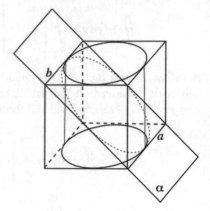

figures. The volume of the cube is $4 \times 4 \times 4 = 16 \times 4$. The volume of the cylinder is $4 \times 4 \times \pi = 16\pi$. The volume of the solid D is half of the volume of the cylinder, $volD = 8\pi$.

Exercise 29.4 Calculate the double integral

$$\iint\limits_A xy\,dxdy$$

where A is the triangle with vertices $O = (0, 0)$, $B = (0, 1)$ and $C = (1, 0)$ (Fig. 29.4). The domain

$$A = \left\{ (x, y) \in \mathbf{R}^2 : a \leq x \leq b; \, p(x) \leq y \leq q(x) \right\}$$

with $(a, b) = (0, 1)$, $p(x) = 0$ and $q(x) = 1-x$, is normal with respect to the x axis. Formula (29.2) yields

$$I = \int_a^b \left[\int_{p(x)}^{q(x)} f(x, y)dy \right] dx = \int_0^1 \left[\int_0^{1-x} xy\,dy \right] dx.$$

Fig. 29.4 The domain of the plane xy: $0 \leq x \leq 1$, $0 \leq y \leq 1-x$

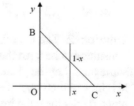

Let us start by calculating

$$\int_0^{1-x} xy\,dy = \left[\frac{xy^2}{2}\right]_{y=0}^{y=1-x} = \frac{x(1-x)^2}{2} = \frac{x^3-2x^2+x}{2}.$$

Therefore,

$$I = \int_0^1 \frac{x^3-2x^2+x}{2}\,dx = \frac{1}{2}\left[\frac{x^4}{4}-\frac{x^3}{3}+\frac{x^2}{2}\right]_{x=0}^{x=1} = \frac{1}{24}.$$

Bibliography

Apostol, T.M.: Calculus II. Blaisdell Publishing Company, New York (1965)

Lax, P., Burnstein, S., Lax, A.: Calculus with Applications and Computing. Springer, New York Inc (1976)

Spiegel, M.R.: Advanced Calculus. McGraw-Hill Book Company, New York (1963)

Stoka, M.: Corso di matematica. Cedam, Padova (1988)

Chapter 30
Differential Equations

30.1 Introduction

We know equations whose solutions are numbers. We studied (see Chap. 22) some methods for calculating a function whose the first derivative is known. We now present the first elements of a wider problem: to determine an unknown function $y(x)$ knowing a relation between $y(x)$ and its first n derivatives, y', y'', ..., $y^{(n)}$, and the independent variable x. Any equation where at least one of these derivatives intervenes, along with x and y, is called a *differential equation*. The number n is called the order of the differential equation.

Differential equations are also frequently met in mechanics, in social and economic phenomena, in engineering, biology, physics and natural sciences. We will outline a short introduction to the topic through examples and some applications.

30.2 Separable Equations

Definition 30.1 A *solution* of a differential equation is a function y that satisfies the equation. Integrating or solving a differential equation means to calculate all the functions that satisfy the equation. The set of the solutions of a differential equation is called the *general integral* of the equation.

Examples of differential equations are $\frac{d^3y}{dx^3} - 4\frac{dy}{dx} + cos2x = 0$ and $dy = (y - x)dx$. The first can be written as $y''' - 4y' + cos2x = 0$. The first equations is of order three, and the second is of order one.

Exercise 30.1 Examine the problem of finding a curve having tangent line at each of its points with slope equal to the double of the abscissa of the point.

Solution The question can be expressed in terms of a differential equation. Indeed, suppose that the curve is the graph of the unknown function $y(x)$; then the required condition leads to the equality

© The Author(s), under exclusive license to Springer Nature Switzerland AG 2023
A. G. S. Ventre, *Calculus and Linear Algebra*,
https://doi.org/10.1007/978-3-031-20549-1_30

$$\frac{dy}{dx} = 2x \tag{30.1}$$

which is a differential equation of the first order. The Eq. (30.1) has the property that the variables x and y are *separable* since the equation may be rewritten as

$$dy = 2xdx \tag{30.2}$$

where y appears only in the left-hand side and x only in the right-hand side; Eq. (30.2) is called a *separable differential equation* and can be solved by the indefinite integration of both sides:

$$\int dy = \int 2xdx$$

which implies $y + c_1 = x^2 + c_2$ and setting $c = c_2 - c_1$, we obtain:

$$y = x^2 + c \tag{30.3}$$

where the numbers c_1, c_2 and c are *arbitrary constants*. For every $c \in \mathbf{R}$ a function is determined by (30.3) which defines the set of the solutions of the Eq. (30.1). The set of functions (30.3) is a family of parabolas having the y axis as the orthogonal axis of symmetry (Sect. 21.8): if $c \leq 0$ each parabola intersects the x axis at the points of abscissas $\pm\sqrt{c}$.

Exercise 30.2 Examine the problem of finding a curve such that the tangent at each of its points has a slope equal to twice the second coordinate of the point.

Solution This condition can be expressed in terms of the differential equation

$$\frac{dy}{dx} = 2y \tag{30.4}$$

We can separate the variables:

$$\frac{dy}{y} = 2dx$$

and calculate the indefinite integrals in the following equality:

$$\int \frac{dy}{y} = \int 2dx$$

Hence, $\ln y = 2x + c$ and, for every $c \in \mathbf{R}$,

$$y = e^{2x+c} \tag{30.5}$$

The set of functions (30.5) is the general integral of Eq. (30.4) and defines a family of exponential curves.

Exercise 30.3 Solve the differential equation

$$y' = 3x^2 \tag{30.6}$$

Solution Equation (30.6) is a first order equation. Put the equation in the form

$$\frac{dy}{dx} = 3x^2$$

separate the variables

$$dy = 3x^2 dx$$

and integrate

$$\int dy = \int 3x^2 dx$$

Therefore,

$$y = x^3 + c \tag{30.7}$$

Equation (30.7) provides the set of solutions of the differential Eq. (30.6), i. e., its general integral. There is a unique solution of (30.6) for each value of the constant c.

Exercise 30.4 Solve the equation

$$\frac{dy}{dx} = 3x^2 \sqrt{y} \tag{30.8}$$

Solution The first-order Eq. (30.8) is separable: indeed, from (30.8) we obtain

$$\frac{dy}{\sqrt{y}} = 3x^2 dx$$

and integration yields

$$\int y^{-\frac{1}{2}} dy = \int 3x^2 dx$$

The general solution of Eq. (30.8) is expressed in the implicit form $2y^{\frac{1}{2}} = x^3 + c$.

Exercise 30.5 Solve the equation

$$\frac{dy}{dx} + 2xy = 0 \tag{30.9}$$

Solution The first-order Eq. (30.9) is separable and assumes the form:

$$\frac{dy}{y} = -2xdx$$

Let's integrate both sides to obtain the general solution:

$$lny = -x^2 + c$$

Exercise 30.6 Solve the equation

$$(1 + y^2)y' = \frac{3}{x} \tag{30.10}$$

Solution The equation may be rewritten:

$$(1 + y^2)\frac{dy}{dx} = \frac{3}{x}$$

Separate the variables and integrate:

$$\int (1 + y^2)dy = \int \frac{3}{x}dx$$

Then

$$y + \frac{y^3}{3} = 3ln|x| + lnc = lncx^3$$

which can be expressed in the form:

$$e^{y + \frac{y^3}{3}} = cx^3$$

which is the implicitly defined general solution of (30.10).

Exercise 30.7 Solve the equation

$$xydy = (1 - y^2)dx$$

Solution The equation is *separable*

$$\frac{y}{1-y^2}dy = \frac{dx}{x}$$

Integrating both sides

$$\int \frac{y}{1-y^2}dy = \int \frac{dx}{x}$$

we get

$$-\frac{1}{2}ln\left|1-y^2\right| = lnx + lnc$$

and from the properties of the logarithm (Sect. 17.9),

$$\frac{1}{\sqrt{1-y^2}} = cx$$

squaring and taking the reciprocals we obtain

$$y = \pm\sqrt{1-\frac{1}{c^2x^2}}$$

which is the required general solution.

30.3   Exponential Growth and Decay

Human populations, bacterial colonies, radioactivity, capital investment projects, epidemics are among the natural, social and economic phenomena represented by quantities that increase or decrease over time in proportion to their value at a given instant. Differential equations are often useful in the study of such phenomena we generically call *quantities.*

We will mention the quantities that, in their temporal evolution, follow an exponential growth or decrease (decay) model: this means that the rapidity of growth or decrease of these quantities is proportional to its value at a given instant.

Let $y(t)$ be the value at instant t of a quantity that follows the model. Let us assume $y(t) > 0$, for every t. As stated above, the rate of change of $y(t)$ is proportional to its value at a given instant t, i. e.,

$$\frac{dy}{dt} = ky$$

where k is a constant called the *growth constant* if $k > 0$ and y is said to *grow exponentially*; if $k < 0$ then k is called the *decay constant* and y is said to *decay exponentially*.

If $k > 0$ then $\frac{dy}{dt} > 0$, then y is an increasing function of time and the model is called a *growth model*. If $k < 0$, y decreases over time and the model is called a *decay model*.

Suppose now that at the instant $t = 0$ the value of the quantity is y_0, i. e., $y(0) = y_0$. In order to know the state of the quantity at any t we solve the problem:

$$\frac{dy}{dt} = ky$$

at the initial state $y(0) = y_0$. We can separate the variables:

$$\frac{dy}{y} = kdx$$

Hence, $\ln y = kt + \ln c$, with c the arbitrary constant to be determined by setting the condition $y(0) = y_0$. So $y = ce^{kt}$ and $y_0 = ce^0 = c$ and, therefore,

$$y = y_0 e^{kt}$$

Exercise 30.8 If a quantity follows an exponential growth model the time required for doubling its initial value is said *doubling time*; if a quantity follows an exponential decay model, the time required for its initial value to be reduced by half is called the *half-life*. The doubling time and half-life are independent of the initial quantity. In fact, in the case of a quantity that follows an exponential growth model, we have:

$$y = y_0 e^{kt}$$

with $k > 0$. The value of y at time t_1 is

$$y_1 = y_0 e^{kt_1} \tag{30.11}$$

Let us find the time interval T so that y doubles, i.e., the value of y becomes $2y_1$ at time $t_1 + T$:

$$2y_1 = y_0 e^{k(t_1+T)} = y_1 e^{kt_1} e^{kT}$$

By (30.11)

$$2y_1 = y_1 e^{kT}$$

and $2 = e^{kT}$. In conclusion, $T = \frac{1}{k} ln2$.

Therefore, it can be seen that the doubling time T is independent of y_0 and t_1. Similarly, if y follows an exponential decay model the half-life time is independent of y_0 and t_1.

Example 30.1 If we hypothesize that a certain human population grows by 2% per year, the doubling time of the population is $T = \frac{1}{0.02} ln2 \cong 34.7$ years.

Bibliography

Burkill, J.C.: The Theory of Ordinary Differential Equations. Oliver and Boyd LTD, Edinburgh (1965)

Ince, E.L.: Integration of Ordinary Equations. Oliver and Boyd LTD, Edinburgh (1961)

Lax, P., Burnstein, S., Lax, A.: Calculus with Applications and Computing. Springer, New York (1976)

Pressat, R.: L'analyse démografique. Presses Universitaires de France, Paris (1961)

Index

Printed in the United States
by Baker & Taylor Publisher Services